VOLUME FORTY FIVE

ADVANCES IN HEAT TRANSFER

VOLUME FORTY FIVE

ADVANCES IN
HEAT TRANSFER

Serial Editors

EPHRAIM M. SPARROW
University of Minnesota, Minneapolis

YOUNG I. CHO
Department of Mechanical Engineering and Mechanics, Drexel University, Philadelphia, Pennsylvania

JOHN P. ABRAHAM
University of St. Thomas, St. Paul, Minneapolis

JOHN M. GORMAN
University of Minnesota, Minneapolis

Founding Editors

THOMAS F. IRVINE, JR.
State University of New York at Stony Brook, Stony Brook, NY

JAMES P. HARTNETT
University of Illinois at Chicago, Chicago, IL

ELSEVIER

AMSTERDAM • BOSTON • HEIDELBERG • LONDON
NEW YORK • OXFORD • PARIS • SAN DIEGO
SAN FRANCISCO • SINGAPORE • SYDNEY • TOKYO
Academic Press is an imprint of Elsevier

Academic Press is an imprint of Elsevier

525 B Street, Suite 1800, San Diego, CA 92101-4495, USA
225 Wyman Street, Waltham, MA 02451, USA
32 Jamestown Road, London NW1 7BY, UK
The Boulevard, Langford Lane, Kidlington, Oxford, OX5 1GB, UK
Radarweg 29, PO Box 211, 1000 AE Amsterdam, The Netherlands

First edition 2013

British Library Cataloguing in Publication Data
A catalogue record for this book is available from the British Library

Library of Congress Cataloging-in-Publication Data
A catalog record for this book is available from the Library of Congress

ISBN: 978-0-12-407819-2
ISSN: 0065-2717

For information on all Academic Press publications
visit our website at store.elsevier.com

Printed and bound by CPI Group (UK) Ltd, Croydon, CR0 4YY

CONTENTS

CONTRIBUTORS

Adrian Bejan
Department of Mechanical Engineering and Materials Science, Duke University, Durham, North Carolina, USA

Suresh V. Garimella
Cooling Technologies Research Center, An NSF IUCRC, School of Mechanical Engineering and Birck Nanotechnology Center, Purdue University, West Lafayette, Indiana, USA

Gad Hetsroni
Department of Mechanical Engineering, Technion—Israel Institute of Technology, Technion City, Haifa, Israel

Khalil Khanafer
Department of Biomedical Engineering, Frankel Vascular Mechanics Laboratory, University of Michigan, Ann Arbor, Michigan, USA

Albert Mosyak
Department of Mechanical Engineering, Technion—Israel Institute of Technology, Technion City, Haifa, Israel

Brian Spalding
CHAM Ltd, 40 High St Wimbledon, London SW 19 5AU, United Kingdom

Kambiz Vafai
Mechanical Engineering Department, University of California, Riverside, California, USA

Justin A. Weibel
Cooling Technologies Research Center, An NSF IUCRC, School of Mechanical Engineering and Birck Nanotechnology Center, Purdue University, West Lafayette, Indiana, USA

PREFACE

The current volume of *Advances in Heat Transfer* contains an eclectic collection of *magna opera* and highly informative summaries of heat transfer topics of current interest. Professor D. Brian Spalding's career has spanned both many subjects in the thermal sciences and many decades of creative activity. Among his legacies is a leadership role in physical modeling and its numerical implementation. His research has generated an unflagging outpouring of original ideas which have served as the motive power for the current advanced status of numerical-based analysis. The life's work of Professor Gad Hetsroni has focused on boiling and two-phase flow. His lengthy involvement with these subjects is brought into focus by his seminal chapter (with A. Mosyak) that also features a novel measurement technique, based on infrared technology, which brings definitive insights into the boiling process.

Professor Adrian Bejan is well known for creating novel and overarching intellectual conceptions. His constructal law of design evolution, enunciated here, is "the law of physics that expresses the natural tendency of all flow systems, bio and nonbio, to morph into configurations that provide greater flow access over time." His chapter is focused on the phenomenon of technology evolution in order to illustrate how the constructal law governs design and evolution in nature.

The thermal management of electronic equipment continues to be of critical importance as invention produces life-altering devices based on electronics. The pursuit of compactness with its concomitant increase in power density has demanded new passive heat spreading technologies that can dissipate extremely high heat fluxes from small hot spots. In their chapter, Professors Justin Weibel and Suresh Garimella have set forth a unique technology that fulfills that need.

The relentless search for uses of nanotechnology and nanofluids is one of the characterizing foci of current thermal engineering endeavors. Professors Khalil Khanafer and Kambiz Vafai have ably documented the most recent advances of nanotechnology for application in thermal storage systems, photovoltaic systems, and solar desalination.

Ephraim M. Sparrow
Young I. Cho
John P. Abraham
John M. Gorman

CHAPTER ONE

Trends, Tricks, and Try-ons in CFD/CHT

Brian Spalding
CHAM Ltd, 40 High St Wimbledon, London SW 19 5AU, United Kingdom

Contents

Abstract

Computational fluid dynamics and its counterpart computational heat transfer are subjects that inspire alarm in precomputer-trained professors and awe in young would-be researchers. One aim of this chapter is to diminish these reactions by clarifying both the laudable and the debatable natures of the subjects. A second aim is to make clear, to those who are not overanxious to follow fashion, that there remains much scope for valuable innovations.

The chapter reviews items selected from approximately half a century of three-steps-forward-two-steps-back actions, and it contains such adumbrations of detail and expressions of personal opinion as its author judges to be conducive to its aim.

Advances in Heat Transfer, Volume 45
ISSN 0065-2717
http://dx.doi.org/10.1016/B978-0-12-407819-2.00001-3

1. INTRODUCTION

Science, whether pure or applied, is no less subject to fashion than other human activities. Following one's predecessors is usually safe and sometimes wise; but it is best done consciously, with the possibility in mind that *not* following may sometimes be better.

The "trends" referred to in the title of the chapter can be fairly called "fashions." Examples will be discussed in respect to computational grids, equation-solving methods, and turbulence models.

Reality *is* and theory *may* be, but it is the latter that scientists prefer to deal with. Just as the theater "magician" persuades his audience to believe in what can surely not be truth, so the inventive scientist seeks to persuade himself, and indeed others too, that his idealizations, though not strictly true, will be *useful*. Lest they be overprotected from criticisms, such artifacts are here called "tricks." Radiation models and techniques for handling awkwardly shaped objects are among the contributions of the computational fluid dynamics/computational heat transfer (CFD/CHT) "tricksters."

Before a novel approach becomes an accepted trick and is honored with the grander title of "model," it appears as a "try-on," by which is here meant that its proposer "wonders if" some new formulation might possibly fit reality better than those in common use. Examples of such musings, in which the author will invite his readers to participate, concern a "mixing length transport" turbulence model, the "population model" approach to turbulent swirling flows, and the "partially parabolic" method.

During the writing of this chapter, the author has been conscious of the serious gaps in his own current knowledge, by exposing which he hopes that some readers will be moved to enlighten him. Should his ignorance prove to be widely shared, however, exposing it may hopefully guide researchers toward avenues that may be profitable to explore.

2. TRENDS

2.1. Computational grid trends

2.1.1 Early choices: Cartesian, cylindrical-polar, and body-fitted

The earliest applications of numerical solution methods to the differential equations of fluid mechanics and heat transfer used grids of Cartesian or cylindrical-polar configuration. Formulating the so-called finite-difference (later "finite-volume") equations linking dependent variables at grid nodes was then rather easy because lines joining such nodes intersected at right angles. Typically, each node was connected with only six neighbors, two in each of the three coordinate directions.

Soon, however, the need to consider flows around curved bodies such as airfoils caused "*body-fitted coordinate*" grids to find favor, even though at least 12 neighbors had to be considered for each node; and still, the equations could be formulated plausibly in more than one way.

2.1.2 Arbitrary polygonal cells

Specialists in the analysis of stresses in solids had meanwhile been taking a different route. Their "finite volumes," which they called "finite elements," were typically tetrahedrons with arbitrary angles between the normal and adjacent sides. This choice allowed their grids to be fitted to bodies of rather awkward shapes, which was probably the reason why fluid dynamicists also began to adopt the idea.

The equations between the variables at the grid nodes now became even more complex and difficult to derive with certainty; but, once done and embodied in computer coding, the difficulties disappeared from view. Reluctance to revisit them discouraged making rigorous tests as to the relative accuracies of the alternative discretization possibilities. (Question #1 to readers: where, if anywhere, have the results of truly comprehensive tests been published?)

Grids of this kind were "unstructured," meaning that geometrically nearby nodes did not necessarily have their values stored in adjacent locations in computer memory. This complicated the task of solving the equations; and for this reason, some computer-code custodians preferred not to follow the fashion, adopting instead a different way of solving the awkward-body-shape problem. They used the "cut-cell" technique.

2.1.3 PARSOL: for "partly solid" cells

In one version of this technique, known as PARSOL [58], the grid was everywhere of the structured Cartesian or cylindrical-polar configuration *except* where cell edges were intersected by the surfaces of solid bodies. Cells having such intersected edges were then split into two parts, one within the body and one outside it. Moreover, this was performed automatically by the computer code; so the bothersome-to-users task of creating an unstructured grid ceased to exist.

Figure 1.1 shows an early example of the application of this technique to the flow of air through a louvered wall. Of course, the Cartesian grid had to

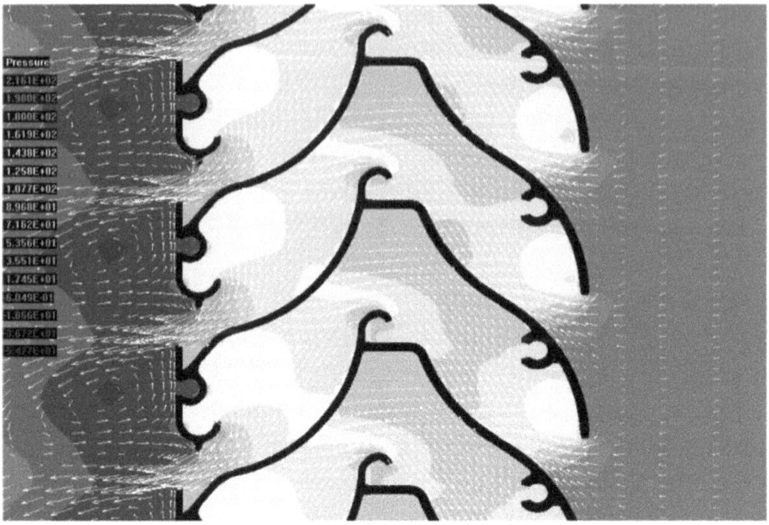

Figure 1.1 PARSOL applied to a louver. (For color version of this figure, the reader is referred to the online version of this chapter.)

be fine enough so that no cell had *two* nonsolid parts in it; but it could not be so fine as properly to represent the boundary layers on the louver surfaces.

2.1.4 Space-averaged rather than detailed-geometry CFD

Early in the present author's career, he had to apply CFD to practical problems in which the detailed geometry of the equipment in question was too intricate to be fitted by any grid that had a cell number small enough for the then-available computers to handle. Specifically, it was necessary to be able, at least to some extent, to simulate the flow of mixtures of steam and water through spaces between the hot water-containing tubes within the shells of nuclear steam generators [1].

Both body-fitted and cut-cell grids were out of the question, because the dimensions of the largest possible grids greatly exceeded tube diameters. Therefore, the heat transfer and frictional interactions between the fluid mixture and the tube bundle were represented via "space averaging." This entailed postulating that coefficients having per-unit-volume dimensions would be able sufficiently to represent the interactions quantitatively; and their local magnitudes were either guessed or computed from believed-to-be plausible formulas. From them were computed the magnitudes per unit shell volume of the heat sinks within the tube-side water and the heat sources in the shell-side mixture.

This was one of the first of the "tricks" alluded to in the title of this chapter. It was accepted as the best that could be done; and it helped steam-generator designers to reduce the flow-induced damage that was then limiting the life spans of their equipment.

2.1.5 IBM: the immersed boundary method

Tube bundles were treated by space-averaged CFD as being "immersed" in the fluids within and outside them. It is interesting therefore that what is called the "immersed boundary method" is becoming popular [2] as a means of avoiding the unstructured grid-creation difficulty. The essential idea is similar to that of space-averaged CFD. It adds such sources or sinks to the finite-volume momentum equations as will reduce to zero the velocities at locations *within* the solid and such as will also ensure that the velocity components at points *just outside* the solid produce vectors parallel to its surface.

As with PARSOL, the grid must be fine enough, when the solid body is thin, for the grid nodes to represent its shape adequately; and the magnitudes of the sources can be computed with various degrees of sophistication. However, the simplicity of the method is such that former enthusiasts for the polygonal cell shape policy appear to be transferring their affections.

Although only now becoming fashionable, its acknowledged roots are old [59]. Figure 1.2 shows a 1995 application to the simulation of air flow within a football stadium [3].

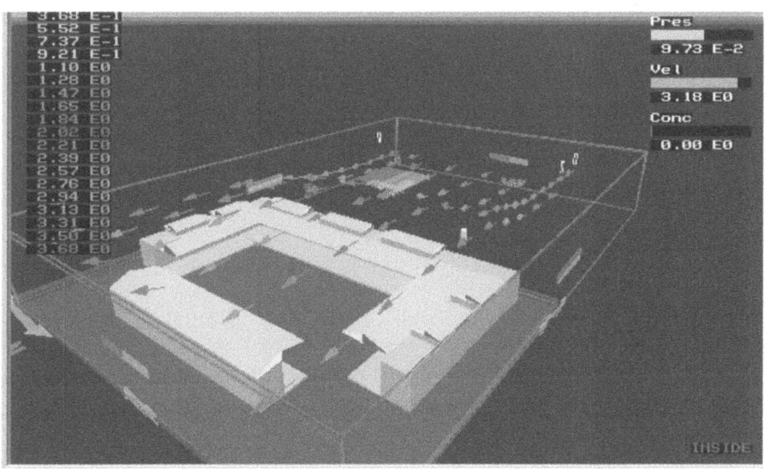

Figure 1.2 Early example of immersed boundary technique. (For color version of this figure, the reader is referred to the online version of this chapter.)

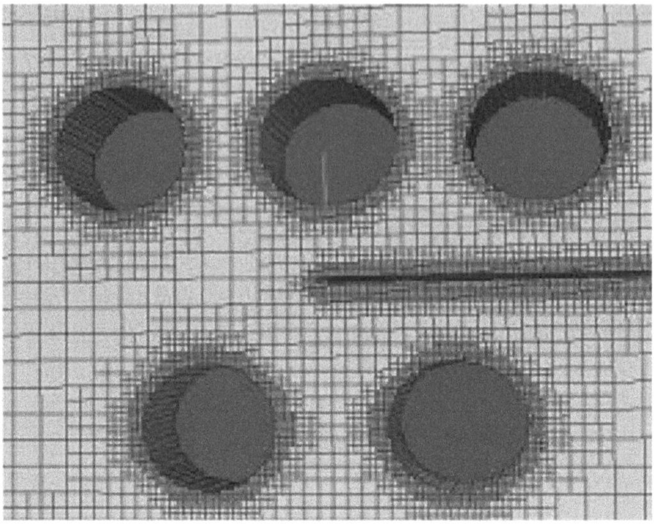

Figure 1.3 Divided Cartesian grid. (For color version of this figure, the reader is referred to the online version of this chapter.)

2.1.6 *Divided Cartesian grids*

It must be admitted that the polygonal cell shape policy does allow the grid to be fine only close to solid surfaces while remaining coarser elsewhere. For this reason, some CFD-code vendors have adopted a compromise solution of the kind illustrated in Fig. 1.3, in which the larger still-Cartesian cells are successively halved in one or all directions, with the smallest cells being closest to the solid surfaces where they are most needed.

To judge from recent CFD publications [60], such grids are becoming more popular than arbitrary polygonal ones, no doubt because the finite-volume equations are easier to formulate.

2.1.7 *The future*

Body-fitted, cut-cell, immersed boundary, and subdivided grids all have their distinct merits; moreover, they are not incompatible with one another. The present author is therefore working on creating grids that combine all features, seeing in such a combination the best that can be envisaged at the present time. The once ubiquitous arbitrary polygonal grid, however, seems unlikely to retain its popularity.

Some success has been obtained with what has been called the X-cell grid, a simple version of which is shown in Fig. 1.4. An interesting feature

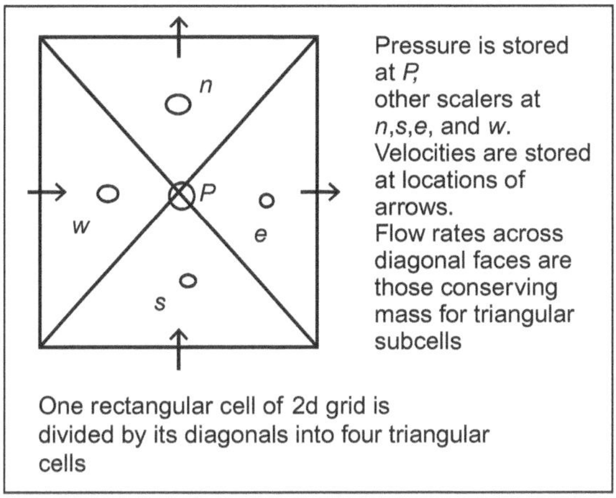

Figure 1.4 The X-cell grid.

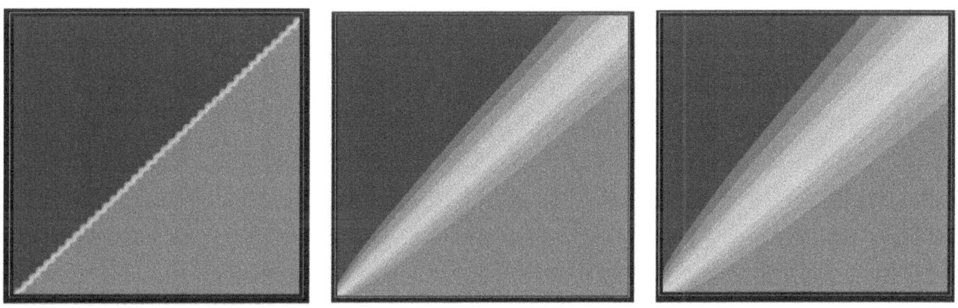

Figure 1.5 Showing the superior numerical-diffusion suppression of X-cells. (For color version of this figure, the reader is referred to the online version of this chapter.)

of this grid type is that the number of control volumes provided for scalar variables such as temperature is four times the numbers of control volumes for mass and momentum conservation. This is not unreasonable because distributions of pressure within fluids, which are deduced from the latter equations, are commonly much more smooth than those of temperature.

More important however than the increase in number is their difference in *shape*; whereas rectangular cells are free from numerical diffusion only when the fluid flows *vertically* or horizontally, X-cell grids are free from it for *diagonally directed* flow. This is illustrated in the three contour diagrams of Fig. 1.5. All of these represent the predicted temperature distribution in a two-dimensional (2D) equal-sided domain, into which colder (blue) fluid flows from the left and hotter (red) fluid flows, with equal absolute velocity, from below. The grid is uniform with 40 rows and 40 columns in diagrams (a) and (b), but it has 80 rows and 80 columns in diagram (c).

The cells of (a) are divided in the X-cell manner; and, as a consequence, the temperature-discontinuity boundary between the two streams remains perfectly sharp. Those of (b) and (c) are not so divided; therefore, the numerical diffusion associated with the conventional upwind differencing causes the interface to become blurred. The blurring is less for case (c), which has the same number of control volumes as case (a); but it is still severe. It is the *triangular shape* of the extra control volumes of X-cell that makes the difference, not their number.

Figure 1.5 admittedly shows X-cell at its spectacularly successful best, because the flow direction is aligned with one of the diagonals. But X-cell is better than the conventional grid of the same number of control volumes *whatever* the flow direction.

Some work has been done on a more advanced version of X-cell in which the velocity components are stored at the same locations as the scalar variables. This gives the grid a so-called collocated character, which has the advantage that the convective contributions to the internode coefficients are the same for both all dependent variables. But there is another advantage too: The pressures are *not* stored at the same location; therefore, the "checkerboard problem" associated with the usual collocated-grid arrangement does not arise!

Attractive though it is, lack of publicity has left this possibility scarcely explored. All that can be reported is that the present author with S. Zhubrin [Ref. 57], several years ago, compared the results of such X-cell-based calculations with results obtained with a body-fitted coordinate grid having an equal number of cells. The flow was the 2D steady laminar flow around and in the wake of a cylinder positioned at right angles to the stream. A comparison of the numerical predictions of the nondimensionalized length of the downstream recirculation zone with the experimental value is conveyed in Table 1.1.

While insufficient in number to be conclusive, these comparisons suggest that X-cell is greatly superior when the grid is coarse.

Table 1.1 Comparison of Numerical Predictions and Experimental Data for Wake Length

NX*NY	Length 1; X-cell	Length 2; BFC	Length 3; exprm
27*13	2.3	1.15	2.75
36*13	2.6	1.25	2.75
60*30	2.8	2.8	2.75

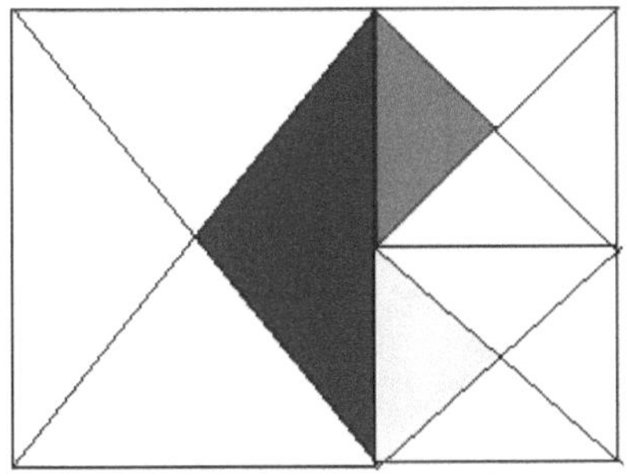

Figure 1.6 X-cell subdivision combined with Cartesian subdivision. (For color version of this figure, the reader is referred to the online version of this chapter.)

The purpose of this chapter is more to point out possibilities than to enumerate certainties. It is therefore appropriate to remark that the X-cell idea is applicable to unstructured Cartesian grids and to structured ones. Figure 1.6 explains clearly enough.

Of course, there is no need to use X-cell subdivision everywhere. For reasons of economy, it would make sense to use it only, so as to preserve realism, where numerical diffusion would otherwise outweigh physical diffusion.

The subject of computational grids will be returned to under "Tricks" in Section 3.3.

2.2. Linear equation solver trends

2.2.1 Point-by-point (i.e., PBP) relaxation methods

Although the presence of convection terms in the finite-volume equations of CFD renders their totality *non*linear, it is common practice to proceed by way of solving a series of equations for a *single dependent* variable. These equations are treated as being temporarily *linear*, by updating their coefficients in what are called "outer iterations," only after all such variables have been attended to.

Parenthetically, it may be remarked that whether this is wise is still doubted by those who remember the SIVA (i.e., simultaneous variable adjustment) method of the early 1970s [4]. It was doing well before it was swept aside by the incursion of SIMPLE (i.e., semi-implicit method for pressure-linked equations) [5]; and surely, SIVA could have been further improved. But decade-long eclipses like this are frequent consequences of science's fashion-following tendency.

The form of the linear equations to be solved is

$$a_{ii}x_i + \sum_j a_{ij}x_j = b_\mathrm{I}$$

where x_i is the value of a dependent variable at node i, x_j is its value at the neighboring nodes j, a_{ii}, and a_{ij} are constant coefficients, and b_I is a source term.

These equations may easily be solved, point-by-point, by updating each x_i in turn, while the x_j's on the right-hand side are treated temporarily as known values, until, that is, they are updated when their turn comes round. Then, as soon as the last value has been updated, a new cycle of updates can be undertaken, so as to determine what adjustments of the first-made updates must be made to accord with the later-made ones.

The updating process just described is often (but not always) called "relaxation." That term will be used in the succeeding text.

The two most common PBP relaxation methods are

- Gauss–Seidel that uses the updated $x_{I'}$'s as soon as they are available and
- Jacobi that delays the updating until each node has been visited once.

The former converges toward the solution more rapidly; but both require many repeated relaxations; and their number increases in proportion to the square or higher power of the number of nodes in the grid.

Therefore, if used in practical calculations, PBP methods must be improved by the application of convergence-*accelerating* devices, of which more will be said later.

2.2.2 General remarks about linear-equation solvers

There are many highly impressive textbooks [6, 7] devoted to the available methods of solving linear algebraic equations. Their authors know vastly more about the subject than does the present one, whose experience nevertheless has highlighted factors that the textbooks fail to emphasize, as follows:

- The *merits of a solver* are to be measured primarily by the brevity of the *computer time* in which it needs to produce a set of x_i *values* that differ by less than a user-assigned tolerance from those that are ultimately found to satisfy all the equations exactly.
- A counterbalancing *demerit* may be (depending on the resources available) the magnitude of the *computer memory* that it requires.
- The *relative* merits of one solver to another depend enormously on the ratios of a_{ij}'s to a_{ii} and to one another.
- They depend very greatly also on the distribution in space of the values of

$$x_{\mathrm{Initial\_guess}} \doteq x_{\mathrm{exact\_solution}}.$$

It would be unnecessary to make such statements were the literature not full of confident *unqualified* assertions regarding the superiority of one method to another. Nor had the author of one highly regarded 500-page textbook provided more than a mere five rather simple cases as tests for the comparison of the methods so learnedly described.

Moreover, the merits of methods are frequently assessed by reference to the magnitudes of the remaining *residuals*, that is, the magnitudes of e_I, defined by

$$e_I = a_{ii}x_i + \sum_j a_{ij}x_j - b_I$$

rather than in terms of the physically meaningful

$$x_{\text{computed}} - x_{\text{exact\_solution}}.$$

It needs to be emphasized that residuals are imperfect measures of the quality of a solution. For example, a particular coefficient a_{ii} may be very large, as occurs when, so as to express one of the boundary conditions, one of the unknowns is being fixed by inclusion of a source term defined as

$$b_i = a_{ii}(x_{\text{fixed}} - x_{ii})$$

wherein x_{fixed} *is the desired value*. Then, even when x_{ii} differs from x_{fixed} by no more than round-off error, the product $a_{ii}(x_{\text{fixed}} - x_{ii})$ can appear as a residual of large size. It is the x_{ii} values that need to be considered, not residuals, and if the absolute value of $x_{ii} - x_{\text{fixed}}$ is less than the tolerance, that is, good enough.

It may be appropriate to voice some further observations concerning the linear equation solver literature at this point, namely, the following:

- It uses a known-only-to-devotees nomenclature, with no deference at all to the solver-*using* community.
- This nomenclature consists largely of surnames of authors: "Krylov subspace," "Lanczos and Arnoldi iterations," "Ritz approximation," "Hessenberg form," "Householder matrix," etc. The "tridiagonal matrix algorithm" (see the succeeding text) is one of the few having memory-assisting significance.
- Although it is recognized that the equation sets frequently arise from the discretization of the differential equations of physics, the physical significances of their solutions are never disclosed.

2.2.3 The Thomas (or tridiagonal matrix) algorithm (i.e., TDMA)

Great importance attaches to the sets of equations that arise when the grid consists of a single chain of interlinked nodes, to which corresponds the reduced equation set:

$$a_i x_i + a_{I-1} x_{I-1} + a_{I+1} x_{I+1} = b_I$$

The reason is that there exists the well-known Thomas algorithm for solving the equations in question exactly *without iteration*. There is no need here to set out the details. It suffices to state that a finite number of operations, proportional to the number of unknowns, proceed from one end of the chain to the other and then back again. At the end of the sequence, all values of x_i are determined.

Of course, grids consisting of a single chain of nodes are rare; but the TDMA can be employed for two- and three-dimensional (3D) grids as well, albeit in an iterative manner. Consideration of how the TDMA then behaves will now be used to explain the influence of coefficient ratios on solver performance noted in Section 2.2.2 in the foregoing. A *2D* example suffices for which the typical equation can be written as

$$a_{I,j} x_{I,j} + a_{I-1,j} x_{I-1,j} + a_{I+1,j} x_{I+1,j} + a_{I-j-1} x_{I,j-1} + a_{I,j+1} x_{I,j+1} = b_{I.j}$$

wherein the subscripts containing i and j indicate node locations in the two coordinate directions.

It has been stated in the aforementioned that the PBP procedures treat the values on the right-hand side of their equations as temporarily known, which enables the left-hand side values to be updated. When the TDMA is applied to 2D problems, *one-half* of the right-hand side values are assumed to be known, that is, those in the second line of the equation; then, *all* the values in the first line can be determined.

Now, the importance of the coefficient ratios can be recognized; if the coefficients $a_{I,j-1}$ and $a_{I,j+1}$ are much smaller than the other a's, the presumption that $x_{I-,j-1}$ and $x_{I,j+1}$ retained their previous iteration values is of no importance whatsoever. This could happen if the domain were very much larger in the *j*-direction than in the *Ii*-direction. The exact solution of the *2D* problem would then be obtained without iteration.

What if the $a_{I,j-1}$ and $a_{I,j+1}$ are much *larger* than the other *a*'s? Then, the changes effected by the TDMA will be small, so that the process would have to be iterated many times to attain convergence. The use of an accelerating procedure would be very desirable.

2.2.4 Acceleration by overrelaxation

When the series of values of $x_{I,j}$, which are produced by a sequence of relaxations, is examined, it is usually found that the values are changing, iteration-by-iteration, in the right direction, but too slowly. It is tempting to "*over*relax," as it is often called, that is, to multiply the increments by some factor greater than unity. But how big should it be?

An obvious answer, which the present author happens *not* to have found in any textbook, is to compute the optimal factor by the following procedure:

- For the sets of $x_{I,j}$'s, both before and after relaxation, compute the residuals e_I.
- Apply increments of *twice the size* resulting from the relaxation and calculate the new residuals, that is, adopt tentatively an overrelaxation factor of 2.0.
- For each of the three sets of the residuals, compute the sums of their squares.
- Assume that the sums of the squares of the residuals vary in a quadratic manner with the relaxation factor (as they must). Hence, deduce what relaxation factor will produce the *minimum* sum.
- Adopt the new $x_{I,j}$'s that correspond to that factor and then make a new relaxation step.

This procedure always works, sometimes spectacularly; and it has been observed that it works better when the sums of the squares of e_I/a_{ii} are minimized rather than those of e_I itself. No claim is being made that it is better than others advocated in the rather large literature concerned with choosing optimal relaxation factors; but, about that literature, it should be remarked that no other front-runner has appeared.

The reason is that *mere* overrelaxation is not enough, for it applies somewhat better corrections at the locations to which less adequate corrections have *already* been applied but *only* to those locations. It takes too narrow a view of what needs to be done. This can be understood by consideration of Fig. 1.7, which illustrates what happens when the Jacobi PBP relaxation is employed.

The problem is that of one-dimensional (1D) heat conduction in a slab of uniform conductivity material, with its faces held at zero temperatures. The *initial guess* is represented by the two upper sloping lines; and the correct solution of the equations is the base of the triangle of which those lines are sides. It is easy to recognize that temperature corrections are needed everywhere; but a finite residual, that is, a heat imbalance, exists *only* at the location corresponding to the apex of the triangle; so, it is only there that the Jacobi relaxation makes any change.

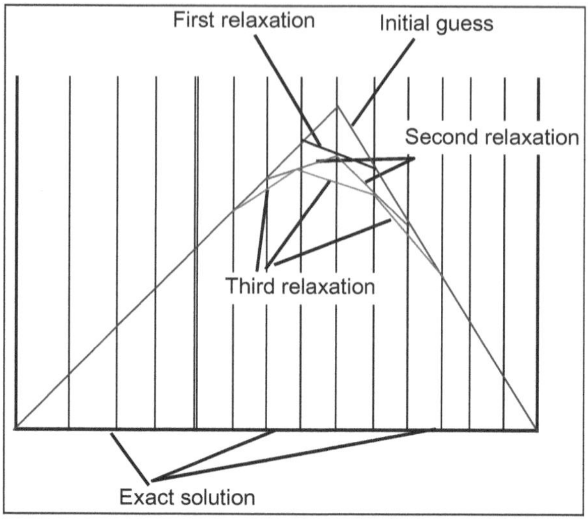

Figure 1.7 Graphic representation of a Jacobi solution process. (For color version of this figure, the reader is referred to the online version of this chapter.)

A Jacobi relaxation, expressed graphically, is a line drawn between nearby pairs of points across the spaces between them. Figure 1.7 shows the sets of lines for the first three relaxations. They have resulted, it might be said, in a "rounding" of the sharp-pointed initial triangle; but the third relaxation curve is still far from the final destination, namely, the zero temperature base.

Overrelaxation merely increases somewhat the cautious adjustments of Jacobi because of its self-imposed restriction of attention to points that currently report errors. What is needed is an acceleration procedure with a wider vision.

The same is true of much more sophisticated relaxation techniques such as Stone's "strongly implicit procedure" [8]. Coupled with TDMA-based initial-guess improvers, such solver systems have for years provided satisfaction without any overrelaxation at all.

However, for whatever reasons, the attention of the linear equation solver specialist shifted long ago from overrelaxation toward acceleration methods of a different kind, now to be discussed.

2.2.5 Conjugate gradient solvers

In 1952, Hestenes and Stiefel [8] introduced the "conjugate gradient" method, which became for a time the leader of fashion. Its strategy was

1. to start with a guessed set of x_i values,
2. to calculate the associated residual e_I,
3. to try another set of x_i's and calculate *their* residuals,
4. to deduce from the two sets of $x_i \sim e_I$ pairs what would be a next try x_i that would probably reduce the sum of the squares of the residuals,
5. then to continue doing this until the sums became small enough.

The method (with various versions of step 5) appears to have had some success; but it was discovered, not surprisingly, that step 1 was the weak point. If one started a long way away from the final destination, the journey lasted appreciably longer.

2.2.6 Preconditioned conjugate gradient solvers

It was therefore decided by someone (who is unclear) (Question to readers #2: can anyone tell me, please?) to start by using a relaxation technique to improve the "initial guess." No great attention was at first paid to which technique should be used; therefore, it was called, rather demeaningly, the *preconditioner*; as though it was something to be used at the start and then discarded.

But it was *not* in fact discarded; it was used again and again after each conjugate gradient "improvement"; and it was at last discovered that some preconditioners were much better than others. How much better, and in what circumstances, is hard to discover from the literature. Therefore, the present author, with Alexey Ginevsky of the Moscow Power Engineering Institute, is working to create a software package for studying the matter, as it might be said, experimentally.

Some preliminary results will now be shown. They relate to a 2D conduction problem in which temperatures at the boundaries of a square domain of uniform conductivity are held at temperature 0.0, while the initially guessed temperature at all other points is 1.0. The grid is a uniform 128 * 128. The following images show temperature contours after each of 10 relaxer-plus-improver iterations, for six different solvers of conjugate gradient type (Fig. 1.8).

Clearly, there are very great differences in convergence behavior. The speed of convergence is successively greater for the first five solvers; but the sixth solver is probably not converging at all, because it is producing values of temperature that lie outside the range zero to one.

The names of the six solvers will not be disclosed here because the investigation, although still in its early stages, has already shown that the relative merits of the solvers are very dependent on the physical problem in question.

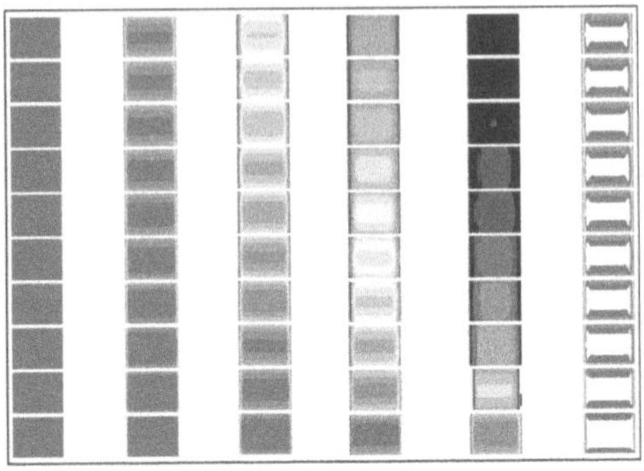

Figure 1.8 Temperature contours produced by six different solvers. First iteration at bottom, 10th at top. Colors mean: red $T=1$; blue $T=0$; white $T>1$ or <0. (For interpretation of the references to color in this figure legend, the reader is referred to the online version of this chapter.)

The purpose of presenting these preliminary results is simply to suggest that contributions to knowledge in this field can be made even by those who are not quite sure what a Krylov subspace actually is. If readers of this chapter wish to take part in the investigation, the software package (called Solvers Simulation Scenario) can be supplied to them.

2.3. Turbulence model trends

2.3.1 Origins

When the present author began his CFD career in the 1960s, the "turbulence model" concept perhaps did not exist, although examples did. Indeed, it may be that it was the publication of the book coauthored with Launder [9] that popularized it. For this reason, the author wishes to state that in his view, the pioneering publications were those of Boussinesq [10] in 1877, Prandtl [11] in 1925, Kolmogorov [12] in 1942, and Prandtl again [13] in 1945. All of these made their contributions before the advent of digital computers. After its advent, the first (and independent) pioneer was Harlow [14].

2.3.2 The effective-viscosity hypothesis

2.3.2.1 Early days

The contribution of Boussinesq was to *guess* that turbulent fluids were similar to laminar ones but possessed a much greater than laminar "effective" viscosity. It was a guess, not indeed very plausible to those who observed turbulent plumes of smoke in the environment; but it proved to be overwhelmingly seductive. If one could only find out what was the effective viscosity at any location, the world of fluid dynamics would indeed be conquered!

Boussinesq's guess was a "try-on"; but it is now sufficiently venerable to be called a "hypothesis" and even a "model." The other just-mentioned pioneers all adopted it; indeed, it can be said that the purpose of their endeavors was to devise a means for predicting what value should be ascribed to the effective viscosity at any point within a turbulent flow.

2.3.2.2 The mixing length hypothesis

The laminar viscosity of a gas is known to be proportional to the mean free path of the molecules and to their average velocity; and it was by analogy with the mean free path that Prandtl [11] in 1925 introduced the concept of the "turbulent *mixing length*." He argued that momentum was transferred between adjoining layers of fluid, having different mainstream-direction velocities, by "parcels" of fluid that traveled a certain distance in a direction normal to that of the mainstream before mingling with and so transmitting momentum to the neighboring fluid.

Whatever the verbal argument employed, the result was the following formula for the effective (kinematic) viscosity:

$$\nu_{\text{eff}} = l_{\text{m}}^2 |du/dy| \tag{1.1}$$

where l_{m} is the mixing length and $|du/dy|$ is the absolute value of the local mainstream velocity gradient. This of course simply transfers the search for one unknown, ν_{eff}, into the search for another, l_{m}, but Prandtl made two proposals with regard to the latter, namely, that

- close to a wall l_{m} was proportional to the *distance from* the wall, y, say, and
- in limited extent turbulent flows remote from walls, such as jets, wakes, and plumes, l_{m} was proportional to the *distance across* the turbulent region.

Values of the constants had to be deduced from experimental data, which was somewhat tiresome; but it could be tolerated if their variation from one situation to another proved not to be too great.

2.3.2.3 Two-equation turbulence models

In the early days of CFD (late 1960s), the mixing length hypothesis was gratefully seized upon, for example, by the many users of the GENMIX computer code [15]; and it proved to be nearly adequate for predicting the so-called "parabolic" flows within pipes and diffusers and in jets, wakes, and plumes. Even there, however, it was not entirely satisfactory, for example, the ratio of l_{m} to jet width proved to be 0.103 for plane jets and 0.075 for axisymmetric ones. The search, therefore, began for some way in which the length scale or some other entity leading to ν_{eff}, could be deduced.

The crucial advance appears to have been that of Kolmogorov [12], in 1942: he proposed that certain time-mean properties of turbulent flows could be deduced from the solution of *differential equations*. The dependent variables of these equations that he used were the

- *kinetic energy* of the fluctuating motion per unit mass, say k, and
- mass-average *frequency* of that motion, say f.

The terms in the equations were those to be found in all of conservation type, namely, time dependence, convection, diffusion, volumetric source, and volumetric sink.

Once k and f were known, Kolmogorov argued the effective viscosity could be deduced from

$$\nu_{\text{eff}} = Ck/f \tag{1.2}$$

where C would be a constant deducible from experimental data.

The work of Kolmogorov did not, in fact, become known to the scientific world until several years later, by which time a different two-equation model had become popular, namely, that of Harlow and Nakayama [14]. Its two dependent variables were the kinetic energy of the fluctuating motion, k, and the volumetric dissipation rate of that energy, ε; and from these, the effective viscosity was to be deduced from

$$\nu_{\text{eff}} = Ck^{3/2}/\varepsilon \tag{1.3}$$

Both Kolmogorov's and Harlow's ideas were hopeful guesses, that is, "try-ons"; but it can be said that the latter's were rather more hopeful than the former's, because Kolmogorov proposed no positive source term for f, whereas Harlow proposed that the rate of generation of ε per unit volume and time should be proportional to k/ε times the corresponding rate of generation of k, that is, to the rate of energy generation by tangential and direct stresses.

The $k-\varepsilon$ hypothesis quickly and deservedly, because of its usefulness, acquired the more honorific appellation "model," to which indeed "classical" soon began to be added. Nevertheless, it is not disrespectful to raise an eyebrow concerning both Harlow's and Kolmogorov's choices of second dependent variable.

Their choice of turbulence energy k as a conserved property is understandable, for is not the first law of thermodynamics an energy conservation law? Indeed, Prandtl himself also proposed [13] a one differential equation model in which the dependent variable was k. But conservation of f or ε? Is it reasonable to suppose that two bodies of turbulent fluid, rapidly mixed together, will dissipate energy at the arithmetic mean of the two individual dissipation rates? Physics knows of no such law.

Interestingly, both Saffman [16] and the present author [17], independently of each other and of Harlow and Kolmogorov, proposed two-equation models with k as the dependent variable of the first equation; but as that of the second equation, they chose W, the sum of the squares of the *vorticity fluctuations*. That quantity, both authors believed, could reasonably be regarded as conserved: for vorticity *is* a conserved quantity.

Arguments were later advanced, and at the time thought persuasive, for preferring the Harlow–Nakayama choice. Years later, however, the same model was independently reinvented by Wilcox [18]; and it is now believed by some to be superior.

The history of turbulence modeling, viewed as a whole, can be seen as an equal-measure mixture of insight and accident. It is, therefore, interesting to speculate as to what might have transpired if Prandtl had happened to decide to make the mixing length l_m itself the dependent variable of his single differential equation. What other terms might he have had invented in order more or less to fit the known experimental data? A speculative answer is supplied in "Try-ons" in Section 4.1.

2.3.2.4 Wall functions

In the early days of CFD, computers had very limited memory; therefore, fine grids could not be afforded. Even for the small 2D parabolic problems dealt with at the start [19], it was recognized that calculating the effective viscosity appropriate to cells close to solid walls presented special problems. The topic appears as "wall-flux relationships" in the index of the foregoing reference. In the book on "elliptic flows" [20], which appeared 1 year later, the entry is "wall functions," which is how it is usually referred to nowadays.

The wall functions used at that time were based upon experimental measurements rendered applicable to more general circumstances by expression in terms of dimensionless quantities. For the simplest possible circumstances, in which the relevant fluid properties (density and laminar viscosity) are constants, as is also the shear stress because of the absence of mass transfer and of pressure gradient, the effective viscosity in the near-wall layer can be expressed as

$$\nu_{\text{eff},n_w}/\nu_{\text{laminar}} = y^+/u^+ \tag{1.4}$$

where y^+ and u^+ are the dimensionless distance from the wall and along-wall velocity in the "law of the wall," for which various empirically based expressions are available. The one that the present author particularly favors, because it covers the whole range in a single formula, [21] is

$$y^+ = u^+ + \left[e^{\kappa u+} - 1 - \kappa u + -(\kappa u+)^2/2 - (\kappa u+)^3/6 - (\kappa u+)^4/24\right]/E \quad (1.5)$$

where $\kappa = 0.417$ and $E = 0.86$.

Some empirically based formulas exist that express the influence of mass transfer through the boundary layer in reducing the effective viscosity, if the flow is *into* the fluid, and increasing it, if the flow is toward the wall, and also of the influences of pressure gradient and of roughness, but not when mass transfer and pressure gradient are *simultaneously* present. Practically nothing is known about the influences of nonunity viscosity and density.

Nearly half a century later, although the increase in computer power has been immense, wall functions are still in use. And the reason is still the same: even if computer power *does* suffice to allow use of arbitrarily fine grids close to walls, knowledge of the physics of low Reynolds number does not have the certainty or generality to make the expenditure worthwhile.

The point will not be expanded upon here; but it will be returned to later, first when "direct numerical simulation" is discussed and later in relation to "urban-terrain" simulation.

2.3.3 Reynolds stress models

Experience of using two-equation models was mixed. Engineers were pleased to have been supplied with computer programs that at least purported to handle turbulent flows at high Reynolds numbers. But the hopes of engineering scientists that a model would be found that fitted a wide range of phenomena, with very little ad hoc "tweaking" of constants, were disappointed. So the question arose: if two equations do not suffice, why not try using more?

Evidence had accumulated that one implication of the effective-viscosity hypothesis was not always correct: The shear stress and the velocity gradient did not always become zero at the same point; and why, anyway, should it be supposed that the effective viscosity, even if it did exist, would be direction-independent? These thoughts gave rise to the notion that the stresses themselves should be the dependent variables of differential equations; and this meant that, since the stress tensor has many components, *many equations* must be solved.

Here, it should be mentioned that the world of engineering science had changed immensely in respect to the number, and indeed nature, of the persons participating in it. Professors, students, theses, journals, and publications all proliferated, and so did the number and power of computers. Therefore, novelties in turbulence modeling, provided that they were not *too mind-testingly* novel, found many enthusiasts; and Reynolds stress models were prominent among them.

The present author has not studied the literature in detail; but to judge from the proceedings of the 2010 8th International Symposium on Engineering Turbulence Modeling and Measurement [22], it cannot be said that Reynolds stress models are as popular as they once were. Other "buzzwords" are at least as prominent in the titles and abstracts of the papers.

2.3.4 DNS

Among the said buzzwords is DNS, standing for direct numerical simulation, that is to say the solution of the unsteady *laminar* Navier–Stokes equations on a sufficiently fine grid and with sufficiently small time steps, for the details of turbulent-flow fluctuations to be accurately simulated. It is not a new line of study having been initiated in the 1970s (see Ref. [23] for a review of early work), and since computers have increased in power by many orders of magnitude since then, it is reasonable to expect that at least *some* useful results would have emerged.

What results could these be? Those referred to in Section 2.3.2.4 in the preceding text as being, regrettably, absent, namely, formulas for the effective viscosity of near-wall regions, as affected by pressure gradient, mass transfer, and nonuniform properties. It would not matter if these formulas were represented as provisional, because grid independence had not yet been demonstrated. At least the *trends* would be instructive. Yet, there is nothing so far as the present author can discover. One could suppose that the practitioners of DNS studies are not greatly interested in the production of results that others are waiting to use.

2.3.5 Large eddy simulation

Large eddy simulation (LES) is an even more prevalent buzzword than DNS. The underlying idea is that DNS requires finer grids and shorter time steps than even the largest of today's computers can accommodate. Very good! Let us use the finest grids and shortest time steps that we can afford; and for the finer-scale and more rapidly fluctuating phenomena, let us use conventional RANS (i.e., Reynolds-averaged Navier–Stokes, even the now-by-some-despised $k - \varepsilon$) modeling. Sometimes "hybrid" methods are employed, whereby the whole space is divided into regions, with RANS used in one and LES in another, the choice being made by the analyst so as to maximize the accuracy/effort ratio.

It is a rational strategy to embark upon, tentatively; and there have been many proofs that it enables realistic simulations to be made of flows, such as those behind a bluff body, which exhibit unsteady wake structures fluctuating periodically from side to side. The present author being an observer from a distance of this field of research makes only two remarks, namely, the following:

- It is evident from comprehensive publications such as Refs. [22, 24] that many variants of LES are being tried without the emergence of one that is widely acknowledged as superior to all.
- No attention appears to be being paid to the easy-to-activate ability of LES to produce the pdf (i.e., probability density function) information needed by heat transfer engineers for deducing volumetric averages of the nonlinear sources characterizing thermal radiation, for example, Ref. [24], admittedly confined to hydraulics applications, does not even include "pdf" in its index.

Thermal radiation is here mentioned because it is an example of a practically important physical process that, from the mathematical viewpoint, is *nonlinear*; for emission, it is proportional to the *fourth power* of the absolute temperature. This entails that a technique such as RANS, which can produce *only* time-averaged temperatures, cannot predict radiation well. LES, on the other hand, because it can compute for what proportion of time the temperature is by various amounts above and below the time-averaged temperature, can calculate the time average of the fourth power of the temperature. Reference [25] contains a full discussion of the possibility here alluded to. Its main message is that the *average* attributes of a population are rarely as interesting as the *departures from the mean*. It is only the strongest gusts of wind that fell the trees.

2.3.6 Population-based models

2.3.6.1 The main idea

That a turbulent fluid is to be regarded as a population is not a new idea. Prandtl, when introducing his mixing length hypothesis in 1925, clearly envisioned "parcels" of fluid having different velocities and directions of motion, which jostled with one another like members of an unruly crowd. The increasing prevalence of the notion in the nine succeeding decades can truly be recognized as constituting a "trend."

Members of human populations can be distinguished by reference to many different attributes: Sex, religion, income, height, and corpulence may all serve; and they have differing importance according to whether one is seeking a wife, a basketball champion, or a sumo wrestler.

The same is true of turbulent fluids. Velocity is the attribute most pertinent to momentum transfer: temperature to thermal radiation, particle density to sifting processes, and chemical composition to combustion. Indeed, it is investigators of chemical reactions who have done most to foster the trend, as the following account will reveal. This account presents the highlights of a much longer review of the subject written in 2010 [26], together with those of some more recent material [27, 28].

Two important characteristics of populations, from the theoretical point of view, are *dimensionality* and *number* of members. If the only attended-to distinguishing feature of a group of men was their height, they would constitute a *1D* population. If it were meaningful to pay attention also to their *weight*, their population would be classed as *2D*; and if the men were sorted additionally in respect to *age*, their dimensionality would be *three*. And so on without limit.

In this chapter, for the sake of concreteness, attention will be confined to *1D* and *2D* populations; and the fluid considered will be a turbulent mixture of gaseous fuel, air, and their products of combustion. As to the number of members, it will be found that valuable information can sometimes be provided by considering populations with as few as *two* members. However, the richness of information increases, as a rule, with the number of members considered, just as is true of the fineness with which the dimensions of the distance and time are discretized in conventional CFD.

2.3.6.2 Graphic representations

Population distributions can be conveniently displayed graphically, whereby it is to be noted that they have their idiosyncrasies. Thus, the single vertical line in Fig. 1.9 represents, by the horizontal position of the single red line of unity height, the local and instantaneous temperature of a single-member population of fluid, that is, of a fluid such as is envisaged by conventional turbulence models.

Figure 1.10, by contrast, represents what might be the same turbulent fluid by means of six lines at arbitrarily selected temperatures, the vertical lengths of which represent proportions of time within which each the fluid is supposed to possess the temperature corresponding to the horizontal position of the line. It is somewhat similar to a pdf, which is a histogram, with

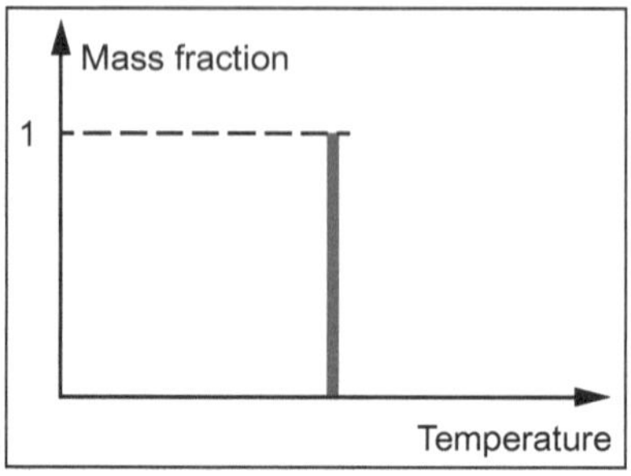

Figure 1.9 One-member. (For color version of this figure, the reader is referred to the online version of this chapter.)

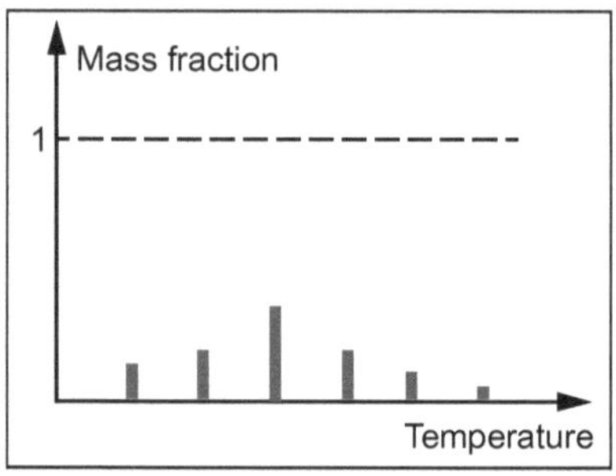

Figure 1.10 Six-member. (For color version of this figure, the reader is referred to the online version of this chapter.)

contiguous vertical strips rather than vertical lines; it differs in that temperature is being thought of as "quantized," in the sense that fluid with between-the-lines temperature is imagined never to exist at all.

Figure 1.10 relates to a *1D* population, for temperature is all that distinguishes one member from another. The strict 2D counterpart would be a plane, having temperature, say, as its abscissa and fuel concentration, say, as it ordinates. Then, the quantized temperature–concentration pairs would appear as dots scattered regularly or irregularly over its surface; and the mass fractions of mixture material associated with each state, corresponding to the lengths of the vertical lines in the 1D diagrams, would have to be represented by the diameters of the dots.

Because of the eyesight strain that such a practice would impose, it is not used. Instead, easier-to-draw-and-read *contour* diagrams will be used, in which practice tends to blur the distinction between population distributions and pdfs that had just been made. However, it is only to the persons writing the corresponding computer programs that the distinction is

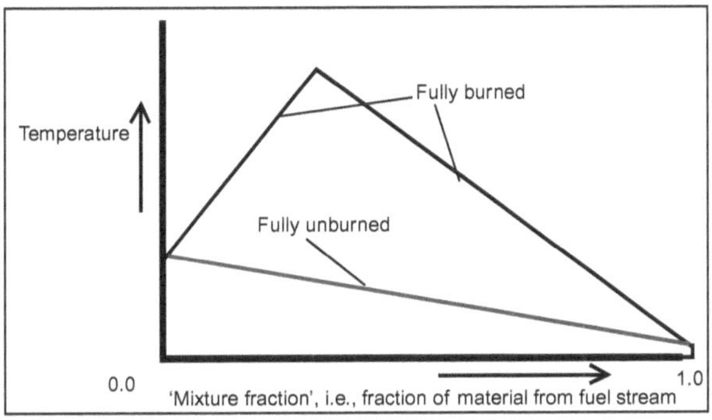

Figure 1.11 Extremes of temperature. (For color version of this figure, the reader is referred to the online version of this chapter.)

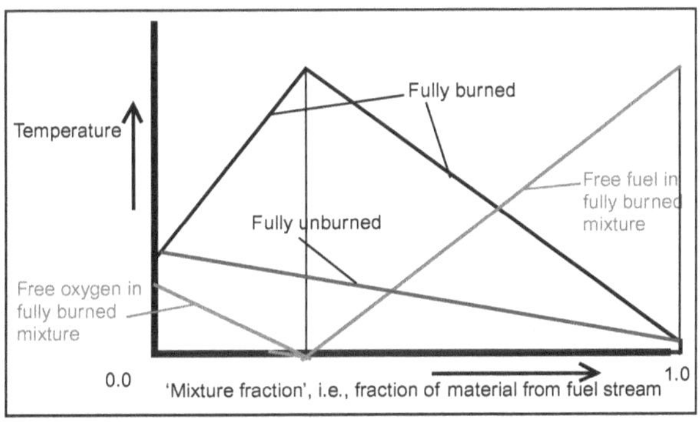

Figure 1.12 Extremes of concentration. (For color version of this figure, the reader is referred to the online version of this chapter.)

important; so it will not be referred to again here. Before the 2D representation is introduced, however, two other diagrams are worth inspecting.

Figure 1.11 shows how the *temperature* of a fuel–air mixture varies with the fuel proportion, when the fuel is (upper, two lines) fully *burned* and (lower, one line) fully *unburned*. The *adiabatic temperature rise* is the vertical distance *between* the upper and lower lines.

Figure 1.12 shows the *free-fuel* and *free-oxygen* values for the fully burned condition. The mixture fraction at which both oxygen and fuel are zero is called "*stoichiometric*."

2.3.6.3 The "TriMix" diagram, a "map" of fuel–air–combustion product states

The "TriMix" diagram, now to be described, is a way of mapping the states that lie *between* the fully burned and fully unburned extremes. Its *vertical dimension* is the *adiabatic* temperature rise resulting from complete combustion of the fuel (to CO_2 and H_2O), hence, the "Tri" in its name. Its *horizontal dimension* is the mass fraction of fuel-derived material or, *in atomic terms*, (1.0 − atomic_nitrogen_fraction/0.768). Points lying outside the triangle correspond to *nonphysical* negative concentrations. They can, therefore, be ignored. The TriMix diagram made its first published appearance in Ref. [27] (Fig. 1.13).

The TriMix diagram will be used as a means of describing the states of turbulent gas mixtures. But, first, its ability to display simple balance-based properties of state will be illustrated by way of five contour diagrams in Fig. 1.14. Any other properties, such as density and viscosity, can also be computed and displayed as well as, if chemical kineticists are to be believed, the rates of chemical reaction.

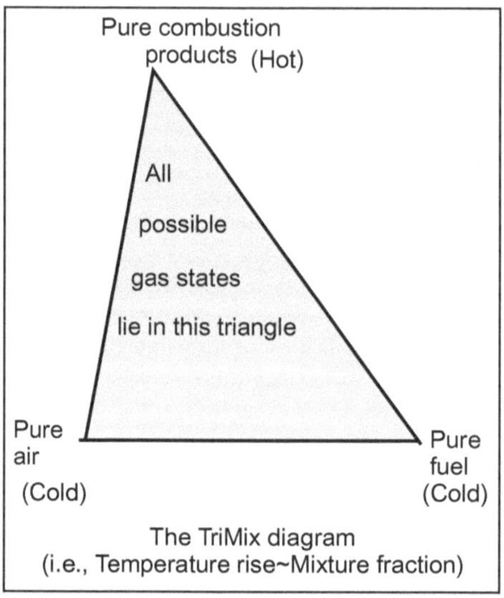

Figure 1.13 The TriMix map. (For color version of this figure, the reader is referred to the online version of this chapter.)

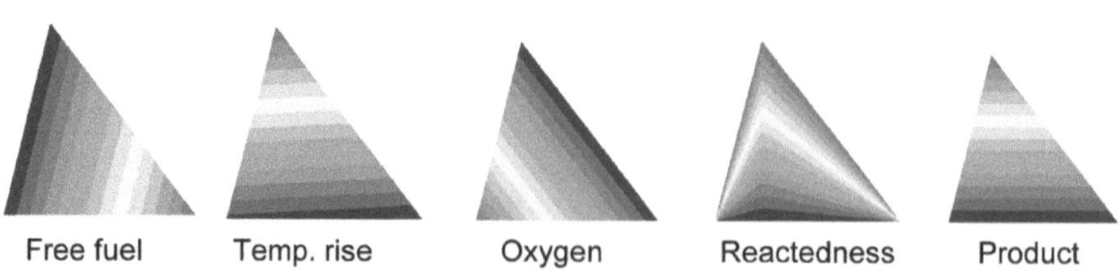

Figure 1.14 Contours displaying gas states on TriMix. (For color version of this figure, the reader is referred to the online version of this chapter.)

There are *three kinds* of reaction to be considered, of which the *rate contours* are shown (red is high rate; blue is low rate) (Fig. 1.15).

Of these reactions, the first is usually desired, being the main producer of energy, whereas the other two are usually undesired, for they produce air pollutants. The designer of a combustor therefore would like to know *where* on the TriMix diagram his hottest gases lie. Preferably, it should be in the stoichiometric-mixture -ratio region; for oxides of nitrogen are generated if the mixture ratio is too lean there and smoke is generated if it is too rich. It should be understood that so far, no particular flame has been considered but knowledge has been assembled about the attributes of *all possible members* of the gases-in-flame population.

2.3.6.4 The combustor-simulation problem

Figure 1.16 shows a TriMix diagram that *could* in principle represent a particular flame or rather a particular location within it, for its contours are those of *the proportions of time* in which the gas at that location is in each of the possible *states represented on the state map*. If he possessed such a diagram for a sufficient number of locations within his combustor, the combustor designer

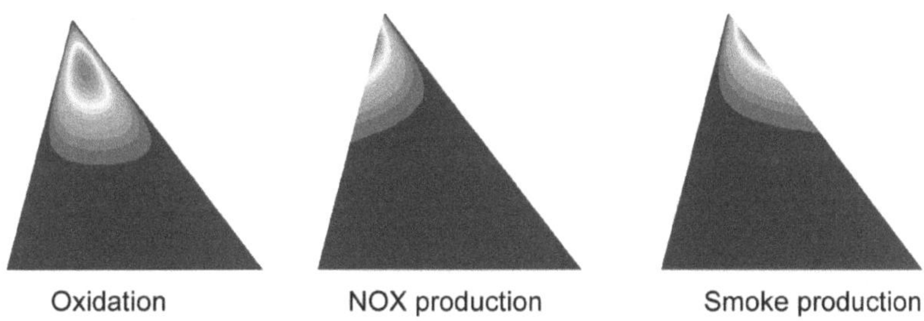

Figure 1.15 Rate contours. (For color version of this figure, the reader is referred to the online version of this chapter.)

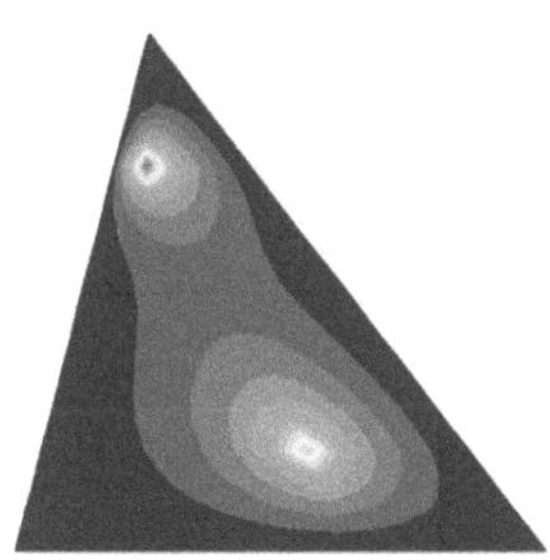

Figure 1.16 2D population distribution. (For color version of this figure, the reader is referred to the online version of this chapter.)

would be able to deduce *total* rates of the three reactions. To provide him with such knowledge is the task of the CFD specialist, who will, of course, need a populational model of turbulence for the purpose.

2.3.6.5 When the turbulent fluctuations are ignored

Very few indeed are the CFD specialists who even know of the existence of such models, let alone use them. It is common still to ignore the fluctuations of concentration entirely, thus in effect presuming that the state of the mixture at a particular location in the combustor is represented by a single point on the TriMix map, as shown in Fig. 1.17.

Two finite-volume equations have to be solved so as to determine the position of the point on the map: one for *mixture fraction* and one for *unburned fuel* fraction. It is better than nothing; but it is not very good. A one-member population is no population at all.

2.3.6.6 EBU: the first two-member population model

The first-ever turbulent-combustion model that took meaningful account of fluctuations appears to have been the so-called eddy break-up" (EBU) model of 1971 [29]. This postulated a population of *two members*, both having the same *fuel ratio*, with *one fully burned* and the other *fully unburned*. The two members were supposed to collide, at rates fixed by hydrodynamic turbulence, forming intermediate-temperature and intermediate-composition material that quickly became fully burned. This model provided a (negative) source term in the finite-volume equation for the unburned fuel fraction, often expressed as

$$\text{source} = -\text{constant} * \text{density} * r * (1 - r) * \varepsilon / k \qquad (1.6)$$

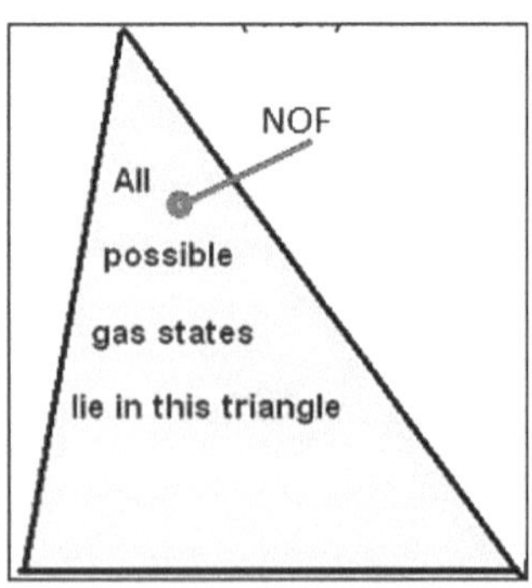

Figure 1.17 The no-fluctuations presumption. (For color version of this figure, the reader is referred to the online version of this chapter.)

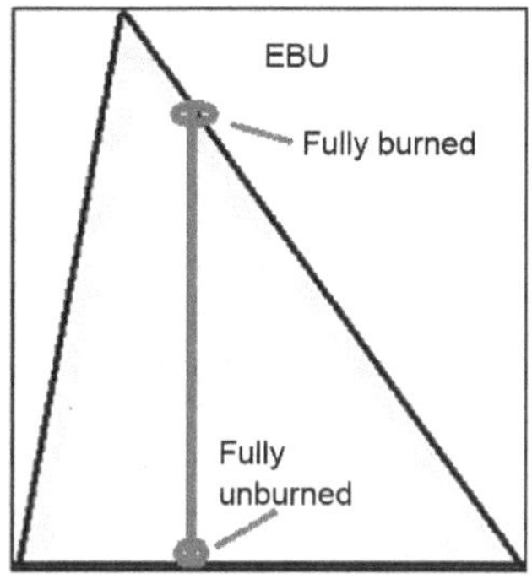

Figure 1.18 EBU on the TriMix map. (For color version of this figure, the reader is referred to the online version of this chapter.)

where *r* is the local reactedness of the mixture, so that r: $(1-r)$ is the ratio of burned to unburned material, and ε *and* k are from the $k-\varepsilon$ model of hydrodynamic turbulence (Fig. 1.18).

The EBU was successful in explaining some puzzling experimental data regarding the almost-constant angle of turbulent premixed flames in the wake of bluff bodies in plane-walled ducts; and the link that it made between the combustion rate and the hydrodynamics of the flow has found its way into almost every subsequent model of turbulent combustion.

2.3.6.7 A two-member model with Navier–Stokes equations for each member

In the 1970s, problems connected with the steam generators of pressurized-water-cooled nuclear reactors stimulated the development of methods of numerical simulation of two-phase flows. This involved the formulation and solution of two sets of interlinked Navier–Stokes equations, one for each phase. Such an algorithm was IPSA (Inter-Phase-Slip Algorithm) [30].

Although conceived with intermingling steam and water in mind, the algorithm could just as easily be applied to the hotter and colder gases that were envisaged in the EBU concept and without its overrestrictive assumption that they must be fully burned or fully unburned. Such a study is reported in Ref. [31], which describes how a shock wave passes along a horizontal pipe containing a combustible gas that is burning slowly.

The wave accelerates the hot-gas fragments more than the cold ones, causing relative motion. The relative motion causes increased entrainment and mixing, which increases the burning rate. This increases the strength of the pressure wave. The result is that the deflagration turns into a detonation (Fig. 1.19).

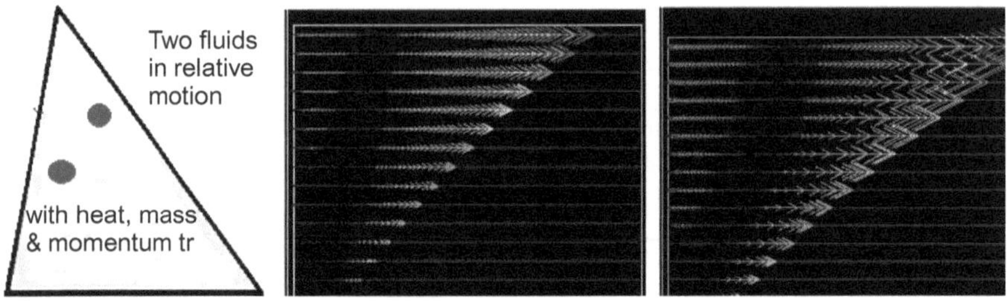

Figure 1.19 Instantaneous TriMix state and successive velocity vectors of colder and hotter members of the two-member population (time increases with vertical position). (For color version of this figure, the reader is referred to the online version of this chapter.)

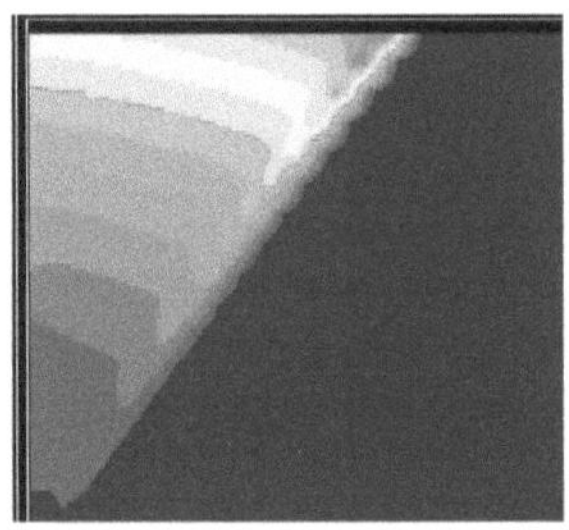

Figure 1.20 Pressure contours, distance horizontal, and time vertical. (For color version of this figure, the reader is referred to the online version of this chapter.)

Figure 1.20 shows the corresponding contours of pressure, which is shared by both gases.

These calculations were performed in 1983; but, for accidental historical reasons, little attention was paid to further development of the model at that time. Now that the ability to predict two-phase flow with interphase slip is more widely spread, it is hoped this chapter may awaken wider interest.

2.3.6.8 A four-member population model

The puzzling facts about turbulent premixed flames in plane-walled ducts, alluded to earlier-mentioned, were

- increasing flow velocity increases flame speed so the flame angle remains constant and
- sufficient increase of velocity *extinguishes* the flame.

EBU, that is, a *two*-fluid model, explained the first, but not the second. The solution [32] (24 years later!) was this: refine the (populational) grid. In other words, use a *four*-fluid model (Fig. 1.21).

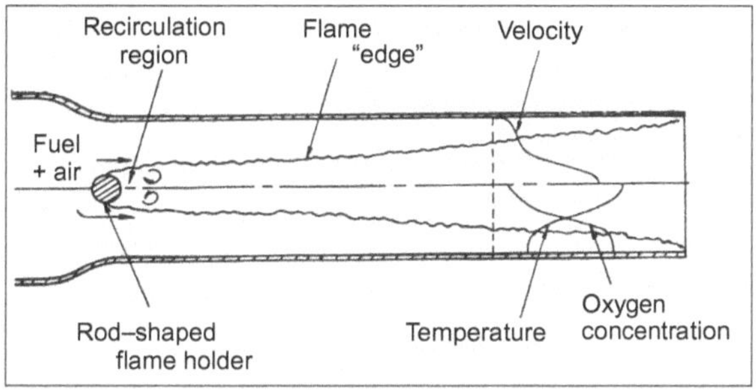

Figure 1.21 The Scurlock [33] experiment that prompted the invention of EBU.

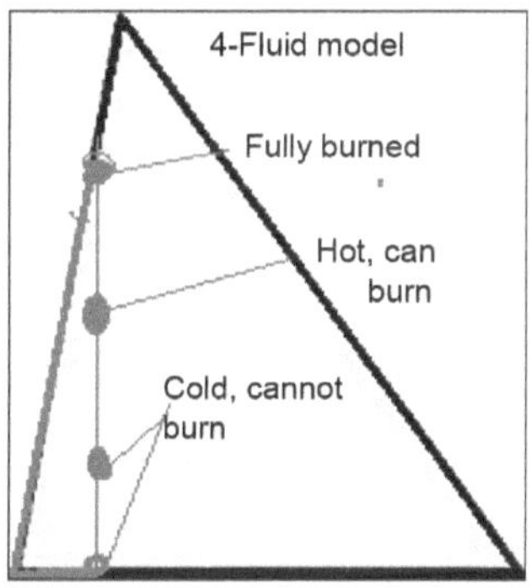

Figure 1.22 TriMix representation of the flour–fluid model. (For color version of this figure, the reader is referred to the online version of this chapter.)

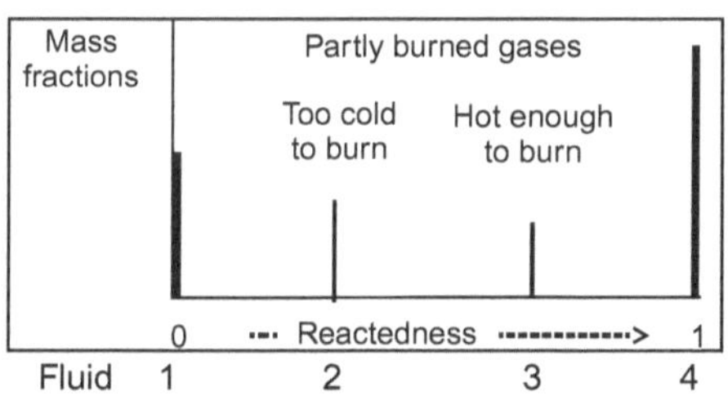

Figure 1.23 Mass fractions of the four population members.

The presence of the *hot, can burn* fluid (see Fig.1.22) allows space for chemical-kinetic limitations to enter. So extinction can be predicted (in principle). The EBU postulate was that fully burned and fully unburned gas fragments collided, at concentration-proportional rates, and the mixture combusted instantly; but with four fluids, there are more pairings possible and, therefore, more varied behavior (Fig. 1.23).

The physical presumptions are that

- fluids 1 and 4 collide at a hydrodynamically controlled rate, like Prandtl's "parcels," producing fluids 2 and 3;
- fluids 1 and 3 also collide to produce fluid 2;
- fluids 2 and 4 collide to produce fluid 3.

This fluid, being hot enough to burn, diminishes in mass fraction at a chemical-kinetically controlled rate, which is why that of fluid 4, the product of combustion, correspondingly increases.

Of course, a four-member population is "too coarse a grid" to permit accuracy, for reaction rates vary with temperature (i.e., reactedness) in a highly nonlinear manner, such as that of Fig. 1.24, wherein the fall to zero at high temperature results from the complete consumption of the reactants.

For such a curve, probably as many as twenty fluids would be needed, if their reactedness intervals were uniform, to achieve acceptable numerical accuracy. But why *not* have 20? Or more?

2.3.6.9 The multimember population

Long though it had taken to move from two to four fluids, the advance to multimember populations proceeded swiftly, both 1D and 2D populations being investigated. Reference [26] provides access to many of the early results, of which only the following set of four will be shown here. In order to point out that *how many* members are needed can be determined by trial-and-error. "Grid-refinement" studies are as practicable (and necessary) for populational grids as they are for spatial or temporal ones.

The same is true of discretized populations. *Grid-refinement* studies, as shown in Fig. 1.25, must be made for a 2D population at one particular geometric location in a flame with reactedness as the vertical dimension and mixture fraction as the horizontal one (TriMix not having been invented at the time).

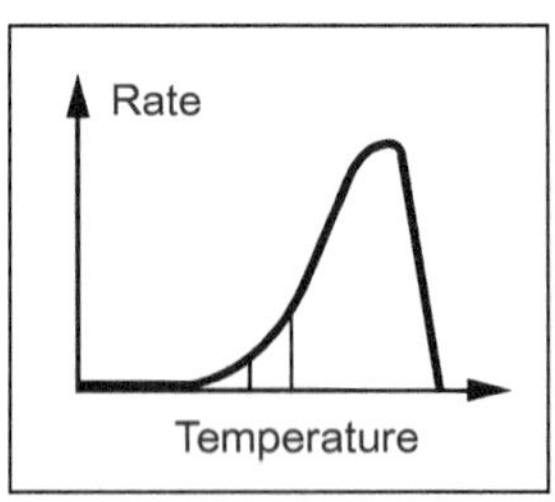

Figure 1.24 Typical variation of reaction rate with temperature.

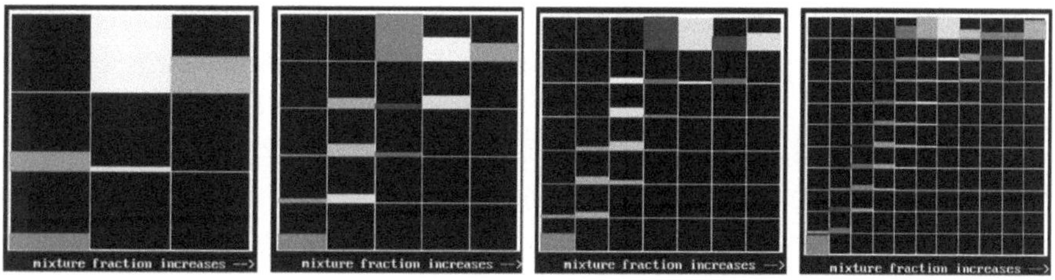

Figure 1.25 Predicted population distributions for 3*3, 5*5, 7*7, and 11*11. (For color version of this figure, the reader is referred to the online version of this chapter.)

At this point, it is appropriate to disclose that there does exist another approach to modeling turbulent combustion that can also be properly called "populational": it is that which is usually referred to as "pdf transport." The concept was first put forward in 1974 by Dopazo and O'Brien [34] in analytical form; then, in 1982, Pope [35] expressed it numerically. However, the chosen numerical method was a version of the Monte Carlo method, the unfamiliarity of which to most CFD specialists has perhaps hindered its acceptance and further development.

Because its associated computer times are rather long, it is unfortunate that the Monte Carlo lacks the time-saving grid-refinement (and grid-coarsening) capability of the discretized-population approach, which is advocated here. The latter provides non-absurd results even with a nine-cell grid, as has just been seen; but the corresponding "nine-particle-group" version of the Monte Carlo method would give no information that could be relied upon at all.

2.3.6.10 Populational and conventional CFD compared

It is now necessary to explain in more detail how such population-distribution diagrams are constructed. Let the colored-area proportion of each box, representing the mass fraction of the population that possesses the corresponding pair of attributes, be represented by the symbol $m_{i,j}$. Then, $m_{i,j}$ obeys a differential equation of the familiar "conservation" form:

$$\partial(\rho\, m_{i,j})/\partial t + \mathrm{div}(\rho_1 v_{i,j} m_{i,j}) = \mathrm{div}(\Gamma_{i,j}\, \mathrm{grad} m_{i,j}) + S_{i,j} \qquad (1.7)$$

in which the four terms represent time dependence, convection, (turbulent) diffusion, and sources in the usual way.

Although the *form* of the equation is familiar, there are several unusual features to be remarked upon:

- The subscripts i,j, which attach to the velocity vector $\boldsymbol{v}$, are a reminder that different population members may possess different velocity components, as was seen in Fig. 1.18 where the same pressure gradient had very different effects on the hotter and colder gases.
- The source term has to express mathematically two quite distinct processes, only one of which, namely, chemical reaction, is recognized by conventional CFD.
- Even this has to be differently expressed, namely, as a *diminution* in the mass fraction of lower-reactedness members of the population and an *increase* in mass fraction of higher-reactedness members.
- The process for which conventional CFD has no counterpart is that expressed earlier as "collision" between more remote members of the population and consequent "production" of material of intermediate attributes. Moreover, a new hypothesis has to be invoked in order that sources and sinks can be evaluated, for example, as a generalization of Eq. (1.6), in which the $r(1-r)$ term can be recognized as the product of the mass fractions of the two members of the population.
- Figure 1.26, extracted from Ref. [36], will enable the reader to envision the possibilities.

These differences from conventional CFD are not such as to present any computer-coding difficulty. *Any* commercial CFD code, therefore, provided that it allows its users to add source terms and make minor modification to built-in diffusion and convection formulations, could be employed for solving the populational turbulence model equations. Why, therefore, is this seldom, if ever, done? The present author has no satisfactory answer to that question.

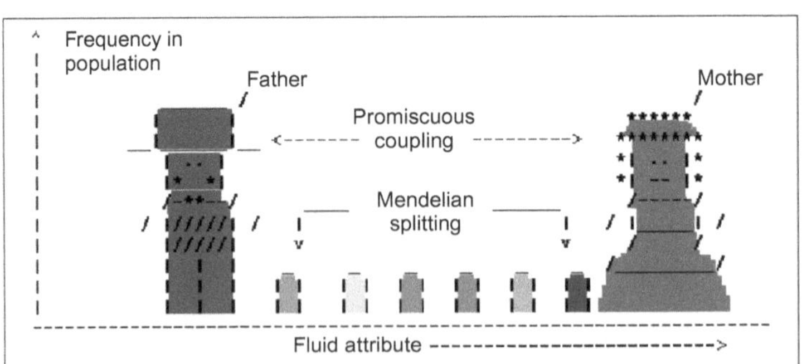

Figure 1.26 Illustration of the "promiscuous-Mendelian" hypothesis. (For color version of this figure, the reader is referred to the online version of this chapter.)

3. TRICKS

3.1. The IMMERSOL radiation model

3.1.1 The magnitude of the radiative problem

Turbulence, *chemical reaction*, *multiphase flow*, and *radiation* are the four main phenomena for which CFD practitioners make use of "*models*," that is, of mathematical idealizations that, although known to fall far short of complete representations, may still, in favorable circumstances, permit useful predictions to be made. Of these,

- the first receives great attention from CFD specialists and enjoys high respect as an engineering science challenge;
- the second is the active concern of, perhaps, an even greater number; and
- the third, though presenting fewer downright mysteries, is the subject of continued and large-scale research.

Radiation, however, although presenting comparable difficulties, has been a less popular subject for research. As a consequence, inability to model radiation realistically is often the *main cause of inaccuracy* in CFD predictions.

This is understandably true of high-temperature processes, such as those in the combustion chambers of engines and furnaces; but it is no less true of lower-temperature ones. Situations in which convective, conductive, and radiative modes of heat transfer may all have similar orders of magnitude include electronic equipment and the living accommodation of human beings.

Radiative heat transfer can be described mathematically with exactness. Perhaps for this reason, it is commonly supposed that enabling a CFD code to add radiation to its predictive capabilities is simply a matter of selecting and attaching to it one or other of the available equation-solving methods. These go under the names of Monte Carlo, discrete transfer, discrete ordinate, zone, etc.

Unfortunately, consideration of how these methods actually perform, when applied to problems of more than modest size, makes plain that they must *all* require very much more computer and elapsed time than anyone can afford. This is so even with neglect of the influences of

- wavelength on absorption and emission,
- impingement angle on the reflectivity of surfaces,
- temperature on the radiative properties of materials,
- chemical composition and "surface finish" of those materials, and
- the complicating presence of turbulent fluctuations of temperature and of multiphase flow.

3.1.2 The action-at-a-distance difficulty

Radiative heat transfer differs in character from conductive and convective heat transfer in that it involves "action at a distance."

Heat *conduction* to a point is influenced by the temperatures of the materials at the immediately *surrounding* locations. Heat *convection* to a point is influenced by the temperature on the immediately *upstream* side. Heat radiation, by contrast, can depend on the temperatures *at all surrounding points*, no matter how far away they are. Admittedly, the more remote points usually have less influence than the nearer ones; but the temperature of the sun affects the heat flux to Earth, that is, remote enough by human standards.

One way of expressing the difference between various modes of heat transfer is to state that the "mean free path of radiation" is often much larger than the dimensions of the domain of study. The "mean free paths of conduction and convection," on the other hand, are usually much smaller, being of the order of the distance between molecules or (in turbulent flow) of the size of the smallest eddies. That of radiation varies inversely with the amount of radiation-absorbing material per unit path length, which is why it is so large in "outer space," where there is no such material.

Where, however, *much* radiation-absorbing material is present, for example, within a furnace, where pulverized-coal particles and finely divided soot absorb scatter and reemit radiation, the mean free path of radiation *can* be much smaller than the apparatus dimensions. Then, radiative transfer can be regarded as similar to heat conduction, but with an increased thermal conductivity.

The "trick," which will now be described, is to treat radiation through *empty* space as though it there too behaved in a conduction-like manner. It is embodied in the so-called [54] IMMERSOL (i.e., immersed solid) model of radiation. Since this, although much used by the present author and his colleagues for many years, has never been adequately described in print, the omission will now be rectified.

3.1.3 IMMERSOL: the main features

3.1.3.1 The dependent variables

IMMERSOL distinguishes three temperatures for a control volume in a medium, which is transparent to radiation, namely,

- T_1, the temperature of the first phase, for example, air;
- T_2, the temperature of the second phase, if present, for example, a cloud of solid particles suspended within the air; and
- T_3, the "radiosity temperature" defined in the succeeding text.

If R stands for the *radiosity*, that is, the sum of all radiation fluxes traversing the control volume, regardless of direction and wavelength, then T_3 is related to it by the equation

$$R = \sigma T_3^4$$

where σ is the Stefan–Boltzmann constant (5.670373×10^{-8} W m^{-2} K^{-4}) and T_3 is measured in degrees Kelvin, as are the other temperatures. Therefore,

$$T_3 = \sigma^{-1} R^{1/4} \qquad (1.8)$$

3.1.3.2 The differential equations

The variables T_1 and T_2 obey differential equations of the familiar conservation kind, distinguished by special source terms, thus,

$$\partial(c_1 \rho_1 T_1)/\partial t + \mathrm{div}(c_1 \rho_1 \boldsymbol{v}_1 T_1) = \mathrm{div}(\lambda_1 \,\mathrm{grad}\, T_1) + S_{1,2} + S_{1,3} \qquad (1.9)$$

and

$$\partial(c_2 \rho_2 T_2)/\partial t + \mathrm{div}(c_2 \rho_2 \boldsymbol{v}_2 T_2) = \mathrm{div}(\lambda_2 \,\mathrm{grad}\, T_2) + S_{2,1} + S_{2,3} \qquad (1.10)$$

wherein

- $S_{1,2}$ and $S_{2,1}$ represent energy transfers per unit volume between phases one and two;
- $\boldsymbol{v}$ represents the velocity vector;
- $S_{1,3}$ and $S_{2,3}$ represent volumetric rates of radiative heat absorption and emission;
- λ_1 and λ_2 represent the sums of the thermal conductivities, laminar plus turbulent, of the respective phases;
- c represents specific heat;
- ρ represents density; and
- t represents time.

T_3 obeys a similar equation but with fewer terms. Specifically, it has a zero on the left-hand side because radiation is not convected in either time or space. Its equation is

$$0 = \mathrm{div}(\lambda_3 \,\mathrm{grad}\, T_3) + S_{3,1} + S_{3,2} \qquad (1.11)$$

3.1.3.3 The source terms

About $S_{1,2}$ and $S_{2,1}$, nothing needs to be said here except that they represent the interphase transfer processes in the usual way. It is the $S_{1,3}$, $S_{2,3}$, $S_{3,1}$, and

$S_{3,2}$ terms that require discussion. The IMMERSOL presumption is that they are related to the three temperatures via the following equations:

$$S_{1,3} = -S_{3,1} = \varepsilon_1' \sigma\left(T_3^4 - T_1^4\right) \quad (1.12)$$

and

$$S_{2,3} = -S_{3,2} = \varepsilon_2' \sigma\left(T_3^4 - T_2^4\right) \quad (1.13)$$

wherein ε_1' and ε_2' are the emissivities of their respective phases per unit length. These quantities are supposedly numerically equal to the absorptivities, which measure the proportion of the radiation flux that is absorbed per meter of its passage through the medium in question.

3.1.3.4 The value ascribed to λ_3

IMMERSOL expresses the thermal conductivity pertaining to the radiosity temperature T_3 in terms of the emissivities per unit length ε' and the scattering coefficients per unit length s' of the two phases in the transparent-to-radiation space and the gap between nearby solid walls W_{gap} as

$$\lambda_3 = 4\sigma T_3^3 / \left\{0.75\left(\varepsilon_1' + s_1' + \varepsilon_2' + s_2'\right) + 1/W_{gap}\right\} \quad (1.14)$$

The origin of this equation, and the meaning of W_{gap}, will be explained in the succeeding text.

3.1.3.5 The boundary conditions

At the walls of, and everywhere within, solid bodies surrounding or immersed in the transparent medium, T_3 is taken as being equal to the temperature T_1 or T_2, according to the phase in question. However, the radiant flux at such walls depends not only on the T_3 gradient in the medium close to the wall but also on the surface emissivity of the wall itself, in a manner that will be discussed in Section 3.1.4.4. At open boundaries of the domain, net radiation fluxes are ordinarily prescribed.

3.1.4 IMMERSOL: the rationale

3.1.4.1 Starting points

In its neglect of wavelength dependency, the IMMERSOL model departs radically from reality; but it does so in a manner that is commonly regarded as acceptable: it employs the widely used "gray-medium" approximation described in many textbooks, for example, in Sparrow and Cess [38].

In that book, and elsewhere, two other accepted concepts are described that IMMERSOL has adopted, namely, those of the *optically thick* and *optically thin* limits. Here, "thick" and "thin" compare the size of the gap between the solid walls enclosing the transparent medium with what can be called the "mean free path" of radiation in that medium, that is, the reciprocal of $(\varepsilon' + s')$. What is distinctive about IMMERSOL is that it is valid both *at* and *between* those limits.

Both extremes arise in practice. Within a large coal-fired furnace, the cloud of burning particles and gaseous combustion products can be regarded as optically thick, for so much solid surface is present per unit volume that rays emanating from the middle of the furnace must be multiply reflected before they escape to the water-cooled walls. The air within the rooms and corridors of inhabited buildings, by contrast, constitutes an optically thin medium; wall-to-wall radiation suffers no impediment.

For optically thick media, there exists a formula that connects the radiative heat-flux vector $\boldsymbol{q}$, in W m^{-2}, with the gradient of the radiosity. It is usually associated with the name of Rosseland [39], and it is

$$\boldsymbol{q} = -(4/3)(\varepsilon' + s')^{-1}\sigma \,\mathrm{grad}\left(T^4\right) \tag{1.15}$$

Here, T is the local temperature of the transparent medium.

If the equation is expressed in terms of an effective thermal conductivity λ_{eff}, involving gradT_3 rather than grad(T_3^4), the expression for λ_{ref} becomes

$$\lambda_{\mathrm{eff}} = (16/3)(\varepsilon' + s')^{-1}\sigma T_3^3 \tag{1.16}$$

At the other extreme, when the medium is so thin as not to participate at all in the radiative heat transfer between two solid surfaces, at temperatures $T_{3\mathrm{hot}}$ and $T_{3\mathrm{cold}}$, say, the heat flux $\boldsymbol{q}$ is well known to obey the formula

$$q = \{1 + (1 - \varepsilon_{\mathrm{hot}})/\varepsilon_{\mathrm{hot}} + (1 - \varepsilon_{\mathrm{cold}})/\varepsilon_{\mathrm{cold}}\}^{-1}\sigma\left(T_{3\mathrm{hot}}^4 - T_{3\mathrm{cold}}^4\right) \tag{1.17}$$

Equation (1.15) is of the flux-proportional-to-gradient kind that CFD codes are well equipped to solve. Equation (1.17) is of the less amenable action-at-a distance kind. The question arises: how can the latter be made more like the former?

3.1.4.2 First steps

Rewriting Eq. (1.17) for the case in which $(T_{\mathrm{hot}} - T_{\mathrm{cold}})$ is small and *in which the wall emissivities are unity* yields

$$q = 4\sigma T^3 (T_{hot} - T_{cold}) \tag{1.18}$$

where T stands for either temperature because they are nearly equal.

Since the temperature gradient equals $(T_{hot} - T_{cold})/W_{gap}$, the effective conductivity that corresponds to Eq. (1.18) is simply

$$\lambda_{eff} = 4W_{gap}\sigma T^3 \tag{1.19}$$

where W_{gap} stands for the distance between the solid surfaces. So, the conductivity increases with interwall distance, as it must do if the heat flux is to be independent of that distance.

It is interesting to compare the value of this conductivity with the thermal conductivities of common materials as seen here:

Atmospheric air	Water at 0 °C	Steel
0.0258	0.569	43.0

wherein the units are W m^{-1} °C^{-1}.

In the same units, and with a wall gap equal to 1 m, the values of λ_{eff} at various temperatures in degrees Celsius are

T_3	20	100	500	1000	1500	2000
λ_{eff}	5.706	11.77	104.8	467.9	1264.1	2663.6

Even taking into account that turbulence may increase the effective conductivity of a fluid by two or three orders of magnitude, it can be concluded from these tables that radiative heat transfer can be significant at room temperature and that at high temperatures, it becomes dominant.

3.1.4.3 Between the "thick" and "thin" extremes

Let now the *reciprocal* of conductivity be considered, that is, the resistivity, λ_{eff}^{-1}, measured in °C m W^{-1}. For the thick medium, Eq. (1.16) yields

$$\lambda_{eff}^{-1} = (3/16)(\varepsilon' + s')(\sigma T^3) \tag{1.20}$$

and, for the thinnest-possible totally empty medium, Eq. (1.19) yields

$$\lambda_{eff}^{-1} = 1/(4W_{gap}\sigma T^3) \tag{1.21}$$

It is, therefore, not unreasonable to suppose that, for intermediate conditions, the two multipliers of σT^3 should be *added* so as to create a more generally valid single resistivity formula, thus

$$\lambda_{\text{eff}}^{-1} = \{(3/4)(\varepsilon' + s') + 1/W_{\text{gap}}\}/(4\sigma T^3) \tag{1.22}$$

This is the source of Eq. (1.14), introduced in Section 3.1.4, and it can be described as the first part of the "IMMERSOL trick." But there is more to come.

3.1.4.4 Wall emissivity as an extra resistance

Equation (1.14) is used in IMMERSOL to calculate the T_3 diffusion fluxes of the finite-volume equations, which CFD codes easily solve. However, something special has to be done for coefficients when one node lies in the transparent medium and the other within a solid, as exemplified by nodes B and A in the Fig. 1.27, in which, for simplicity, the transparent medium consists of a single phase.

As is usual in CFD codes, the conductivities pertaining to the cell are stored at each grid node. Therefore, the radiative heat flux crossing the boundary between cells B and C will be deduced from the formula:

$$\text{flux}_{\text{B to C}} = (T_{3,\text{B}} - T_{3,\text{C}})/\{(x_{\text{I}} - x_{\text{B}})/\lambda_{3,\text{B}} + (x_{\text{C}} - x_{\text{I}})/\lambda_{3,\text{C}}\}$$

where x is the horizontal coordinate.

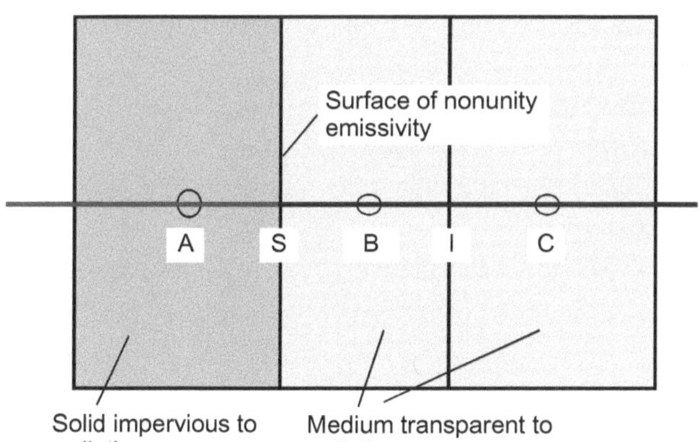

Figure 1.27 Computational cells near-solid surface. (For color version of this figure, the reader is referred to the online version of this chapter.)

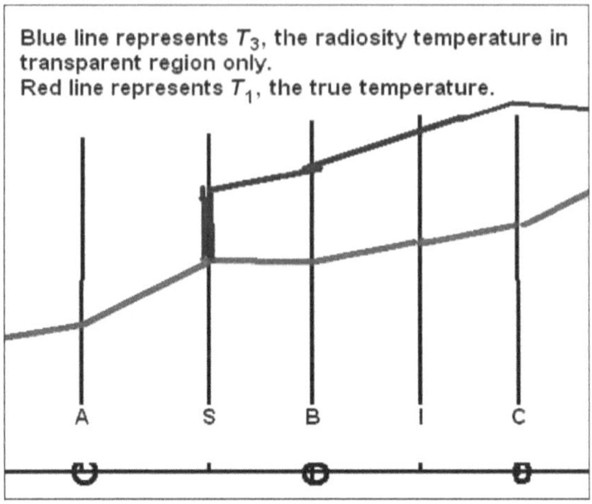

Figure 1.28 T_1 and T_3 profiles near a solid surface. (For color version of this figure, the reader is referred to the online version of this chapter.)

The calculation of the radiant flux at the S interface, however, requires more careful study because the surface emissivity can cause a discontinuity of T_3 gradient there, as is illustrated in Fig. 1.28, which shows the postulated profiles of *both* T_3 and T_1 because of their inescapable interaction.

Here, it is postulated that T_3 and T_1 are equal to each other within the solid; but, whereas the latter has a finite gradient everywhere, the former may have an infinite one at the interface between the phases. The fluxes of energy in question are as follows:

- Conduction from A to S, namely, $(T_{1,A} - T_{1,S})\lambda_{1,A}/(x_S - x_A)$
- Conduction and convection from S to B, namely, $(T_{1,S} - T_{1,B})\lambda_{1,B}/(x_C - x_S)$
- Radiation from S to B, namely, $(T_{3,S} - T_{3,B})/\{(x_C - x_S)/\lambda_{3,B} + (1 - \varepsilon_S)/\varepsilon_S\}$ wherein the term involving ε_S is inserted so as to conform with Eq. (1.17) in the preceding text. This is the second part of the "IMMERSOL trick."

Requiring the fluxes to be in balance at the surface S enables the equal-by-definition values of T_3 and T_1 there to be evaluated. The necessary formula is as follows:

$$\begin{aligned} T_{1,S} &= T_{3,S} \\ &= \{T_{1,A}\lambda_{1,A}/(x_S - x_A) + T_{1,B}\lambda_{1,B}/(x_B - x_S) \\ &\quad + T_{3,B}/\{(x_B - x_S)/\lambda_{3,B} + (1 - \varepsilon_S)/\varepsilon_S\}/[\lambda_{1,A}/(x_S - x_A) \\ &\quad + \lambda_{1,B}/(x_B - x_S) + 1/\{(x_B - x_S)/\lambda_{3,B} + (1 - \varepsilon_S)/\varepsilon_S\}] \end{aligned} \tag{1.23}$$

3.1.5 IMMERSOL: conclusions

The equations now assembled have transformed the problem of numerically simulating radiative heat transfer into one that can be solved by any computer code capable of handling conductive and convective heat transfer. The fields of T_3 can now be computed, and the radiant heat flux vector $\boldsymbol{q}$ any point can be deduced from its gradient via

$$\boldsymbol{q} = -\lambda_3 \,\mathrm{grad}\, T_3, \tag{1.24}$$

with λ_3 calculated from Eq. (1.14).

That is not to say that its predictions will *always* be in close agreement with experimental reality. Only for partly or wholly transparent layers between parallel uniform-temperature walls is that to be expected. But, although further research is needed, limited experience has shown that it never makes physically unreasonable predictions; and its computational expense is small.

So small is it indeed that IMMERSOL can be used when wavelength dependency is too great to be ignored. This arises, for example, when short-wavelength solar radiation provides a source of heat, that is, redistributed by way of long-wavelength low-temperature infrared radiation between the terrestrial objects on which it impinges. Thus, to use T_3 as the measure of the latter radiation and T_4, say, as a measure of the former would be much more realistic than to ignore the wavelength dependence entirely, as is commonly done. To split the wavelength ranges into 10 or more bands would not significantly strain computer resources.

The author is, however, not aware of any practical exploitation of this possibility. As mentioned in the preceding text, radiative-transfer research is not very fashionable.

3.2. The wall-distance trick

3.2.1 How to calculate W_{gap}

The seriously interested reader of Section 3.1 will have perceived a lacuna in the argument, for nothing has been said about how the vitally important W_{gap} quantity is to be calculated or indeed about what it *means* for spaces that are cluttered with solid objects of various shapes and sizes. A "data center," that is, a large hall filled with computer cabinets and air-cooling equipment, is a case in point. Radiative heat transfer plays a significant part in lowering the temperature of the "hot spots," in which the center manager must watch out for. But what meaning has W_{gap} in such a cluttered space?

There is an answer and again it has a somewhat "tricky" character: as does IMMERSOL itself, it produces answers that are exact in simple circumstances and never unreasonable in others. In order to whet the reader's appetite, attention is drawn to three images (Figs. 1.29–1.31), the origin of which will be explained.

Let the task be to apply the IMMERSOL model for calculating the rate of radiative heat transfer between the two boxes of Fig. 1.29, the walls of which are held at different temperatures. The radiation will supposedly pass through the also empty duct that connects them, the walls of which will be supposed insulated. The emissivities of duct and box walls will be supposed known.

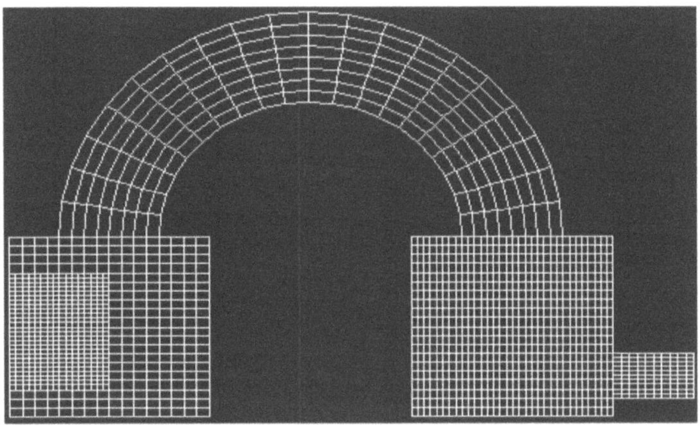

Figure 1.29 Computational grid for two boxes and a connecting duct.

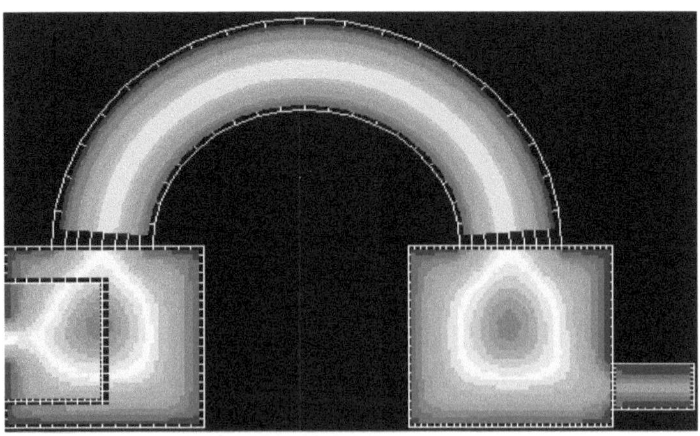

Figure 1.30 Computed contours of distance from a solid wall. (For color version of this figure, the reader is referred to the online version of this chapter.)

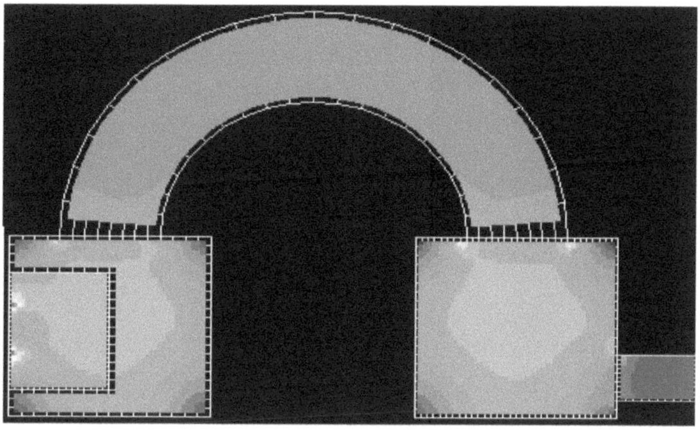

Figure 1.31 Computed contours of gap between walls. (For color version of this figure, the reader is referred to the online version of this chapter.)

Since IMMERSOL has reduced the task to the level of a heat-conduction one, it is easy; but it requires W_{gap} values to be known for every point in order that local conductivity can be computed.

Figure 1.30 shows the distribution not of W_{gap} but of a related quantity, W_{dis}, the distance of each point within the boxes or duct from the nearest solid wall. This is not a quantity that IMMERSOL uses; but it is calculated at the same time as W_{gap}.

Understandably, the lowest values are at the walls themselves; and the highest values are at the centers of the boxes. The corresponding values of W_{gap} appear in Fig. 1.31.

Equally understandable, W_{gap} appears to have a uniform value inside the connecting duct, for its walls are indeed parallel. Within the boxes, it is less uniform; but its highest value is about twice as large as that within the duct. These are plausible findings, the origins of which lie in the numerical solution of a Poisson equation now to be described.

3.2.2 The L *equation*

The trick that IMMERSOL has exploited comes from the turbulence modeling field. There, it is often desired to calculate the distance of a point in a fluid from the nearest solid wall; and a convenient way of doing so is to solve the following differential equation:

$$\mathrm{div}\,\mathrm{grad}L = -1 \tag{1.25}$$

with the boundary condition $L=0$, wherever solid is present. Since the original publication [40] is somewhat obscure, the idea that it expresses will be spelled out at length.

The equation is similar to that for temperature within a uniformly conducting medium, having a uniform heat source, and in contact with solids and other surfaces at which the temperature is held at zero. When it has been discretized and expressed in customary finite-volume form, it is easily solved by the linear equation solver of any CFD package, whether the geometry is 1D, 2D, or 3D.

The variable L is not itself the distance from the wall, even though it is proportional to that distance at locations very close to a wall. Its dimensions are indeed those of length-*squared*. However, the wall distance can be deduced from the solution for L, as can also a plausible estimate of the effective distance between walls. The method is to derive, by considering a simple geometry, namely, that between two parallel walls, relationships between

- the distance from the wall W_{dis} and
- the distance between walls W_{gap}, on the one hand, and
- the local value of L and
- the local value of its gradient, on the other

and thereafter to *presume* that the relationships have general validity.

3.2.3 The parallel-wall situation

Let the distance measured from one wall be y, and the distance to the opposite wall y_1. Then the 1D form of Eq. (1.25), namely,

$$d^2L/dy^2=-1, \tag{1.26}$$

can be integrated to give

$$dL/dy=-y+A, \tag{1.27}$$

where A is a constant, and then further to give

$$L=-y^2/2+Ay+B, \tag{1.28}$$

where B is another constant.

Insertion of the boundary condition $L=0$ at $y=0$ and $y=y_1$ yields

$$B=0, \text{ and } A=y_1/2$$

with the result

$$L = y(y_1 - y)/2 \tag{1.29}$$

$$L' = y_1/2 - y \tag{1.30}$$

where L' stands for dL/dy.

Elimination of y_1 from the two equations yields

$$L = y(L' + y/2) \tag{1.31}$$

which is a quadratic equation, easily soluble for y.

From its solution, follow with W_{dis} substituted for y and W_{gap} for y_1:

$$W_{dis} = \left(L'^2 + 2L\right)^{1/2} - L' \tag{1.32}$$

$$W_{gap} = 2\left(L'^2 + 2L\right)^{1/2} \tag{1.33}$$

It is these equations that are employed generally, L and L' being obtained for each point in 2D or 3D space from the numerically computed solution of Eq. (1.25).

They have been found to give plausible results in all situations; but of course, neither W_{dis} nor W_{gap} has an unequivocal meaning when the walls exhibit corners, whether concave or convex.

It is interesting in this connection to consider a pipe of circular cross-section, for which it is again easy to obtain an analytical solution for L. The earlier-mentioned equations then show that the expression for W_{dis} is exactly correct in the immediate vicinity of the wall; and it rises to a maximum equal to radius divided by the *square root of 2* at the center of the pipe.

W_{gap} on the other hand varies between pipe radius times $2^{1/2}$ at the center to the radius times $2^{-1/2}$ near the pipe wall. One can therefore conclude that even extreme departures from the presumed parallel-plane conditions lead to not wholly implausible results.

3.2.4 Concluding remark

Before leaving the topic of wall-distance and wall-gap calculation, it is necessary to make clear that the L-equation method has no physical basis whatever. Its status is that of a "lucky guess," its only justification being that "it works." Those who are reluctant to use so dubious a trick should ask themselves: "What else can I do?"

3.3. The cut-link trick

3.3.1 Introduction

It has already been mentioned in Section 3.1, under "Trends," that grids of arbitrary polygonal cells are giving way in respect to popularity to variants of the "cut-cell" or "immersed boundary method" [53] kind. The present author has for several years been using one variant of the former, of which the details have never been fully published; moreover, he is at present in the process of developing a simpler but more powerful variant. It therefore accords with the purposes of this chapter to describe the first variant, to explain why it is now being superseded, and to record in more detail than is usual what the main features of the new variant actually are.

The first variant was called "PARSOL," because it handled cell PARtially filled with SOLid material. The new one is called "SPARSOL," which stands for Structured PARSOL. Strictly speaking, both variants have some unstructured features, but SPARSOL has fewer than PARSOL, as will be explained. The main difference is that PARSOL was a true "cut-*cell*" technique in that rectangular cells, of which some *edges* were cut by the surfaces of solid bodies, were regarded as divided into *two* "*subcells*." In general, these were of nonrectangular shape. Balance equations were formulated by treating both these subcells as control volumes. The number of equations to be solved therefore increased, which required somewhat troublesome changes in the solver.

SPARSOL, by contrast, considers cut *links* rather than cut *edges*, a link being the line joining two cell-center nodes. It therefore creates no new control volumes or corresponding equations; so the solver requires no change. What *it* does is to make such changes to the *coefficients* and perhaps also to the *source terms* of those equations as will best express the interactions between the fluid and the solid materials.

"*Best* express" is the phrase used, not "*perfectly* express"; and it is preferred advisedly. Perfection is never to be expected of a *finite*-volume (or *finite*-difference or *finite*-element) method. "Near enough" is all that can be hoped for; and if it is conjoined with simplicity, it is very good indeed.

3.3.2 The pros and cons of PARSOL

The major "pro" of PARSOL was of course that it removed entirely the grid-generation problem, which so troubled the arbitrary unstructured-grid users. Defendants of the latter practice argued that theirs was more accurate—and probably, they were right. But how great the difference in accuracy was and whether it was worth the effort were never systematically put to the test. (Question to readers #3: Is the last statement true? And, if not, where can a systematic comparison be found?)

When first introduced, PARSOL was used only for hydrodynamics problems. Therefore, no finite-volume equation for the in-solid subcell had to be solved; and the only changes to the coefficients of the finite-volume equations were those that accounted for the reduced volume areas and distances of the in-fluid subcell when compared with the uncut whole cell. An account *was* taken of the size and orientation of the (often) inclined interface between the solid and the fluid; and the wall functions used in calculating the velocity components and turbulence quantities were duly modified. That pressure gradients were no longer aligned with velocity directions was not considered however; nor was the fact that sometimes the true thickness of the fluid boundary layer was *not*, as was tacitly presumed, much larger than the near-wall cell size.

It was when PARSOL started to be used for conjugate heat transfer problems that the in-solid subcells had to be used as supernumerary finite volumes. The changes made to the solver were at first excessively explicit, which entailed that the convergence of the temperature equation was sometimes inordinately slow. Later, more implicitness was built-in, with consequent increased speed of convergence, and still was added simultaneous solution for the radiosity temperature, that is, the T_3 of IMMERSOL.

This last addition increased the complexity of the solver, which occasionally led to divergence or at least to physically dubious solutions. It was while seeking to interpret and correct these deficiencies that it was recognized that it was the extra-control-volume feature that was the basic cause of the trouble, thence sprang the search for a better alternative, with SPARSOL as the result.

There was however an independent and even stronger reason for making the change, namely, that PARSOL could not cut its cells into more than two parts: one solid and one fluid. If it was to handle thin solid objects that cut the grid obliquely, only a very fine grid could be used; and this incurred serious computer-time penalties. SPARSOL is free from this crippling restriction.

3.3.3 Detecting the link intersections

3.3.3.1 The problem

Whether they are the cut edges of PARSOL or the cut links of SPARSOL, both procedures require intersections of lines with surfaces to be detected, their locations stored, and associated geometric quantities to be computed.

Detection can be done in more than one way. Having seen no discussion of this important matter in the literature, the author will provide one, based on his own experience.

First, it must be explained that the shapes of the surfaces of solid objects are always supplied in one of two ways: via *formulas* or via *facets*. The position, shape, and size of a sphere, for example, can be completely and compactly specified by the formula:

$$(x - x_0)^2 + (y - y_0)^2 + (x - x_0)^2 = r^2 \tag{1.34}$$

where x, y, and z are points on the surface; x_0, y_0, and z_0 are the Cartesian coordinates of the center; and r is the radius. Just four parameters will suffice. For such shapes, it is easy to determine the location of intersections of their surfaces with any straight line, whether an edge or a link, by way of algebra.

Far more often, however, the same sphere will be described by way of facets, as illustrated in Fig. 1.32, which shows a somewhat crudely facetted sphere within its bounding box. The information needed to describe it is voluminous, consisting mainly of the Cartesian coordinates of each facet vertex. The facet method is nevertheless often preferred for the use of formulas because it can be used for objects of *any* shape. A procedure must be therefore devised for determining the intersection, if it exists, of each facet with each grid line. Such a procedure will be described; it can be used for two- or 3D Cartesian or cylindrical-polar grids, whether structured or unstructured. Extension to body-fitted grids is also feasible.

3.3.3.2 The 2D projection method (2DPM)

In this, the first-used method, now superseded, the facets of the object were projected on to a plane normal to one of the coordinate directions, as triangles or quadrilaterals.

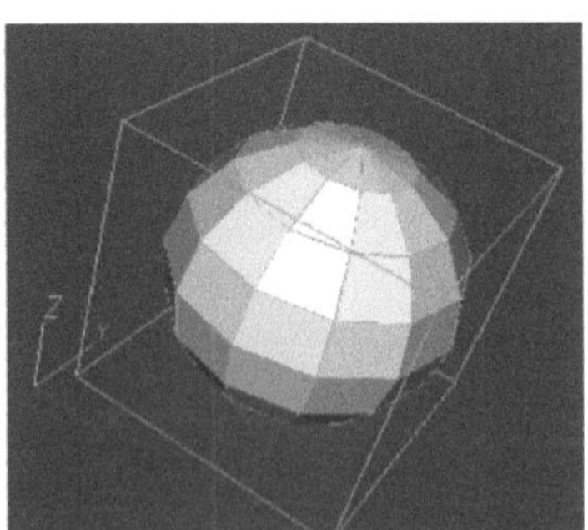

Figure 1.32 Facetted sphere. (For color version of this figure, the reader is referred to the online version of this chapter.)

Cell edges appeared in the projection plane as points, and the existence of an intersection was evidenced by the cell-edge point lying within the projected facet area. This was detected by calculating the areas of triangles having facet-edge projections as base and the grid-line projection as apex. The determination of whether the grid-line projection lay inside or outside the facet projection depended on the signs of these areas. Thereafter, the normal-to-plane location of the intersection could be calculated, and the geometric properties of the cut-cell computed and stored for use.

The disadvantages were the slowness of the area calculation, the hit-or-miss nature of the decision as to whether points lay inside or outside, and the obscurity of the intersection calculation. Moreover, as implemented, no advantage was taken of the economies that can be made when the flow situation to be simulated is two- rather than 3D. The 2DPM coding was thus found, after intensive study, to have several drawbacks, of which the most serious were as follows:

- It could not be relied upon always to detect intersections between facets and cell edges, because of the lack of control of "tolerances," that is, the differences of distance between what was and what was not an intersection.
- It could not directly treat the commonly occurring 2D flow situations, but had to convert them into pseudo-3D ones, which was at best uneconomical and at worst contributed to the "missed-intersection" phenomenon.
- Even when intersections were correctly detected and their positions computed, the excessive amount of computation involved imposed a serious delay on the start-up of the true CFD calculations. This was the most serious of the three.

3.3.3.3 The 2D section method (2DSM)

It was for these reasons that the now-preferred 2D *section* method was invented. It is so named because it finds first the *straight-line segment* that a facet makes *if* it intersects one of the planes containing cell edges (for PARSOL) *or cell centers* (for SPARSOL). Then, it seeks intersection *points* that the segment may make with the grid lines corresponding to the two other coordinate directions. Figure 1.33 illustrates this.

The procedure is as follows:

- Choose a first constant-grid coordinate plane. If the grid is cylindrical-polar, of course, the grid coordinate cannot be the radius, for that is not a plane. Then, lines corresponding to the not-chosen grid coordinates can be imagined as inscribed on the chosen plane either as straight lines or (in the cylindrical-polar case) as circles.

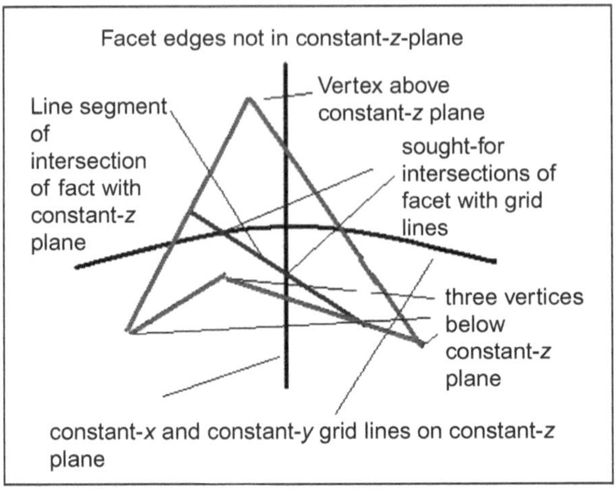

Figure 1.33 Illustrating the 2DSM. (For color version of this figure, the reader is referred to the online version of this chapter.)

- Check the coordinates of the vertices of the facet to determine whether they all have equal values of the normal-to-plane coordinate. If they do, no intersections can exist, so the facet can be discarded; otherwise, proceed as follows.
- Examine each edge of the facet in turn and consider the coordinates of the vertices at each of its ends. If the normal-to-plane values of all of these are less than that of the plane or all of them exceed that value, no intersection is possible so the facet can be discarded; otherwise, proceed as follows.
- For each pair of vertices lying on opposite signs of the chosen plane, calculate the in-plane coordinates of the location of the vertex-joining edge with the chosen plane. This lies at one end of a facet-with-plane intersection *segment*.
- By comparison of the coordinates of these pairs with the coordinates of the "inscribed" grid lines, determine whether the segment intersects one or more of these. If not, discard the facet; otherwise, proceed as follows.
- Use the appropriate algebraic expression, quadratic, or linear according to whether the coordinate in question is or is not the radius, to calculate the coordinates of the intersection point.
- Unless the problem is 3D rather than 2D, choose a *second* constant-coordinate plane and repeat the process, with one difference: it is necessary, having found a facet-plane intersection segment, to seek its intersections only with the grid lines corresponding to those of the first chosen *plane*. The reason is that intersections with other grid lines have already been found.

- The final step is to record which part of the intersected grid lies inside the object by making use of the convention connecting the order of listing of the facet vertices with the side of the facet on which the object itself lies.
- When this has been done for all the facets of the current object, repeat the process for the next object and continue thus until all the objects have been dealt with.

3.3.3.4 Other aspects of facet-grid-line intersection detection

The just-described 2DSM works well when it is presented with a well-ordered set of objects and associated facets. However, architects and others often present, as objects around which they require flow to be computed, collections of facets, produced by Computer-Aided-Design packages, which are far from being well ordered. From the *architect's* viewpoint, they are satisfactory, for the buildings and other objects have the right *visual appearance*, but commonly encountered defects are the following:

- Adjacent facets may bear conflicting information about which side is "in" and which side is "out."
- Some facets are absent, so that the surface of a solid object appears to have holes in it.
- Some needless facets are supplied, having identical vertex coordinates (listed but in a different order), for example, those representing the floor of one room and the ceiling of the room below it, when the flow in neither room has to be computed.
- Far too many facets are supplied, for example, those representing the individual steps in a staircase, for the computational grid to be take account of.

It is desirable therefore to pass such CAD-package output through a *facet-fixing* program before it is delivered to the CFD code; but even after this, some difficulties may be encountered. Specifically, the facets supplied may show that some of the objects *overlap*, in the sense that two or more objects lay claim to the same locations in space. Alternatively, two objects may *touch*, in the sense that their facets both intersect the same grid line at the same location, but from opposite sides. These problems will be discussed in SPARSOL terms, which is to say that the grid lines in question will be internode links rather than cell edges.

The "overlapping" problem can be solved by imposing the rule "last claimant wins." There can be only one material at any particular grid node, for example; and the "last-prevails" principle allows material settings to be made without inquiry as to whether a previous one has been made at the node in question.

The "touching" problem can be solved by the same rule. All that is necessary is to arrange that no decisions about the magnitudes of internode coefficients are made until all material settings have been completed.

One final problem should be mentioned: a link between grid nodes can be cut twice by facets belonging to the same body. This can occur when the thickness of an object is small compared with internode distances. In such circumstances, it might be said that the grid ought to be refined, but with structured Cartesian or cylindrical-polar grids, the needed refinement near the body results in *not*-needed refinement elsewhere. The more economical solution is to move one of the nearby grid nodes so that it lies within the object; then, one of the two intersections moves from the cut internode link to another, which was not previously cut. Thus, the double-cut difficulty is removed. But this is just one aspect of geometry adjustment, discussed in the succeeding text.

The overlapping, touching, and double-cutting problems are not confined to objects that are defined by facets; they can arise also with objects defined by formulas. After the intersections have been found, it is immaterial which method was used to define the shapes.

3.3.4 Changing coefficients in SPARSOL

3.3.4.1 The problem

The coefficients connecting the solved-for variables at neighboring grid nodes typically represent the influences of diffusion (laminar and turbulent) and convection. In the immediate vicinity of a solid surface, it is usually the former influence that predominates; therefore, the present discussion will consider diffusive influences exclusively. Moreover, for concreteness, the dependent variable considered will be temperature; and a single phase will be supposed present at each grid node. The material property in question is thus the thermal conductivity, λ.

The commonly used formula for the heat transfer coefficient, C_{LM}, between points L and M in Fig. 1.34 is, because all the material between the points is solid,

$$C_{\mathrm{LM}} = \lambda_{\mathrm{solid}} A / (x_{\mathrm{M}} - x_{\mathrm{L}}) \tag{1.35}$$

wherein x denotes the horizontal position and A is the surface area of the face between the cells. That for C_{MR}, on the other hand, is

$$C_{\mathrm{MR}} = A / \{(x_{\mathrm{MR}} - x_{\mathrm{M}})/\lambda_{\mathrm{solid}} + (x_{\mathrm{R}} - x_{\mathrm{MR}})/\lambda_{\mathrm{fluid}}\} \tag{1.36}$$

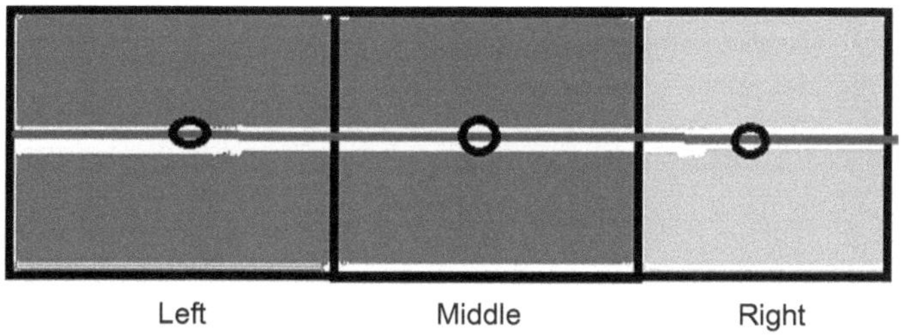

Figure 1.34 Cells wholly filled with solid (darker) of fluid (lighter). (For color version of this figure, the reader is referred to the online version of this chapter.)

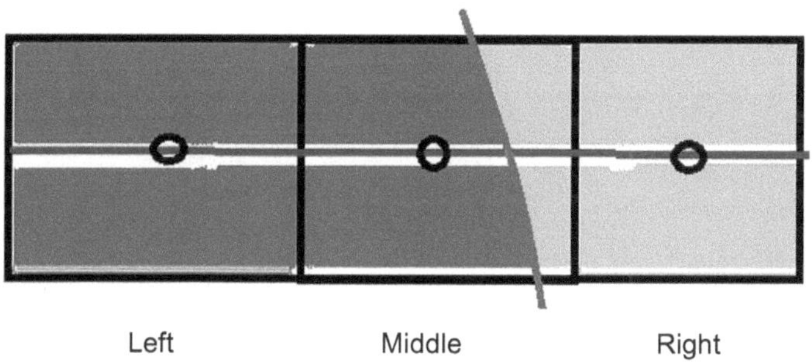

Figure 1.35 Solid object cuts link MR in nonstandard position. (For color version of this figure, the reader is referred to the online version of this chapter.)

wherein x_{MR} denotes the horizontal position of the interface between M and R. This has the value $W_M/2$, where W_M is the horizontal width of cell M.

3.3.4.2 Changing the distances

Consider now the situation in Fig. 1.35.

The equation to fit this situation is obviously

$$C_{MR} = A/\left\{ \left(x'_{MR} - x_M\right)/\lambda_{solid} + \left(x_R - x'_{MR}\right)/\lambda_{fluid} \right\} \tag{1.37}$$

wherein x'_{MR} now denotes the horizontal position of the location at which the object surface cuts the internodal link MR. This location is known as a consequence of the operations described in Section 3.3.4.1. Therefore, two actions to be taken after finding the intersections are first to determine which nodes lie in the solid and second to recalculate the distances from the nodes to the interfaces.

A major defect of PARSOL was its inability to handle thin objects. How SPARSOL deals with them is shown in Fig. 1.36.

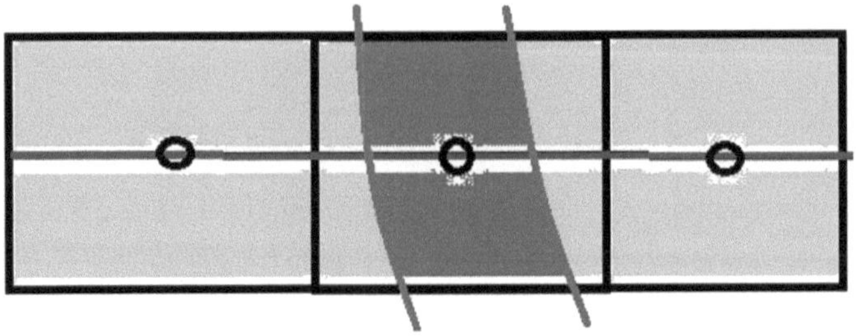

Figure 1.36 Thin object cuts two neighboring links. (For color version of this figure, the reader is referred to the online version of this chapter.)

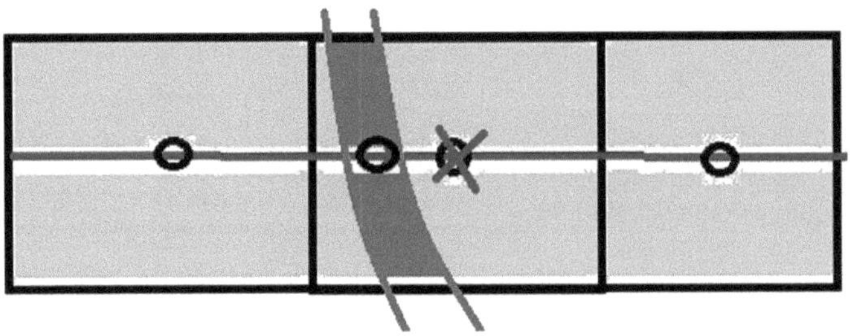

Figure 1.37 Thin object cuts one link. (For color version of this figure, the reader is referred to the online version of this chapter.)

Figure 1.36 differs from Fig. 1.35 in that link LM is intersected and link MR. This simply entails that the equation

$$C_{LM} = A/\{(x'_{LM} - x_L)/\lambda_{fluid} + (x_M - x'_{LM})/\lambda_{solid}\} \tag{1.38}$$

must be used also, the quantity x'_{LM} being modified from its standard value.

What if the two surfaces of the thin body cut the *same* internode link as shown in Fig. 1.37?

More than one strategy could be chosen; but the one here illustrated is move the node. The equations are therefore

$$C_{MR} = A/\{(x'_{MR} - x'_M)/\lambda_{solid} + (x_R - x'_{MR})/\lambda_{fluid}\} \tag{1.39}$$

and

$$C_{LM} = A/\{(x'_{LM} - x_L)/\lambda_{fluid} + (x'_M - x'_{LM})/\lambda_{solid}\} \tag{1.40}$$

wherein x'_M replaces x_M. This is all that needs to be said.

3.3.4.3 Changing the areas

It is, however, not only the distances that require to be changed in order to express the influence of nonstandard intersections of internode links. Consider, for example, the situation illustrated in Fig. 1.38.

Earlier, the intersected horizontal (red) links were considered; but attention is now turned to the nearby not-intersected vertical (green) links; and the question is: how should these be modified so as to account for the nonstandard locations of the red-link intersections? The answer is obvious: by way of the *areas* in the coefficient formulas. Obviously, the coefficient between the LowLeft and HighLeft nodes, for which λ_{fluid} is the correct conductivity, is associated with a *less*-than-standard area.

Correspondingly, the coefficient between LowRight and HighRight, for which λ_{solid} is the correct conductivity, is associated with a *greater*-than-standard area.

These area changes can be deduced from the red-link-intersection locations resulting from the actions described in Section 3.3.4. How? By linear interpolation. Are there interpolation formulas that will take into account the putative curvature of the object surface also? "No" is the answer, to which might be added: "Of course not. Do not expect too much. If you want more accuracy, use a finer grid."

In the previously mentioned examples, it has been only the red links that have been intersected. What should be done if only the green links had been intersected is obvious: y-distances should be changed rather than x-distances; but what should be done of both are intersected as illustrated in Fig. 1.39.

The safest answer to this question is to do nothing special at all. The LowLeft-to-LowRight and LowRight to HighRight links are not

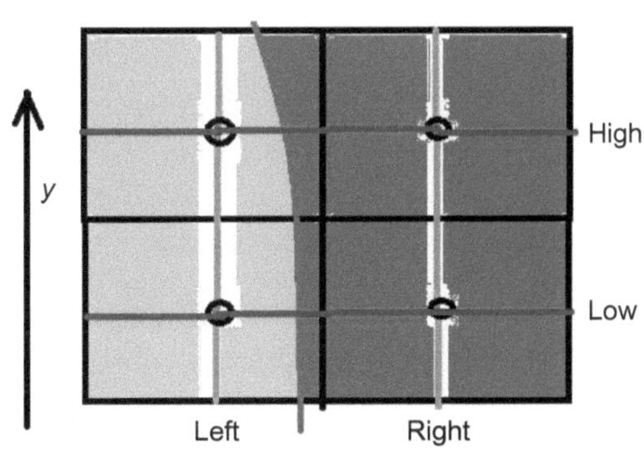

Figure 1.38 How intersected links influence areas. (For color version of this figure, the reader is referred to the online version of this chapter.)

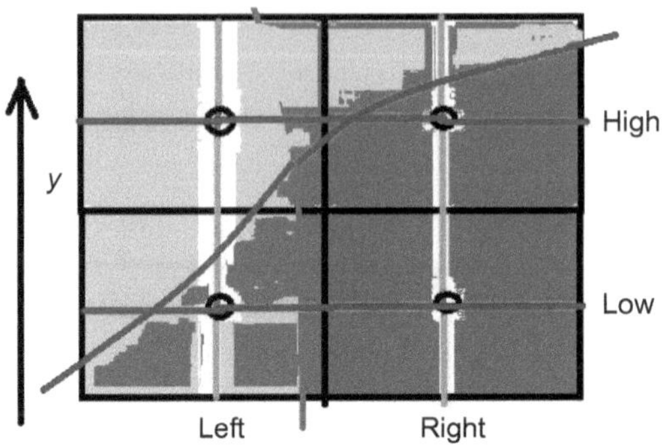

Figure 1.39 Nodes with two links. (For color version of this figure, the reader is referred to the online version of this chapter.)

intersected; but they do have slightly reduced areas. The other links are singly intersected; so they should have their coefficients computed from Eqs. (1.39) and (1.40). More complex rules could be thought of; but it is better to accept the "do not expect too much" advice.

3.3.4.4 Adding fluid-side resistances

It was mentioned in the preceding text that if the flow is turbulent, the thermal conductivity to be used is the sum of the laminar and turbulent values. However, close to walls, the turbulent contribution varies steeply with distance from the wall; and with structured grids of economically tolerable fineness, this often means that the only way to determine an appropriate "effective" value is by use of a "wall function." No difference in principle arises when links are cut by the surface of bodies at nonstandard locations; but of course, the changed distances have to be taken into account.

3.3.5 Modifying sources

Source terms, in balance equations for mass, momentum, energy, and chemical species, are often proportional to the volume of the cell or more precisely to the volumes of each of the materials within the cell. The same is true of the terms representing variations with time when as unsteady-state computation is in progress.

PARSOL modified the volumes associated with cut cells, diminishing these by the amount that was ascribed to the "subcells" that it created. SPARSOL also modifies volumes associated with each of its nodes. Since the number of these nodes remains constant, some of its volumes are diminished and others increased. The magnitudes of these changes are computed from the link-intersection locations in straightforward ways, too obvious to be spelled out here.

Momentum sources play a *special role, for the presence of solids* within the fluid commonly forces the velocity at nodes within them to have the value prescribed or calculated for the object. In SPARSOL, these are best specified by way of cell-wise linearized sources, thus

$$\nu\_\text{source} = \text{large\_number}(\nu\_\text{object} - \nu\_\text{local})$$

This has the result of introducing such a large momentum source, if the local velocity differs at all from the object velocity that ν_local is forced to equal that velocity very closely. Very often, the object is not moving, so that ν_object equals zero, as is shown in Fig. 1.40, where SPARSOL is being used to represent the flow deflection caused by an airfoil at a high angle of incidence.

Of course, with such a coarse grid, the velocity distribution close to the airfoil surface cannot be well represented; therefore, the calculated frictional force on the object will be very far from correct. This is the reason why those concerned with the simulation of flows around aircraft, for which the frictional component of drag is very significant, often go to great trouble in order to place finely divided cells in the boundary-layer region. Figure 1.41 illustrates this trend.

The computational expense is very great. Therefore, two remarks are worth making:

Figure 1.40 SPARSOL's use of momentum sources to represent an airfoil. (For color version of this figure, the reader is referred to the online version of this chapter.)

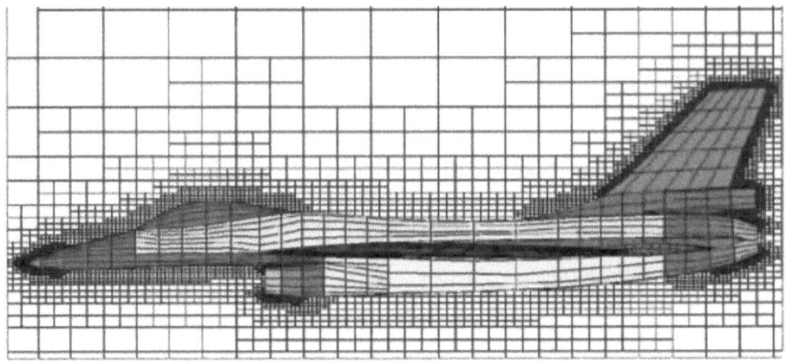

Figure 1.41 Unstructured-grid refinement near-solid surface. (For color version of this figure, the reader is referred to the online version of this chapter.)

- First, it is improbable that the grid is *ever* made fine enough for numerical accuracy, for turbulence models require energy sources to be computed and these depend upon the squares of velocity gradients. Very fine grids are needed if these are to be accurate.
- Second, there is a much simpler and cheaper way of achieving the required fineness. It is to calculate only the pressure distributions in the air just outside the boundary layer by means of a 3D *elliptic*-flow solver (which may even be of the highly economical potential-flow kind). Then flow within the boundary layer can be calculated on as fine a grid as necessary by a *parabolic-flow* solver.

The second feature may, like IMMERSOL, LTLS, and SPARSOL, also be regarded as a "trick," additionally "unfair" because it robs the heavy-weight-computing enthusiast of the excuse for demanding ever more number-crunching power. Yet, the argument in its favor is compelling. The essential characteristic of boundary layers is that they have one *predominant direction of flow*. Then, if the mathematical solution procedure itself "goes with the flow," 3D problems can be solved with a 2D grid; and one does not need much memory for that.

Of course, the 3D and 2D calculations have to be linked. The parabolic solver has to report to the elliptic one the "displacement-thickness" distribution over the aircraft's surface, in order that the elliptic solver can modify its first guess, which may have been to assume that the thickness was zero everywhere. So back-and-forth iteration is necessary. But it is not to be expected that *many* iterations will be needed; unless, that is, the conditions are close to those leading to boundary-layer separation. Aircraft-safety rules preclude that, in any case.

The idea of coupling potential-flow with boundary-layer calculations is far from new; and a fairly recent NASA "2D-airfoil-challenge" exercise [37] showed that its accuracy is comparable with that achievable by all-elliptic 3D Reynolds-averaged Navier–Stokes solvers. But there, the boundary-layer methods employed were all of the preCFD integral-equation kind; and these are restricted to 2D. No one seems to have thought of using *numerical* parabolic solvers, which, of course, are not so restricted. Was not, after all, the now-widely used SIMPLE algorithm [5] first introduced for 3D boundary-layer calculations?

3.3.6 Concluding remarks about SPARSOL

SPARSOL, as just described, is simply a variant of what has come to be known as the immersed boundary method. Whether it is superior or inferior to others can be determined only by extensive tests that are beyond the present author's competence. All that can perhaps be claimed is that, because of the publisher's generous space allocation, it is at least the most completely described. (Question to readers #4: is this true?)

4. TRY-ONS

4.1. A differential equation for mixing length

4.1.1 What ludwig prandtl might have done

Prandtl was nearing the end of his life when he published his one-equation turbulence model in 1945; so perhaps he did not have time to recognize that a more advantageous dependent variable than k might have been chosen. Since his time, other choices have been made, in particular that of Nee and Kovasnay [41] in 1968.

The novelty introduced by them was the proposal that the effective viscosity, ν_{eff}, as well as appearing as a coefficient in the diffusion terms, should *itself* appear as the dependent variable of the differential equation. And why not? If Harlow and Kolmogorov could treat dissipation rate, ε, and frequency, f, as conserved properties, why not ν_{eff} as well?

Had Ludwig Prandtl followed the same line of thought, then the k-equation would not been needed. He might well have chosen his *mixing length* as the conserved property; then the "*mixing length transport model*" might be among those used by engineers today.

4.1.2 The spalart–allmaras viscosity-transport model

Before speculating further about what Prandtl might have done, it should be mentioned that, long after it had become widely accepted that only turbulence models with two or more differential equations were worthy of study, the Nee–Kovasnay innovation was revived and developed by Spalart and Allmaras [42, 43]. Their one-equation model has been shown often to perform and more sophisticated ones, at least in the aeronautics-related problems for which it has been tested. Extensions to supersonic flow have been successfully made [44].

Inspection of the now-extensive literature on the subject has evoked the following thoughts in the present author's mind:

- Spalart's remark in Ref. [44], "turbulence modeling *can* stagnate," is a wise warning, which he at least is heeding.
- Also notable is his remark: "No deep reason was seen why two equations were indispensable, although this remains a widespread position."
- He has also felt free to formulate his one equation in unconventional ways, using vorticity rather than strain rate in source terms and allowing the turbulent-diffusion terms to conserve not ν_{eff} but $\nu_{\text{eff}}^{1.62}$.
- It is with such a free-from-preconception attitude that any alternative one-equation model should be considered.

4.1.3 The "mixing length transport try-on"

Much is known about the distributions of mixing length in turbulent flows, including that, at high Reynolds numbers, it increases with flow-direction distance x, raised to a power, equal to

- unity, in a plane mixing layer, a plane jet, or a circular-section jet, each having its own proportionality constant;
- one-half in a plane wake, for example, behind a circular cylinder; and
- one-third in an axisymmetric wake, for example, behind a sphere.

Moreover, it tends to be 0.41 times the distance from the wall, as a solid wall is approached; and downstream of a grid of parallel rods, it tends to be 0.103 times the distance between the rods [45]. Within a long pipe of circular cross-section, it obeys the Nikuradze [46] formula:

$$l_{\text{m}}/R = 0.14 - 0.08(1 - y/R)^2 - 0.06(1 - y/R)^4 \tag{1.41}$$

where y is the distance from the wall and R is the pipe radius.

In the semilaminar region very close to a wall, the formula of van Driest [47] is believed to prevail. It is

$$l_{\text{m}}/y = 0.41\{1 - \exp(-y^+/26.0)\} \tag{1.42}$$

All this constitutes perhaps a richer body of empirical information than the ν_{eff} transport modelers start from.

The task is to devise a differential equation of the form

$$\partial(\rho l_{\mathrm{m}})/\partial t + \mathrm{div}\left(\rho_1 \boldsymbol{v}_{i,j} l_{\mathrm{m}}\right) = \mathrm{div}(\Gamma\,\mathrm{grad} l_{\mathrm{m}}) + S \tag{1.43}$$

which, when solved numerically, will yield mixing length distributions that accord with these simple-situation findings; then, it can be reasonably hoped, the solutions to the equations with different boundary conditions will also fit experimental findings reasonably well.

The crucial question is: how should the source-and-sink term, *S*, be expressed in terms of velocities and other variables, so as best to fit the data? A reasonable first guess for high Reynolds numbers would be

$$S = const1 * strain\_rate * (l_{\mathrm{m}} - 0.41 * W_{\mathrm{dis}} - 1.03 * profile\_width) \tag{1.44}$$

wherein

- l_{m} is the local mixing length, to be used with Eq. (1.1);
- *const1* is a dimensionless constant or a function of dimensionless properties of the local flow, chosen so as to fit known experimental data regarding mixing length distributions, such as those cited earlier;
- *strain_rate* would be deduced from local velocity gradients in a conventional manner;
- W_{dis} is distance from the wall computed by way of the "thick" of Section 3.2; and
- *profile_width* would be introduced so as to reflect the influence of such geometric factors as the spacing of the rods in an upstream grid.

Equations (1.42) and (1.43) are *only* first guesses; and a combination of physical intuition and numerical resourcefulness will be needed to translate them into finite-volume equations and convergent solution procedures. For example, at the upstream edge of an ideal mixing layer, the strain rate is infinite and l_{m} is zero; for the furthest upstream control volume, an approximation must be employed for their finite product that makes physical sense.

4.1.4 How "const1" might be determined: the "reverse-engineering" approach

Publications about turbulence models commonly report what empirical auxiliary functions are to be used, but not how they were arrived at. In the present case, those functions have not yet been determined; so it is proper to propose how they *could* be. The approach suggested is here called "reverse engineering," for it starts with given facts and tries to work out how they came about.

The facts to start with are the experimentally determined velocity profiles. For the simple situations listed in the preceding text, these can be approximately represented as piecewise polynomials. Then, the "reverse engineering" starts by *imposing* the corresponding velocities, if one's CFD code allows it, at the corresponding nodes of the computational grid, by supplying linearized momentum sources to the cells surrounding them.

Thus, if the source is specified as, say, $1 \times e^6 (u_{experimental} - u)$, it will be large enough to make the calculated velocity differ little from the experimental one; and its magnitude, if printed out, will provide significant Information. Thus, if the effective viscosity of the computer code has been set equal to zero, the printed-out sources disclose magnitudes of the shear stresses that are present in the real flow.

The next move is to calculate, from the printed-out source differences, what are the shear forces at the boundaries of the velocity cells. This can be done by working from the free-stream boundaries of the layer, where the sources are zero, toward the center. From the shear forces and the known velocity differences, the effective viscosities for each internode link can be computed. If the code permits the link-by-link insertion of effective viscosities, doing this and observing whether the printed-out sources indeed now become small, is a useful test of the accuracy with which the whole operation has been conducted.

Once the effective viscosities are known, the corresponding mixing lengths for each link can be computed. Then, *solution* for l_m can be activated, the just-computed "experimental" values being fixed by linearized-source terms, in the same way as was done for velocity. The then-printed-out l_m sources provide information about the cell-by-cell values of the *whole* of the right-hand side of Eq. (1.43), that is, the contributions of both the S and the Γ terms. Disentangling the two contributions will involve (cautiously) making some presumptions.

Fortunately, not one but several simple situations exist to which this "reverse-engineering" process can be applied; and the relative importance of the S and the Γ terms is unlikely to be the same in each. Therefore, disentanglement may not prove to be too difficult. PhD students indeed are likely to relish the challenge.

In addition to the simple situations already cited, there exist others for which reliable data are available. The "backward-facing-step" is one that has been used for the critical assessment of two-equation models [48]. It is of interest because it exhibits free-shear-layer *and* near-wall *and* 2D recirculation effects. Whether the one-equation "mixing length transport model" could handle all three simultaneously is a question of great interest. The Spalart–Allmaras model does *not* appear to have been tested for this flow.

4.1.5 Concluding remarks about mixing length transport

The purpose of the foregoing paragraphs is not to persuade readers that l_m *is* a better dependent variable to use than ν_{eff} or k, but merely that it *might* be, and that nobody knows. Not only *can* turbulence modeling research stagnate, but also it *has stagnated.* It is hoped that some readers of this chapter will see and seize some of the many opportunities for further progress that still exist.

4.2. The population approach to swirling flow

4.2.1 The problem

Swirling turbulent flows are of great practical importance. They are employed

- in gas-turbine combustors in order to promote mixing of fuel and air and thereby to increase thrust per unit volume and combustion efficiency;
- in the large mechanically stirred reactors of chemical industry for similar purposes;
- in hydro- and aerodynamic cyclones in order to promote *un*mixing, that is, the preferential separation of elements of the mixture experiencing different body forces.

Such flows are also regarded as of such scientific interest that special conferences of theoreticians are devoted to them. Perusal of their proceedings, however, reveals an astonishing fact, namely, that although the preferential-separation process experienced by materials *carried by* the swirling flow is considered that the *turbulence itself* is subject to preferential separation receives no direct attention. Instead, attempts are made to "tweak" the constants and functions of turbulence models of the kind that are used for *non*swirling flows, but with meager success.

4.2.2 A "try-on" solution

In a recent unpublished presentation at such a conference, [55] to which he now seeks to give wider publicity, the author proposed that the analogy with two-phase flows should be exploited. Figure 1.42 illustrates how a stream comprising a uniform mixture of water droplets of water suspended in vapor-phase steam behaves when it flows through a curved duct. The flow is from bottom to top. The colored contours denote volume fractions, yellow signifying "high" and light blue "low."

Clearly, and understandably, the greater centrifugal forces on the droplets have caused the water to migrate to the larger-radius region, forcing the air to move to the inside of the bend. The point of the demonstration was to argue that, if the force differential had been caused by differences of *velocity* rather than of density, the effect would qualitatively have been the same.

In connection with Fig. 1.20, it was seen that the use of a two-member population of a single-phase mixture, when Navier–Stokes equations are solved for each of the members, can simulate qualitatively the transition from deflagration to detonation. May it not be therefore that the same model could at least throw some light on the "velocity-sifting" phenomenon that gives swirling turbulent flows their special character? And that a multi-member model might do still more?

For multimember populations, to solve Navier–Stokes equation sets for each member would be excessively costly; and it would be disproportionate in view of the guesswork that would be needed concerning the friction forces opposing the sifting process. Nevertheless, first steps with such guesswork applied to a 17-member model were reported in the just-mentioned presentation, the application being to an imaginary flow in the space between concentric cylinders, rotating about their common axis at different velocities, as shown in Fig. 1.43.

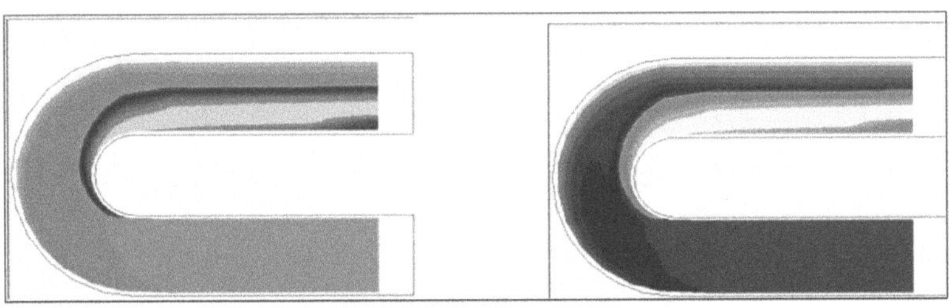

Figure 1.42 Computed volume fractions of water (left) and steam (right). (For color version of this figure, the reader is referred to the online version of this chapter.)

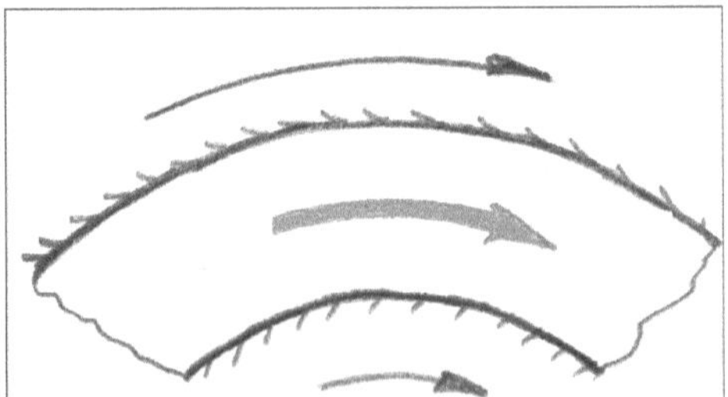

Figure 1.43 Flow between rotating coaxial cylinders. (For color version of this figure, the reader is referred to the online version of this chapter.)

It was postulated that at some entrance plane, on the left, the velocity population was extremely orderly, each member being located at the radius that corresponded to a linear velocity distribution from inner to outer radius. Turbulent mixing was then postulated as occurring, with a diffusivity corresponding to experimental data in fully developed plane channel flow. The consequent intermingling of the members was then computed.

The contour diagrams in Fig. 1.44 were computed for cylinders of infinite radius, that is, for *no* curvature. They showed for 3 of the 17 fluids how turbulent mixing causes the mass fractions to spread with distance downstream. This diffusion process is opposed by the collision/engulfment process that tends to even out the local probability density functions; and the fact that the contour lines become horizontal on the right indicates that the two opposing phenomena finally balance (Fig. 1.45).

Thereafter, calculations were carried out with finite radius, and in order to throw light on the role of the centrifugal forces, two cases were considered. In the first, it was the larger-diameter cylinder that had the higher velocity; in the second, it was the smaller-diameter one. One would expect the first to diminish the intermingling effect and the second to increase it. This expectation is borne out by the corresponding fluid-population distributions shown in Figs. 1.46 and 1.47, respectively.

The calculations just described were made to show that a multimember population, with longitudinal velocity as the population-distinguishing attribute, could be made at little expense and that the results are qualitatively plausible. The exercise was of the "try-on" character; and the conclusion no more than: "Yes, I think it might work."

Perhaps that is more or less what Ludwig Prandtl thought after his first experiments with the mixing length model.

4.3. Hybrid CFD "Try-on"

4.3.1 The general idea

The word "hybrid" is often used nowadays in the turbulence modeling literature to describe the practice of employing different models of turbulence in different parts of the same field of flow. An example is the use of a steady-state RANS model close to a wall and an unsteady-state LES one elsewhere [49]. The "try-on" now to be proposed can be regarded as an extension of the idea: it involves using not only different formulations of the governing equations in the different regions but also different methods for solving them.

The idea is of course not new: and it was in common use by aerodynamicists long before CFD existed. Even then, aerodynamicists could predict

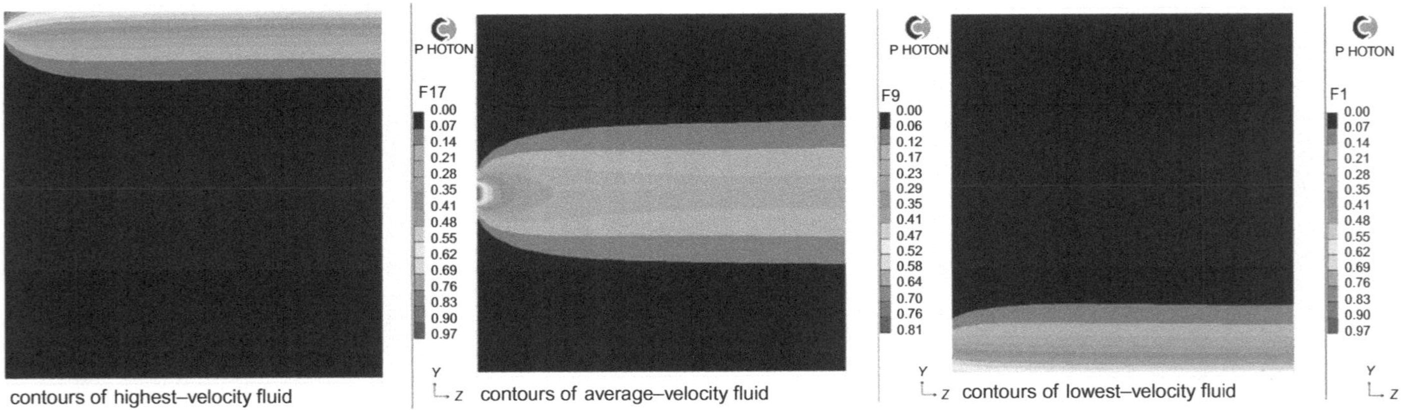

Figure 1.44 Contour diagrams of mass fractions of highest-, middle-, and lowest-velocity fluids, with flow from left to right and radius vertically upward. (For color version of this figure, the reader is referred to the online version of this chapter.)

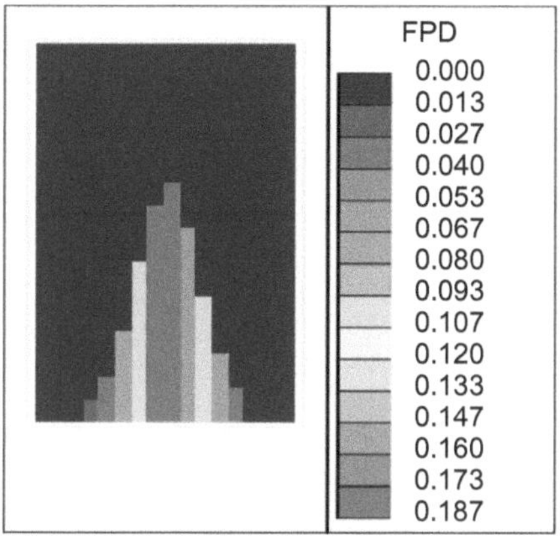

Figure 1.45 Computed fluid-population distribution far downstream for a location midway between the moving surfaces, for the case of zero curvature. Fluid 9, the middle-velocity population member, has the highest mass fraction, namely, 0.187. (For color version of this figure, the reader is referred to the online version of this chapter.)

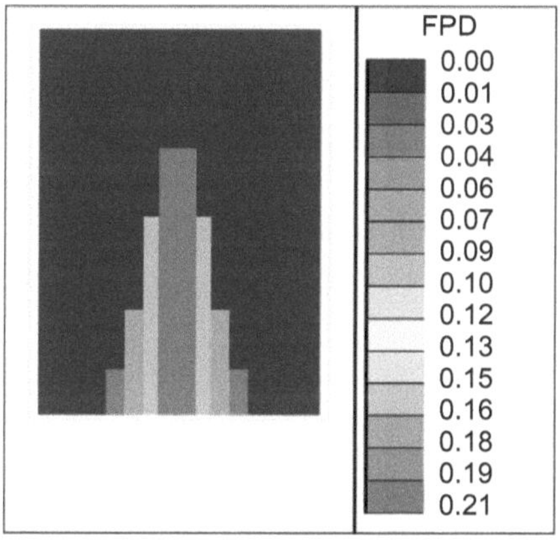

Figure 1.46 Computed fluid-population distribution far downstream for a location midway between the moving surfaces, for the case of faster-moving outer cylinder. Fluid 9, the middle-velocity population member, still has the highest mass fraction, and it has risen to 0.21. (For color version of this figure, the reader is referred to the online version of this chapter.)

their lift and drag. They used a combination of *potential-flow* theory with *boundary-layer* theory, proceeding iteratively:

- First source–sink distributions were sought that caused streamlines to coincide with the shape of the airplane. This led to distributions of *pressure* over the surface.

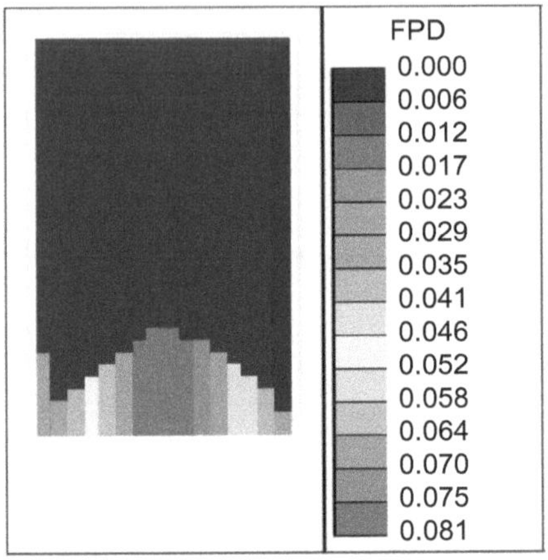

Figure 1.47 Computed fluid-population distribution far downstream for a location midway between the moving surfaces, for the case of faster-moving inner cylinder. Fluid 9, the middle-velocity population member, still has the highest mass fraction; but it has fallen to 0.081. (For color version of this figure, the reader is referred to the online version of this chapter.)

- Then, they used boundary-layer theory to calculate the "displacement thickness" of the layer, that is, the extent to which the airplane was bigger than at first supposed.
- Then, they repeated the first step with a new specification of airplane shape; and the second with the consequentially new pressure distribution.
- And so on until the changes of displacement-thickness distribution became small enough to ignore.

Of course, the boundary-layer theory was primitive, being of the 2D integral-profile kind. However, the principle was sound, and it still is. A recent application of the method to a 2D airfoil was referred to in Section 3.3.5; and the accompanying remarks will now be expanded upon.

4.3.2 The partially parabolic method extended

When, at Imperial College in the 1970s, the SIMPLE method, having been invented for 3D parabolic-flow problems, was recognized as applicable to elliptic problems also, both 2D and 3D, it was the latter that attracted the most attention. However, computers were still small, and any memory-saving device was welcome. One such device was the so-called partially parabolic method [50–52]; and the efficacy of the method was demonstrated by reference to turbulent flows in curved ducts and around the sterns of seagoing vessels. This exploited the fact that, the downstream-to-upstream terms in the momentum equations being negligible, the equations became parabolic in nature; so they could be solved, once the pressure distribution was known, by the marching-integration procedure of Ref. [5], which required only *2D* storage. Only the mass conservation equation, from which the pressure was computed, exhibited significant downstream-to-upstream influences; so it alone required a 3D grid.

The method attracted no attention outside Imperial College; and the present author's interests became deflected in the direction of two-phase flows in nuclear steam generators and in gas-turbine combustion chambers. To neither of these was the partially parabolic method applicable, for they exhibited no predominant direction of flow. Easily achievable developments were therefore not then pursued; but three such advantageous developments will now be outlined, as follows:

1. In the all the work carried out in the 1970s, the *same* grid was used for the pressure as for the other variables; but there was no need for this, and advantage in respect to realism could have been attained by using much finer cells in the parabolic grid than in the elliptic one. This is affordable, because the parabolic grid requires only 2D storage.

   To allow for differences in the main flow-direction step sizes, the pressure gradients used in the momentum equation would have to be interpolated; but, since pressure varies much more gradually than other variables, little diminution of accuracy is to be expected.
2. In the early work, the volumes of space traversed by the elliptic and the parabolic calculations were also the same; but in the applications envisaged in the present "try-on," which include flows around aircraft and missiles, the parabolic calculation would be confined to the regions in which velocity gradients were significant, namely, close to the wall, and in the jets and wakes; for elsewhere, the flow is inviscid. This would generate further economies.
3. There would also be no need to have only one parabolic grid. Indeed as exchanges of information between elliptic and parabolic solutions proceeded, the stagnation point on the airplane nose might well shift from iteration to iteration. So a different parabolic grid would be needed each time. But why not? Grid generation for parabolic flows is very easy. Moreover, it is probable that *several* parabolic grids would be better than one; and the upstream boundary condition for farther-downstream ones would be reduced by interpolation in the values at the outlet surfaces of upstream ones.

It remains to ask: Why is this hybrid "mix-and-match" method not used? Are the advantages of economy and realism here envisaged not worth having? Or not recognized? Or recognized, but deemed illusory? Or is it that the possibility of solving 3D boundary-layer equations numerically is not widely understood?

4.3.3 Simulating automobile aerodynamics

Early CFD calculations of flow around automobiles made by the present author used computational grids such as that in Fig. 1.48. They extended over a much larger volume than that occupied by the vehicle; and, despite the fact that in the greater part of the volume the flow was inviscid, the 3D Navier–Stokes equations were solved throughout (Fig. 1.49).

It was a foolish practice of course; but even now, it is customary to employ a single grid, albeit of unstructured form. Therefore, it appears reasonable to ask: Why not use instead

- a 3D elliptic potential-flow solver for the inviscid flow, as has been proposed in Section 4.3.2 for the airplane;
- a 3D Navier–Stokes parabolic solver for the roof and sides; and
- embedded 3D Navier–Stokes solvers for the wakes of the car body, the wheels and the wing mirrors?

As compared with the airplane problem, the iterative interfacing will be somewhat more complex; and it is probable that more iterations will be needed before the interactions between the regions—elliptic-inviscid, parabolic-viscid, and elliptic-viscid—have procured mutually agreeable solutions of the interlinked equations. It is not *certain* that the final result of the simulation will be the same as that of a single fine-grid elliptic

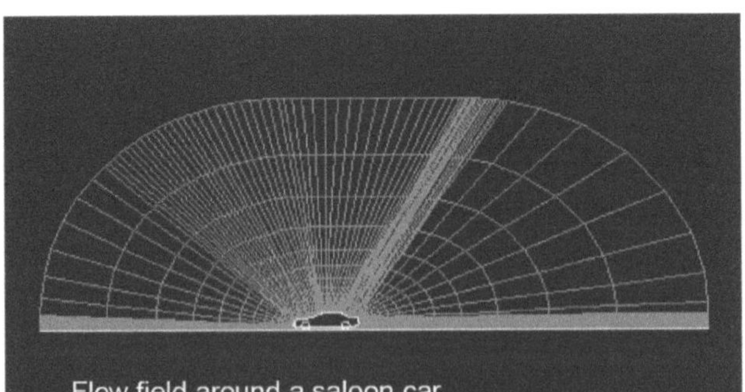

Figure 1.48 The whole grid. (For color version of this figure, the reader is referred to the online version of this chapter.)

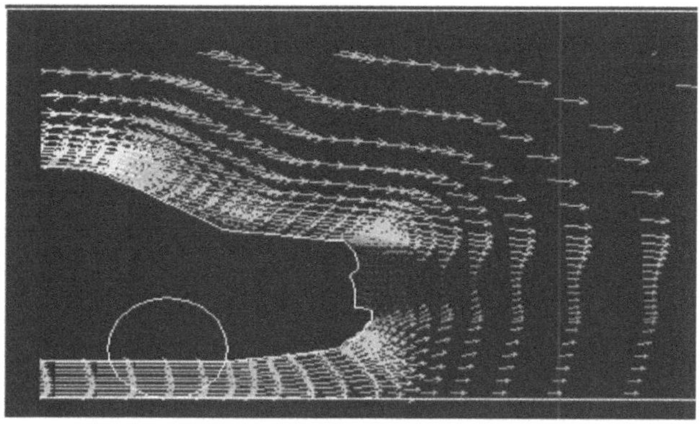

Figure 1.49 The wake. (For color version of this figure, the reader is referred to the online version of this chapter.)

Navier–Stokes solution for the whole space or that the total time-and-memory computer resource will be smaller. But, it seems highly probable and certainly worth a try.

4.3.4 Environmental applications

For the design of wind farms and the investigation of atmospheric-pollution phenomena, it is necessary to calculate the fields of velocity, temperature, and concentration in the atmospheric boundary layer, in spaces that extend several kilometers in the horizontal directions but have much smaller vertical heights. Such problems are well suited to solution by the partially parabolic method; although the wind directions are different near the ground at higher altitudes, the differences are not so great that a single direction of "marching integration" cannot be found for which all normal-to-plane velocities are negative. In other words, a "predominant direction of flow" can exist.

Probably no part of the domain can be regarded as inviscid; so the problem is more akin to the curved-duct and ship's-stern problems, which were solved already in the 1970s. So much the better! Nevertheless, it would be possible to profit from the iteration-between-linked-regions technique that has been outlined in the foregoing Sections 4.3.2 and 4.3.3. Finer-grid partially parabolic regions can be embedded inside coarser-grid ones, in the same way as elliptic regions.

Lastly, let flow over an urban terrain be considered, for example, the campus of the university of Delft, shown in Fig. 1.50 [56]. It is customary to employ 3D elliptic solvers for simulating such flows, with the finest grid that can be afforded (1 m in the case illustrated).

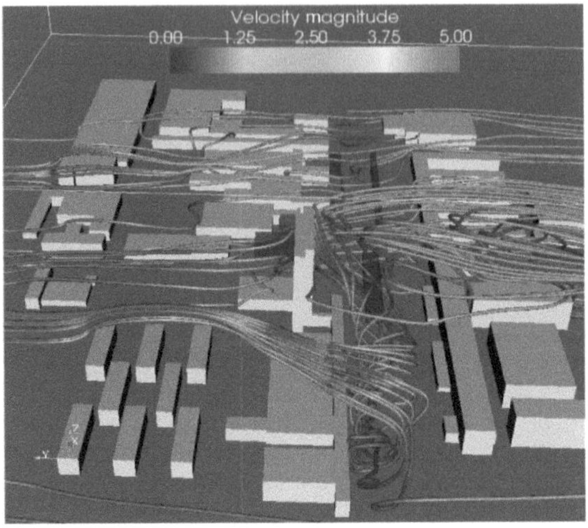

Figure 1.50 Urban-terrain simulation. (For color version of this figure, the reader is referred to the online version of this chapter.)

Inspection of the streamlines confirms the expectation that the prevailing wind enforces a predominant direction of flow in most of the space but that recirculation regions do exist in the wakes of the buildings. It follows that the original partially parabolic method can*not* be used; but the extended one, which allows the embedding of recirculation regions as was outlined in connection with flow around cars, *can* be employed.

Of course, the number of recirculation regions may be rather large; but

- the flows within them need not be simulated simultaneously;
- visiting them in the order upstream-before-downstream will maximize speed of convergence;
- usually only a few of them are of interest to the user, so it makes sense to select them for most frequent and finest-grid treatment.

Generally speaking, the hybrid-CFD approach of the present "try-on" will allow, it is suggested, choices to be made that are optimal in respect to both economy and fitness for purpose.

4.3.5 Generalizing wall functions

In natural- or urban-terrain simulations, it will often be useful to split the whole atmosphere into at least an upper and a lower region. The surface separating them, which might be either horizontal or parallel to the undulating ground, should be high enough to ensure that no flow occurs in the direction opposite to the predominant one. Then, the original partially parabolic method can be used for the upper region; and any embedded elliptic volumes are contained within the lower one.

Of course, the two regions must exchange information, between iterations, regarding the pressures above and below the boundary and the fluxes of mass and of horizontal-direction momentum across it; and, as far as the upper region is concerned, this might as well be cast in wall-function form.

There is no need to go further; but far-seeing researchers might perceive that, once very many such situations have been analyzed, it may be possible to recognize quantitative connections between the aforesaid fluxes and some averaged properties of the below-dividing-surface contents. Thus, the effective shear-stress coefficient might be expected to depend on the amount of solid material, its surface area, and the typical solid-element size. One day indeed, a large-scale research program might be instituted of which the final deliverable would be a comprehensive set of properly parameterized formulas for the "effective roughness" of forests and cities. These could be used by practicing engineers, town planners, and environmentalists who had not time themselves to make massive CFD calculations.

Alas, no such pragmatic research programs are visible on the current scientific scene; but, when funding agencies are more wisely directed, they may appear in the future.

5. CONCLUDING REMARKS

The earlier-mentioned miscellany of fact and speculation, of history and prophesy, of argument and opinion, and of "broad-brush" and "nitty–gritty" has been launched with the motives expressed in the abstract. Readers who reach the end, will now look on CFD/CHT, it is hoped, as less daunting and awesome than they thought; and also as less *finished*. If some of them are caused to think "I believe that I could do better than *that*, the author will be well pleased."

REFERENCES

[1] A.K. Singhal, D.B. Spalding, Mathematical modelling of multiphase flow and heat transfer in steam generators, in: Multiphase Transport: Fundamentals, Reactor Safety, Applications, vols. 1–5, Hemisphere Publishing, Washington, DC, 1980, pp. 373–406.

[2] C.S. Peskin, Flow patterns around heart valves: a digital computer method for solving the equations of motion, PhD thesis, Albert Einstein Coll. Med, 1972.

[3] S. Kang, An improved immersed boundary method for computation of turbulent flows with heat transfer, PhD thesis, Stanford University, 2008.

[4] L.S. Caretto, R.M. Curr, D.B. Spalding, Two numerical methods for three-dimensional boundary layers' computer methods, Appl. Mech. Eng. 1 (1) (1972) 39–57.

[5] S.V. Patankar, D.B. Spalding, A calculation procedure for heat, mass & momentum transfer in three-dimensional parabolic flows, Int. J. Heat Mass Transf. 15 (1972) 1787–1806.
[6] G.H. Golub, C.F. van Loan, Matrix Computations, fourth ed., John Hopkins, 2013.
[7] Y. Saad, Iterative Methods for Sparse Linear Systems, SIAM Publications, Philadelphia, 2003.
[8] M.R. Hestenes, E.L. Stiefel, Method of conjugate gradients for solving linear systems, J. Res. Natl. Bur. Stand. B 49 (1952) 409–436.
[9] B.E. Launder, D.B. Spalding, Mathematical Models of Turbulence, Academic Press, London and New York, 1972.
[10] J. Boussinesq, Theorie de l'ecoulement turbulente, Mem. Pre. Par. Div. Sav. 23 (1877), Paris.
[11] L. Prandtl, Bericht ueber Untersuchungen zur ausgebildeten Turbulenz, ZAMM 3 (1925) 136–139.
[12] A.N. Kolmogorov, Equations of motion of an incompressible turbulent fluid, Izv. Akad. Nauk. SSSR Ser. Phys. VI (1–2) (1942) 56.
[13] L. Prandtl, Ueber ein neues formelsystem fuer die ausgebildete turbulenz, Nachr. Akad. Wiss. Goettingen (1945).
[14] F.H. Harlow, P.I. Nakayama, Turbulence transport equations, Phys. Fluids 10 (11) (1967) 2323–2332.
[15] D.B. Spalding, GENMIX: A General Computer Program for Two-Dimensional Parabolic Phenomena, Pergamon Press, Oxford, 1977.
[16] P.G. Saffman, A model for inhomogeneous turbulent flow, Proc. R. Soc. London A317 (1970) 417–433.
[17] D.B. Spalding, The prediction of two-dimensional steady turbulent elliptic flows, in: ICHMT Seminar, Herceg Novi, Yugoslavia, 1969, See also D. Brian Spalding, A two-equation model of turbulence, VDI Forschungsheft 549 (1972) 5–16.
[18] D.C. Wilcox, Multiscale model for turbulent flows, AIAA J. 26 (1988) 1311–1320.
[19] S.V. Patankar, D. Brian Spalding, Heat and Mass Transfer in Boundary Layers, first ed., Intertext, London, 1967, 2nd ed. 1970.
[20] A.D. Gosman, W.M. Pun, A.K. Runchal, D.B. Spalding, M. Wolfshtein, Heat and Mass Transfer in Recirculating Flows, Academic Press, London, 1969.
[21] D.B. Spalding, A single formula for the law of the wall, Trans. ASME E J. Appl. Mech. 28 (1961) 455–458.
[22] M.A. Leschziner, et al., (Ed.), Proceedings of ETMM8, Marseille, France, 2010.
[23] U. Schuhmann, G. Groetzbach, L. Kleiser, Direct numerical simulation of turbulence, in: W. Kollmann (Ed.), Prediction Methods for Turbulent Flows, von Karman Institute, 1980.
[24] W. Rodi, G. Contstantinescu, T. Stoesser, Large-Eddy Simulation in Hydraulics, CRC Press, 2013.
[25] D.B. Spalding, PPAs & PPBs, facilitating the comparison of experimental & DNS-, LES-, PANS-, PDF-transport and MFM models of turbulence, in: M.A. Leschziner, et al., (Ed.), Proceedings of ETMM8, Marseille, France, 1, 2010, p. 343.
[26] D.B. Spalding, Turbulent mixing and chemical reaction; the multi-fluid approach. http://www.cham.co.uk/phoenics/d_polis/d_lecs/mfm/mfm00.htm, 1995–2010.
[27] D.B. Spalding, Benjamin Franklin and computational fluid dynamics: the population approach to turbulence, ASME J. Heat Transf. Comput. Fluid Dyn. 0022-1481, 135 (2013) 011005-1–011005-12.
[28] D.B. Spalding, Combustion and turbulence; old models and new, in: Huw Edwards Award Lecture, Institute of Physics, 2012.
[29] D.B. Spalding, Mixing and chemical reaction in steady confined turbulent flame, in: Proceedings 13th International Combustion Symposium, The Combustion Institute, 1971, pp. 649–657.

[30] D.B. Spalding, Numerical computation of multiphase flow and heat-transfer, in: C. Taylor, K. Morgan (Eds.), Contribution to Recent Advances in Numerical Methods in Fluids, Pineridge Press, Swansea, 1980, pp. 139–167.

[31] PHOENICS Library Case W979. http://www.cham.co.uk/phoenics/d_polis/d_applic/d_react/2flrpl.htm.

[32] D.B. Spalding, Models of turbulent combustion, in: Proc. 2nd Colloquium on Process Simulation, Helsinki University of Technology, Espoo, Finland, 1965, pp. 1–15.

[33] G.C. Williams, H.C. Hottel, A.C. Scurlock, in: Third Symposium on Combustion, Williams and Wilkins, 1949, p. 21.

[34] C. Dopazo, E.E. O'Brien, An approach to the auto-ignition of a turbulent mixture, Acta Astronaut. 1 (1974) 1239.

[35] S.B. Pope, A Monte Carlo, Method for the PDF equations of turbulent reactive flow, Combust Sci. Technol. 25 (1981) 150–174.

[36] D.B. Spalding, Enlarging the frontiers of computational heat transfer, in: Advances in Computational Heat Transfer, CHT-8, Marrakech, 2008.

[37] S.M. Klausmeyer, Comparative results from a CFD challenge over a 2D three-element high-lift airfoil, in: NASA Technical Memorandum 112858, 1997.

[38] E.M. Sparrow, R.D. Cess, Radiation Heat Transfer, Wadsworth Inc., California, 1966.

[39] S.G. Martyushev, M.A. Sheremet, Characteristics of Rosseland and P-1 approximations in modeling nonstationary conditions of convection-radiation heat transfer in an enclosure with a local energy source, J. Eng. Thermophys. 21 (2012) 111–118.

[40] D.B. Spalding, Calculation of turbulent heat transfer in cluttered spaces, in: Proc. 10th Int. Heat Transfer Conference, Brighton, UK, 1994.

[41] V.W. Nee, L.S.G. Kovasnay, The calculation of the incompressible turbulent boundary layer by a simple theory, in: Proc. AFOSR/IFP Conference on Computation of Turbulent Boundary Layers, 1968.

[42] P.R. Spalart, S.R. Allmaras, A one-equation turbulence model for aerodynamic flows, La Recherche Aerospatiale 1 (1994) 5–21.

[43] P.R. Spalart, Strategies for turbulence modelling and simulations, Heat Fluid Flow 21 (2000) 252–263.

[44] S. Deck, P. Duveau, P. D'Espiney, P. Guillen, Development and application of Spalart–Allmaras one-equation turbulence model to three-dimensional supersonic complex configurations, Aerosp. Sci. Technol. 6 (2002) 171–183.

[45] R. Gran Ollson, Gescwindigkeitsuns temperatur verteilung hinter einem gitter bei turbulenter stroemung, ZAMM 16 (1936) 257–267.

[46] J. Nikuradze, Gesetzmaessigkeit der turbulenten Stroemung in glatten Rohren, VDI Forschungsheft 356 (1932) 20.

[47] E.R. Van Driest, On turbulent flow near a wall, J. Aerosp. Sci. 23 (1956) 1007.

[48] F.T.C. Thangam, C.G. Speziale, Turbulent separated flow past a backward-facing step; a critical evaluation of two-equation turbulence models, in: NASA Contractor Report 187532 AD-A233 478; ICASE Report No. 91-23, 1991.

[49] J. Frohlich, D. von Terzi, E. Severac, R. Vaibar, Hybrid LES/RANS simulations—state of the art and perspectives, in: Proceedings ETMM8, Marseille, France, 2010.

[50] V.S. Pratap, D.B. Spalding, Numerical computations of the flow in curved ducts, Aeronaut. Q. 26 (1975) 219–228.

[51] A.M. Abdelmeguid, N.C. Markatos, K. Muraoka, D.B. Spalding, A comparison between the parabolic and partially-parabolic solution procedures for three-dimensional turbulent flows around ships' hulls, Appl. Math. Model. 3 (4) (1979) 249–258.

[52] A.K. Singhal, D.B. Spalding, A two-dimensional partially-parabolic procedure for axial-flow turbomachinery cascades, in: Aero. Res. Council Report Number R&M 3807, 1978.

[53] R. Mittal, G. Iaccarino, Immersed boundary methods, Annu. Rev. Fluid Mech. 37 (2005) 239–261.
[54] http://www.cham.co.uk/phoenics/d_polis/d_enc/enc_rad3.htm.
[55] D.B. Spalding, in: Lecture at Third International Symposium: Heat Transfer & Hydrodynamics in Swirling Flows, Moscow Power Engineering Institute, 2008.
[56] S. Kenjeres, B. ter Kuile, Modelling and simulations of flow, turbulence and dispersion in urban environments with non-homogenous vegetation, in: Proceedings of ETMM8, Marseille, France, 2010, p. 170.
[57] D.B. Spalding, S.V. Zhubrin, X-cell: a new algorithm for fluid-flow simulation in PHOENICS, 1996, http://www.cham.co.uk/phoenics/d_polis/d_enc/x-cell.htm.
[58] http://www.cham.co.uk/phoenics/d_polis/d_enc/parsol.htm.
[59] http://www.cham.co.uk/phoenics/d_polis/d_applic/applic.htm.
[60] http://www.cham.co.uk/phoenics/d_polis/d_enc/x-cell.htm.

CHAPTER TWO

A Study of Micro-scale Boiling by Infrared Techniques

Gad Hetsroni, Albert Mosyak
Department of Mechanical Engineering, Technion—Israel Institute of Technology, Technion City, Haifa, Israel

Contents

Abstract

This is a critical review of the application of infrared (IR) techniques in micro-scale boiling phenomena. Investigations performed within the last decade were used to analyze the specific features of micro-scale IR measurements. The method is illustrated by various

Advances in Heat Transfer, Volume 45
ISSN 0065-2717
http://dx.doi.org/10.1016/B978-0-12-407819-2.00002-5

examples of heat transfer determination using image analysis of the heater surface temperature. The good results obtained in the last decade by widespread use of IR technique in experimental studies of convective problems have proved the IR technique to be an effective tool in overcoming several limitations of the standard sensors.

1. INTRODUCTION

1.1. Capability of infrared thermography heat transfer measurements

The heat transfer results in micro-channels are significantly different from those obtained in larger size channels [1–5]. The use of IR techniques in experimental studies of convective heat transfer problem has proven to be an effective tool in overcoming several limitations of the standard sensors both for temperature measurements and flow visualization [6,7].

The use of quantitative IR thermography requires resolution of several problems. These problems include accurate characterization of the IR system performance and its calibration: the use of optics to increase the spatial resolution, the choice of the most appropriate heat flux sensor, its characterization especially with regard to the lateral thermal conduction effects, etc. [8–10].

Measuring heat fluxes in thermo-fluid dynamics requires both a thermal sensor (with its related thermophysical model) and temperature transducers. In standard techniques, where temperature is measured by thermocouples, resistance temperature detectors (RTDs), etc., each transducer yields either the heat flux at a single point or the space-averaged one. For example, cartridge heaters may be embedded in the copper block and one thermocouple inserted halfway along the micro-channels may be used to infer the micro-channel base temperature whereas the second thermocouple inserted opposite at given distance in the copper block allows calculating heat flux using 1D heat conduction equation. Hence, in terms of spatial resolution, the sensor itself has to be considered as zero-dimensional. This limitation makes measurements particularly troublesome whenever the temperature, and/or heat flux, fields exhibit high spatial variations. Instead, the IR camera constitutes a truly 2D transducer, allowing for accurate measurements of surface temperature maps even in the presence of relatively high spatial gradients. The sensed intensity will differ across pixels from the net radiated intensity from the source due to variations in the configuration factor between the physical measurement area and the corresponding pixel array. Accordingly, the heat flux sensor also becomes 2D, as long as the necessary corrections are applied. A first analysis of heat transfer measurements by IR thermography and a review of some of their applications were presented by Carlomagno and de Luca [9]. The basic characteristics of the method were evaluated also by Hetsroni *et al.* [11]. Those authors found the dependence of the frequency response on the physical properties of the sensor material, its geometry, as well as on the frequency of temperature fluctuations.

When compared to standard techniques, the use of an IR camera as a temperature transducer in convective heat transfer measurements appears advantageous from several points of view. In fact, since an IR camera is fully 2D (up to 1 M pixels), in addition to producing a whole temperature field it permits an easier evaluation of errors due to radiation and in-plane conduction. Further, it is nonintrusive (e.g., not allowing conduction through thermocouple or RTD wires), it has high sensitivity (down to 0.02 K), and low response time (down to 20 μs). As such, IR thermography can be effectively employed to study micro-scale boiling phenomena with both steady and transient techniques [12a,b].

1.2. Methodology of IR measurement in microsystems

The measurement of heat transfer in single- and two-phase flow by the hot-foil IR technique has been thoroughly studied by Hetsroni and Rozenblit [13] (experimentally) and Hetsroni *et al.* [11] (theoretically). This method is based on the measurement of local temperatures on the air-outside surface of a thin heater (a constantan foil 0.050 mm thick), while the other side of the heater is subjected to a flow of liquid and where the heat transfer coefficient between the wall and inner flow is to be determined (Fig. 2.1). The foil was attached to the window and was coated on the bottom (i.e., not on the water side) with a black matt paint of about 0.02 mm thickness. Constant heat flux was achieved by supplying DC power. The transient heat conduction problem was solved numerically for film heater with black paint layer. The mean temperatures between the film sides did not exceed 0.12 °C.

By measuring the bottom temperatures of the heated foil, heat flux and mean fluid temperature it is possible to compute the heat transfer coefficient from the foil to the flowing fluid. In case of simulated computer chip, a thin heater is deposited on the surface of the wafer. This model of IR temperature measurement with an opaque face layer is shown in Fig. 2.2. Thin-film opaque heater is deposited on the surface of wafer facing IR radiometer.

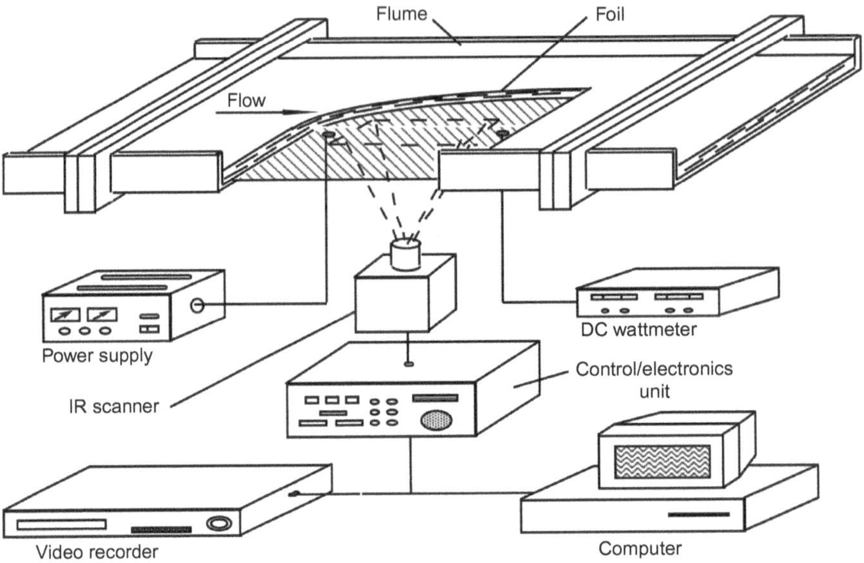

Figure 2.1 A schematic diagram of the measurements.

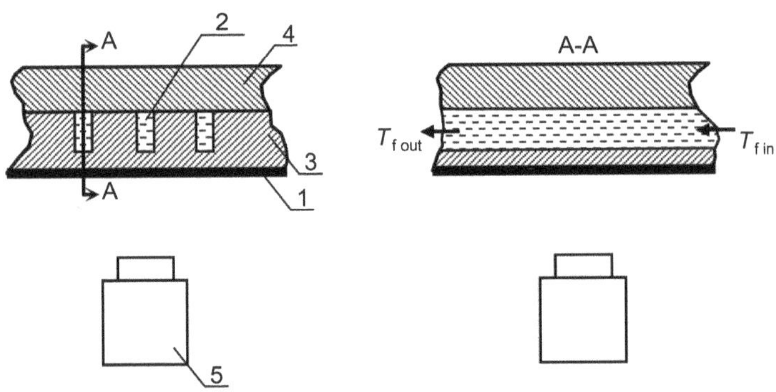

Figure 2.2 Simplified model of the IR measurement with an opaque face layer. 1. Face layer (thin film heater), 2. micro-channel, 3. wafer, 4. cover, 5. IR video.

The face layer in this case must be thin and its material should have high enough thermal conductivity. According to this method the face layer should be heated by supplying electrical current to determine the heat transfer coefficient. A transparent cover made from sapphire makes possible to perform IR measurements of thin fluid layer adjacent to the cover. In this case the IR radiometer should be directed toward sapphire window. The temperature of the heater measured by IR cameras is slightly higher than the temperature of the inner micro-channel wall.

In case of simulated computer chip, a thin heater is deposited on the surface of the chip which is facing the IR radiometer while micro-channels are etched in the other side, covered by a sheet of quartz. In these micro-channels, water (or some other coolant) is circulating. The temperature of the surface of the water can also be measured by the IR radiometer. According to this method, the thin heater is heated by supplying a DC electrical current. The temperature is measured on the surface of heater by IR.

Hetsroni *et al.* [14] and Tiselj *et al.* [15] estimated numerically that the aforementioned deviation is about 0.15–0.8 K in the range of heat fluxes, materials, and dimensions typical to microsystems. Therefore, the temperature measured by the IR camera must be corrected according to the calculated estimation for given topology and material of the micro-device.

Measurement of the temperature field of a micro-object by an IR camera has a number of problems. The small diameter of the device causes a substantial amount of IR radiation from the background. If the background has a temperature different from that of the small-sized object, the surface temperature of the device will be measured with an uncertainty due to the background radiation. The problem of careful accounting of background influence on the object temperature measurement received attention in the handbook by Zissis [16].

IR lenses with different focal distance make possible to focus IR camera on very small area of heat sink. The IR microscopic lenses drastically confine the field of view of the IR camera. For example, only a few millimeters along the capillary tube can be observed. It is also difficult to identify the position of the measuring point along the tube. Moreover, using microscopic lenses does not completely eliminate the influence of the background radiation. That is the reason why a special method was used, similar to the one of the disappearing filament in color temperature measurement [17]. The method consists of compensating the background by controlling its temperature to a level close to the measured temperature of the capillary tube surface. This method was used for investigation of the heat transfer in small-sized channels where the heat transfer coefficient exhibited an unusual behavior.

The basic radiation model of the measurement of a temperature field in a micro-device is presented by Hetsroni *et al.* [18]. According to this model, in general, the micro-scale object is located between a face layer (nearest to the IR radiometer) and foreground layer. The micro-scale object may represent either any part of the wall surface of the micro-device or the liquid in the micro-channel for cooling such devices. In the latter case, the liquid temperature is measured close to its surface. Both the face layer and the foreground layer may be present or omitted in this model according to the characteristics of the micro-object studied. The transparency to an IR irradiation of the face layer and the object may vary from transparent to completely opaque. Each layer and the object itself in such model may have a different emissivity and temperature.

If the face layer is either absent in this model or is transparent to IR radiation, the detector of the IR camera may collect the sum of the emitted and reflected energy coming from the object. We should add to this sum the energy transmitted through the object from the foreground layer (e.g., the heater) if the object itself is transparent to IR radiation. If the face layer is opaque, we can judge the thermal regime of other parts only by an indirect method. The detector of the IR camera may collect not only some background radiation reflected by the object, but that which falls directly within the field of view of the IR camera.

1.3. Microscale phenomenon of boiling

1.3.1 Pool boiling

Performance of boiling equipment is limited by the transition from nucleate to film boiling. The latter is a regime of diminished heat transfer, characterized by the absence of liquid contact on the heater surface and can be accompanied by the ultimate physical destruction of the heater depending on its material of construction. The distraction of the heater is often termed as *burnout*. At the transition from nucleate boiling to film boiling (called boiling crisis), the boiling heat flux achieves its maximum value. This maximum is characterized as the critical heat flux (CHF). Basic considerations led to the creation of a schematic of processes involved in boiling and their relations to burnout, as illustrated in Fig. 2.3 [19].

At high values of heat flux, the flow is past the point where a bubbly regime can be sustained. According to Kutateladze [20], this kind of transition occurs at a heat flux of the order of 100 kW/m^2. The vapor flow is accommodated by "expanding" the two-phase region sufficiently to allow an increasing amount of disengagement from the liquid in the pool. The characteristic scales of such a flow pattern are much greater than that of the capillary length, $\delta = \{\sigma/g(\rho_l - \rho_v)\}^{0.5}$, and the detailed behavior depends very much on the geometry and dimensions of the fluid domain. One can thus refer to a "scales separation" phenomenon in high heat flux, saturated

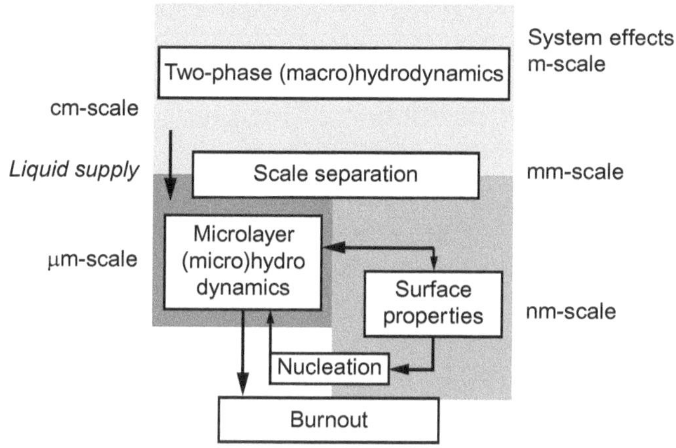

Figure 2.3 Schematic of processes involved in boiling and their relations to burnout [19].

pool boiling. One implication is that the heater and the extended liquid micro-layer on it operate independently of any significant influence of the hydrodynamics of the liquid outside of the micro-layer. A subsequent implication is that since the presence of the liquid pool is only incidental, the burnout phenomenon can be studied in isolation by focusing on the heater-micro-layer system alone. In conjunction with the above-identified scale separation, Theofanous *et al.* [21] examined the IR records to facilitate understanding of the flow regimes that govern micro-hydrodynamics.

1.3.2 Boiling in micro-channels

The study of heat transfer in microsystems such as metal capillary tubes requires also taking into consideration the effect of direct radiation from the background in the field of view of the IR camera. Incoming radiation from extraneous elements to the IR sensor cannot be neglected when obtaining temperature measurements of small-sized devices. A new method, described by Hetsroni *et al.* [22], was applied to decrease a systematic error caused by background radiation. The proposed method is based on compensating the background radiation by controlling its temperature to the level equal to the temperature of the capillary tube. This is achieved by recording the IR data against a background whose temperature was maintained at a given value by a thermostat. In this document, we describe and compare with previous predictions new experimental results and physical approaches developed by studies of micro-scale boiling using the IR technique.

2. BOILING INCIPIENCE

2.1. Models for prediction of incipient boiling heat flux and wall superheat

The onset of bubble nucleation usually requires that the temperature of the heated surface exceeds the saturation temperature of the liquid, corresponding to the pressure. A bubble nucleus will grow if the temperature of the fluid at the distance from the wall equal to the bubble height is greater than the superheat requirement, which depends on the pressure. There has been a great deal of analysis of bubble nucleation and a number of semiempirical models of boiling incipience have been proposed during the past few decades [23,24]. Two general approaches were used to describe the parameters corresponding to the onset of nucleate boiling (ONB): the first is based on an analysis of the behavior of a single bubble on a rough surface where the temperature of the surrounding liquid exceeds the saturation temperature. The second approach considers vapor bubbles generated in the liquid (homogeneous nucleation) or in a cavity on a surface (heterogeneous nucleation) based on the classical kinetics of nucleation.

The first approach developed by Hsu [25] is widely used to determine the ONB in conventional size channels and in micro-channels [26–32]. These models consider the behavior of a single bubble by solving the 1D heat conduction equation with constant wall temperature as a boundary condition. It should be stressed that such models are valid only when the medium is continuous. Measurements by Ye *et al.* [33] showed that bubble nucleation in water may be considered as continuous when the radius of the cavity $r_c \geq 10$ nm.

2.2. Comparison between models and experiments

There is a number of experiments directed at studying the incipience of nucleate boiling in heated channels of hydraulic diameter $d_h = 0.04$–20 mm. The overall characteristics of the experiments are presented in Table 2.1 [34]. Measurements of hydrodynamic and thermal parameters corresponding to ONB were performed in circular, annular, rectangular, and trapezoidal channels where water, refrigerant-22, and fluorocarbon R-113 were used as working fluids. ONB was recorded by various means: analysis of the pressure drop–mass flux characteristic curve; relation between the heat flux and the wall superheat at the starting point of the sharp drop in wall temperature ($\Delta T_{S,ONB} = T_{W,ONB} - T_S$, where T_W is the temperature of

Table 2.1 Onset of nucleate boiling

Author	Channel geometry	Hydraulic diameter (mm)	Mass flux (kg/m^2s)	Fluid	Heat flux (kW/m^2)	Pressure (MPa)
Bergles and Rohsenow [35]	Cylindrical tube	2.387	(2.1–19.4) $\times 10^3$	Distilled water	1890–18900	0.261
Sato and Matsumara [26]	Cylindrical tube	–	–	Water	100–750	0.1
Unal [37]	Cylindrical tube	4–20	132–2818	Water; refrigerant-22	20–1920	0.1–15.8
Hino and Ueda [39]	Cylindrical tube	7.0	158–1600	Fluorocarbon R-113	11.78–46.90	0.147
Kennedy *et al.* [43]	Cylindrical tube	1.17, 1.45	800–4500	De-ionized and de-gassed water	0–4000	0.344–1.034
Hapke *et al.* [80]	Cylindrical tube	1.5	100–500	De-ionized water	50–200	0.1
Yarin *et al.* [34]	Cylindrical tube	1.07	49–146	Distilled water	62–162	0.1
Stoddard *et al.* [124]	Annular channel	0.724–1.0	85–1428	Fully de-gassed water	124–1000	0.344–1.034
Su *et al.* [125]	Annular channel	1.0, 1.5	45–180	Pure water	40–210	0.2–3.5
Qu and Mudawar [29]	Rectangular channel	0.231×0.713	130–1440	De-ionized water	200–2000	0.12
Liu *et al.* [32]	Rectangular channel	0.275×0.636	309–883	De-ionized water	100–730	0.1
Lee *et al.* [40]	Trapezoidal channel	0.413	170–899	De-ionized water	1.49–500	0.161

Characteristics of experiment [39].

the heated wall, corresponding to nucleate boiling, and T_S is the saturation temperature); and visual detection of bubble growth and departure from the heated wall. Experiments were carried out in the range of mass flux $m=45$–$2100\ \mathrm{kg/m^2\,s}$, heat flux $q=1.49$–$18{,}900\ \mathrm{kW/m^2}$, and pressure $P=0.1$–15.8 MPa.

2.2.1 Wall superheat

Bergles and Rohsenow [35] studied the wall superheat that corresponded to the nucleate boiling of distilled water in a stainless steel tube of inner diameter $d=2.387$ mm. The measurements of $\Delta T_{S,ONB}=T_{W,ONB}-T_S$ corresponding to the ONB were performed at different values of inlet temperature, T_{in}, in the range of inlet velocity $U_{in}=3.74$–19.42 m/s, and Reynolds numbers corresponding to inlet parameters $Re=9\times10^3$–2×10^5. Wall temperature measurements were made at the heated length to the tube diameter ratio of $L/d=29$ and 48. Experiments were restricted to low heat fluxes. This usually limits subcooled boiling tests to the region of low wall superheat. During the subcooled nucleate flow boiling of a liquid in a channel, the bulk temperature of the liquid at ONB, T_B, is less than the saturation temperature. At a given value of heat flux, the difference $\Delta T_{sub,ONB}=T_S-T_B$ depends on L/d. The experimental parameters are presented in Table 2.2.

Table 2.3 shows that the ratio $q_{ONB}/\rho U_{in}c_p$ changes relatively weakly in the range of wide variations of heat flux and inlet flow velocity. The wall superheat significantly depends on heat flux. The data of Bergles and Rohsenow [35] are shown in Fig. 2.4. This dependence is close to $\Delta T_{S,ONB}\sim q_{ONB}{}^{0.5}$. The results presented by Sato and Matsumura [26] and the results of flow boiling in the 1.07-mm tube with average roughness of $k_s=0.29\ \mu$m obtained in the range of the Reynolds number $Re=50$–150 by Yarin *et al.* [34] are also shown in Fig. 2.4. The values of $\Delta T_{S,ONB}T_{S,ONB}$ of Sato and Matsumura [26] are higher than those obtained at the same heat flux by Bergles and Rohsenow [35] and Yarin *et al.* [34]. According to studies by Hsu [25], Sato and Matsumura [26], Davis and Anderson [27], Kandlikar *et al.* [34], and Liu *et al.* [32], one can conclude that $\Delta T_{S,ONB}\sim q_{ONB}^{0.5}$ and

$$q_{ONB}=M\frac{k_L h_{LG}\rho_G}{\sigma T_S}\Delta T_{S,ONB}^2 \tag{2.1}$$

In this equation, M is constant over a wide range of $q_{ONB}=10^3$–$10^6\ \mathrm{W/m^2}$, and k_L, ρ_G, σ are the thermal conductivity of liquid, the vapor

Table 2.2 Liquid subcooling at ONB point [34]

Author and method of ONB detection	Heat flux q_{ONB} (MW/m^2)	Inlet flow velocity U_{in} (m/s)	Pressure P (MPa)	Relative heated length L_{ONB}/d	Fluid subcooling $\Delta T_{sub.ONB}$ (K)
Unal [37] High-speed photographic technique	0.38	2.121	13.9	1250	4.3
	0.45	2.121	15.8	1250	4.1
Bergles and Rohsenow [35] The dependence of the wall excess temperature on the heat flux	9.774	19.2	0.261	29	108.3
	9.774	19.2	0.261	48	66.6
	6.67	9.54	0.261	29	103.3
	5.67	9.54	0.261	48	50.0
	2.145	3.74	0.261	29	92.2
	2.145	3.74	0.261	48	56.6
Kennedy *et al.* [43] The dependence of the pressure drop on the mass flux	1.5–4.0	1.0–4.0	1.034	137	5–12
	1.5–3.0	1.3–2.6	1.034	110	13–19
	1.5–3.0	3.5	0.69	137	4.0
	1.5–3.0	1.3	0.69	110	4.0
	1.5–3.0	2.0	0.344	137	5.0
	1.5–3.0	4.3	0.344	110	5.0

Table 2.3 Parameters of experiments by Bergles and Rohsenow [35]

	Inlet Temperature		Subcooling				Parameter	
Inlet Velocity	$L_{ONB}/d=29$	$L_{ONB}/d=48$	$L_{ONB}/d=29$	$L_{ONB}/d=48$	Heat Flux	Wall Superheat	$L/d=29$	$L/d=48$
U_{in} (m/s)	T_{in} (°C)		$\Delta T_{sub,ONB}$ (K)		q_{ONB} (MW/m^2)	$\Delta T_{S,ONB}$ (K)	$q_{ONB}/\rho_{ONB}U_{in}c_p$ (K)	
19.42	8	40	108.3	66.6	9.974	27	0.126	0.126
9.54	13	52	103.3	50	5.67	19.4	0.149	0.146
2.145	22	47	92.2	56.6	3.71	12.2	0.144	0.144

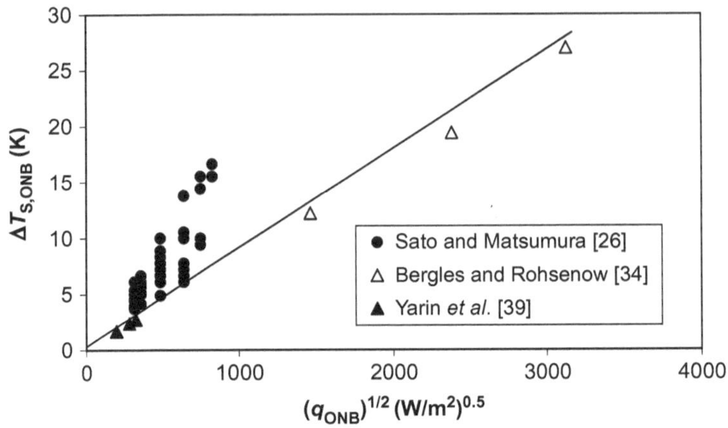

Figure 2.4 Dependence of wall superheat on heat flux.

density, and the surface tension, respectively. For example, M is 1/12 in Hsu [25], 1/8 in Sato and Matsumura [26], $1/8(1+\cos\theta)$ in Davis and Anderson [27], where θ is contact angle, and 1/(9.2) in Kandlikar *et al.* [36].

Experiments by Bergles and Rohsenow [35] corresponded to boiling incipience in relatively short channels when subcooling of the working fluid at the ONB point was about $\Delta T_{sub,ONB}=50$–100 K, where $\Delta T_{sub,ONB}=T_S-T_{B,ONB}$ and $T_{B,ONB}$ is the bulk (mean mass) fluid temperature at the channel cross-section where ONB occurs. Unal [37] studied the incipience of boiling at relatively high values of L_{ONB}/d_h. The experimental determination of the ONB in this study has been carried out in such a way that the number of bubbles appearing on the developed films had been counted for different inlet temperatures at a constant heat flux q, pressure P, and mass flux G. It was observed that not all the bubbles were attached to the wall of the test section, but some were in the bulk of the flow. In these experiments, the value of $\Delta T_{sub,ONB}$ did not exceed 4.3 K.

Heat transfer characteristics during flow boiling of water in a $d=1.5$ mm tube were studied by Hapke *et al.* [38]. The measured roughness was on the order of magnitude of 5 μm. The axial distribution of the external wall temperature was measured using the thermographic method. The temperature reached a maximum at the point after the single-phase liquid exceeded saturation temperature. A wall excess temperature (above saturation) occurred at this point. The local temperature varied near the initial point when the boiling started. Amplitudes of 2 K with a frequency of approximately 2 Hz were achieved. The results measured were compared with those presented by Sato and Matsumura [26] and by Bergles and Rohsenow [35]. A satisfactory agreement existed only for heat fluxes of about $q=50\ \mathrm{W/m^2}$. However, it was necessary that the wall had quite a high excess temperature to initiate the nucleate boiling in the experiments by Hapke *et al.* [38] compared to those reported by Sato and Matsumura [26] and by Bergles and Rohsenow [35]. A mass flux dependence of the wall excess temperature was reported by Hapke *et al.* [38]. Figure 2.5 shows the dependence of ΔT_S on the parameter q_{ONB}/m.

Surface temperature measurement of a heated capillary tube ($d_{out}=1.5$ mm, $d_{in}=1.07$ mm, root mean square (rms) roughness of 0.4%) by means of an IR technique was carried out by Hetsroni *et al.* [22]. The IR camera that was used has the spectral band of 3.4–5 μm. It is cryogenically cooled and has a temperature range of measurement from −10 to 450 °C. This camera has a thermal imaging and measurement system with full-screen temperature measurement and built-in storage and analysis capabilities. It uses a 256 × 256 platinum silicide focal plane array detector, which provides a superior image without the use of mechanical scanning.

Incoming radiation from extraneous sources to the IR sensor cannot be neglected when obtaining temperature measurements of small-sized devices. Schematically, the method is depicted in Fig. 2.6. The surface temperature of the small-sized object is determined from the IR image, which is recorded by IR camera against the background. The background temperature was also measured by the IR camera. Both the object and the background screen were made of the same stainless steel and were painted by the same matt black paint, so that the object and background had equal emissivity ε.

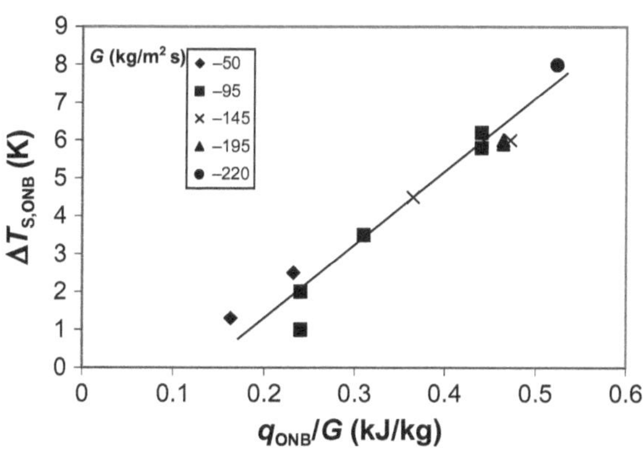

Figure 2.5 Dependence of wall superheat $\Delta T_{SONB}=T_{W,ONB}-T_S$ at ONB point on the parameter q_{ONB}/G observed by Hapke et al. [38] in the $d=1.5$ mm tube.

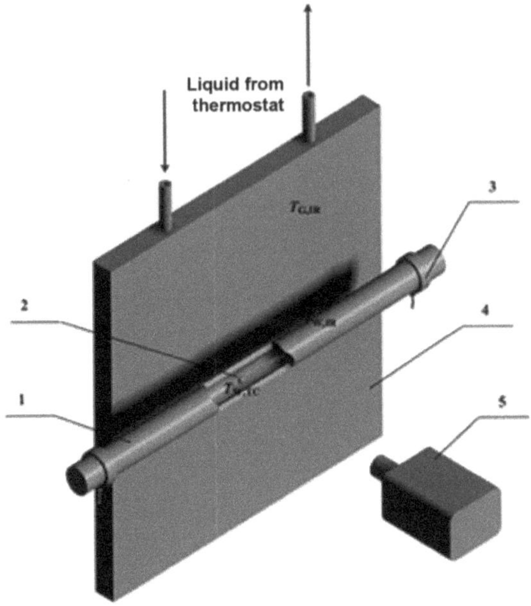

Figure 2.6 Scheme of infrared measurement of surface temperature capillary tube and calibration method [7]. 1. Calibration section, 2. thermocouple, 3. electrical contacts, 4. screen (background), 5. IR video camera.

A thermocouple array was used to obtain the value of background temperature. The shielded thermocouples were spaced by caulking on the background screen surface and provided the temperature distribution on the surface. This temperature distribution showed that the screen's temperature was uniform along the tube. The method could be considered as image compensating method. In other words, the capillary tube becomes almost indistinguishable against the background, like a "disappearing filament" used in some methods of temperature determination using a pyrometer. A series of experiments were then conducted to evaluate the precision of the temperature measurements on the heated surface of the capillary tube, using the method described earlier. Figure 2.7 shows an IR image and surface temperature profile of the capillary tube. For single-phase flow, $x < 65$ mm, the dependence of the surface temperature on the distance from the inlet is linear. At the distance of $x = 65$ mm and $T_W = 110$ °C, ONB occurs.

Qu and Mudawar [29] performed experiments to measure the incipient boiling heat flux q_{ONB} in a heat sink containing 21 rectangular microchannels, 231 μm wide and 713 μm deep. Tests were performed using de-ionized water with inlet liquid velocities of 0.13–1.44 m/s; inlet

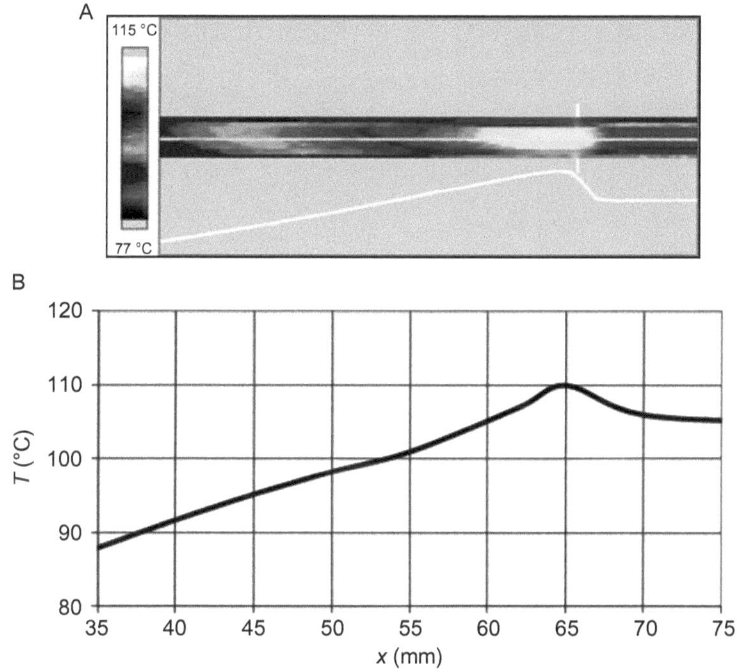

Figure 2.7 Infrared image a) and surface temperature profile b) of the capillary tube (d_{out} = 1.5 mm, d_{in} = 1.07 mm) Hetsroni et al. [7]. (For color version of this figure, the reader is referred to the online version of this chapter.)

temperatures of 30, 60, and 90 °C; and outlet pressure of 0.12 MPa. Using a microscope, boiling incipience was identified when the first bubbles were detected growing at and departing from the micro-channel wall. The authors conclude that bubble behavior at incipient boiling in micro-channels is quite different from that in large channels. At incipient boiling, a small number of nucleation sites were observed close to the exit of several micro-channels. The detachment size was comparable to that of the micro-channel cross-section for lower velocities and decreased progressively with increasing velocity. Results of these experiments are presented in Fig. 2.8. Figure 2.8 shows that the wall superheat is directly proportional to the heat flux.

Sato and Matsumura [26] and Bergles and Rohsenow [35] have proposed an equation for the incipient boiling condition in the case that surface cavities of all sizes are available for nucleation. Hino and Ueda [39] studied incipient boiling of fluorocarbon R-113 in a stainless steel tube of d = 7 mm at mass flux m = 158–1600 kg/m^2 s. Inlet subcooling $\Delta T_{sub,in} = T_S - T_{in}$ ranged from 10 to 30 K. In the case where the upper limit of available cavity

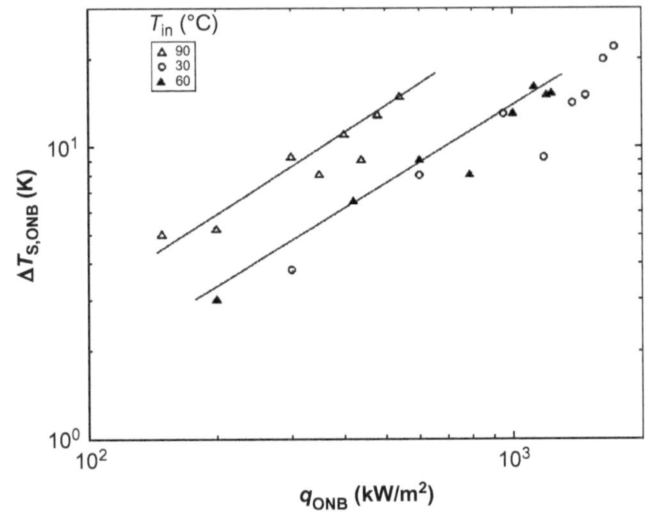

Figure 2.8 Dependence of wall superheat on heat flux. Experiments performed by Qu and Mudawar [29] in rectangular parallel micro-channels 231 μm wide and 713 μm deep.

sizes was restricted to a radius of r^*_{max}, the incipient boiling condition was expressed as follows:

$$q_{ONB} = \frac{k_L}{r^*_{max}} \Delta T_{ONB} - \frac{2\sigma k_L T_S}{h_{LG}\rho_G \left(r^*_{max}\right)^2} \tag{2.2}$$

Figure 2.9 shows the relation between the heat flux and the wall excess temperature at the ONB position obtained by Hino and Ueda [39] in the range of r^*_{max} = 0.22–0.34 μm. Experimental points show that the wall superheat at the ONB position was practically independent of the mass flux and the inlet subcooling. The lines shown in this figure represent the values of Sato and Matsumura [26] and Bergles and Rohsenow [35]. The wall excess temperatures reported by Hino and Ueda [39] were much greater than those predicted by Eq. (2.1).

The data for ONB in trapezoidal micro-channels of d_h = 41.3 μm presented by Lee *et al.* [40] are shown in Fig. 2.10. Figure 2.10 illustrates a comparison of the data reported by Lee *et al.* [40] and the predictions of Eq. (2.2) with various values of r^*_{max}. For the experimental data points in Fig. 2.10, the saturation temperature corresponds to the local pressure at each of the ONB locations. The local pressure is estimated by assuming a linear pressure distribution in the channel between the inlet and exit ones. To this we can

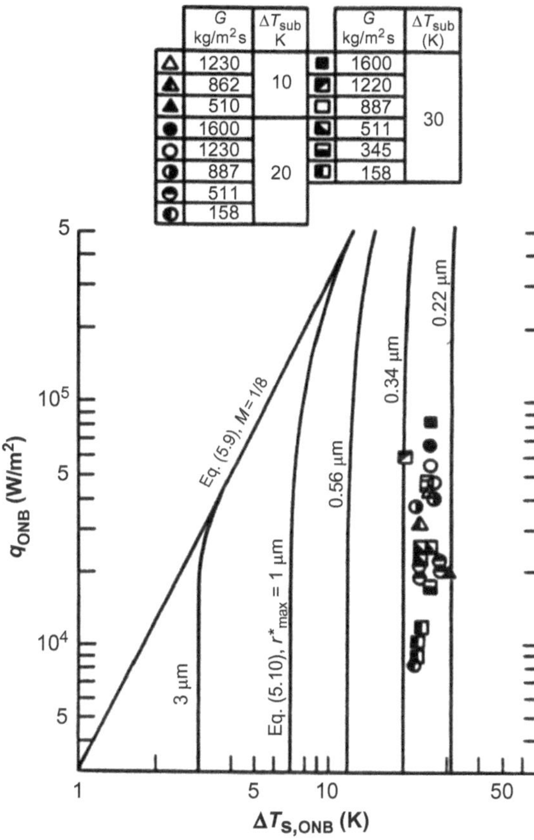

Figure 2.9 Relation between heat flux and wall superheat at the position of incipient boiling [39].

add that this assumption was not discussed by the authors. The system pressure may vary from case to case. For Fig. 2.10, an average system pressure of 161.7 kPa over various different cases was employed. As for the wall temperature, it is assumed that the channel wall temperature is uniform as the channel is relatively short and the wall material, silicon, has relatively high thermal conductivity. The figure indicates that most of the cavity sizes ranged from 1.5 to 4 μm. This is consistent with the maximum roughness on the side wall of the channels used in experiments. Experiments showed that for a given value of r^*_{max}, the wall excess temperature $\Delta T_{S,ONB}$ does not depend on mass flux.

Boiling incipience in parallel micro-channels with low-mass-flux subcooled water flow was studied by Mosyak *et al.* [41]. The test module

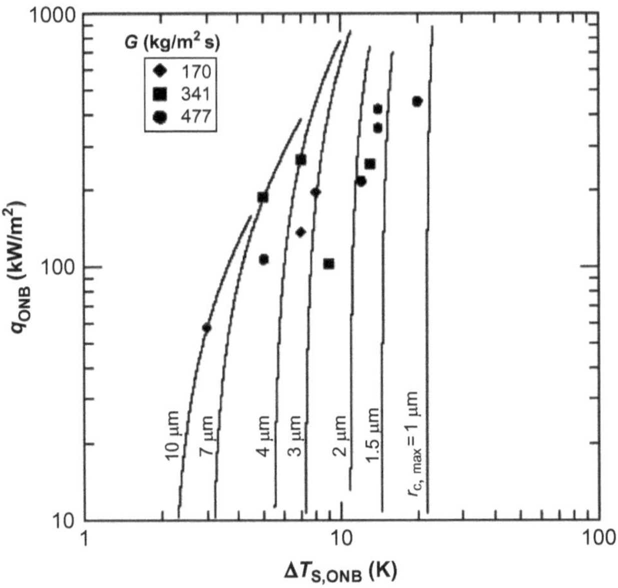

Figure 2.10 Relationship between heat flux and wall superheat in micro-channels of $d_h = 41.3$ μm [40].

was fabricated of 14 mm × 20 mm and 1.25-mm-thick aluminum plate and placed into the heat sink housing. Twenty-five micro-channels were machined on one side of this plate using a precision sawing technique. The micro-channels were 200 μm wide, 580 μm deep, and a wall thickness of 200 μm. A thermal high-speed imaging radiometer was utilized to study the temperature field on the electrical heater and the working fluid temperature distribution along the micro-channels. The IR camera was suitable for temperature measurements in the wavelength range of about 5 μm; it has a sensitivity of 0.1 K and a typical resolution of 256 pixels per line. The measurement resolution was 0.03 mm. A radiometer permitted the obtainment of a quantitative thermal field in the line mode, an average temperature in the area mode, and a temperature of a given point in the point mode.

To provide additional insights into the role of surface roughness on ONB in micro-channels, roughness parameters were studied extensively. Surface parameters may be characterized by rms roughness R_q, center-line average roughness R_a, maximum peak height $R_p = \max(y_i)$, maximum valley depth $R_v = \min(y_i)$, and maximum height of the profile $R_t = R_p - R_v$ [42].

Measurements carried out along the channel at intervals approximately of 0.1 mm are shown in Fig. 2.11 [41]. As can be seen, the surface is

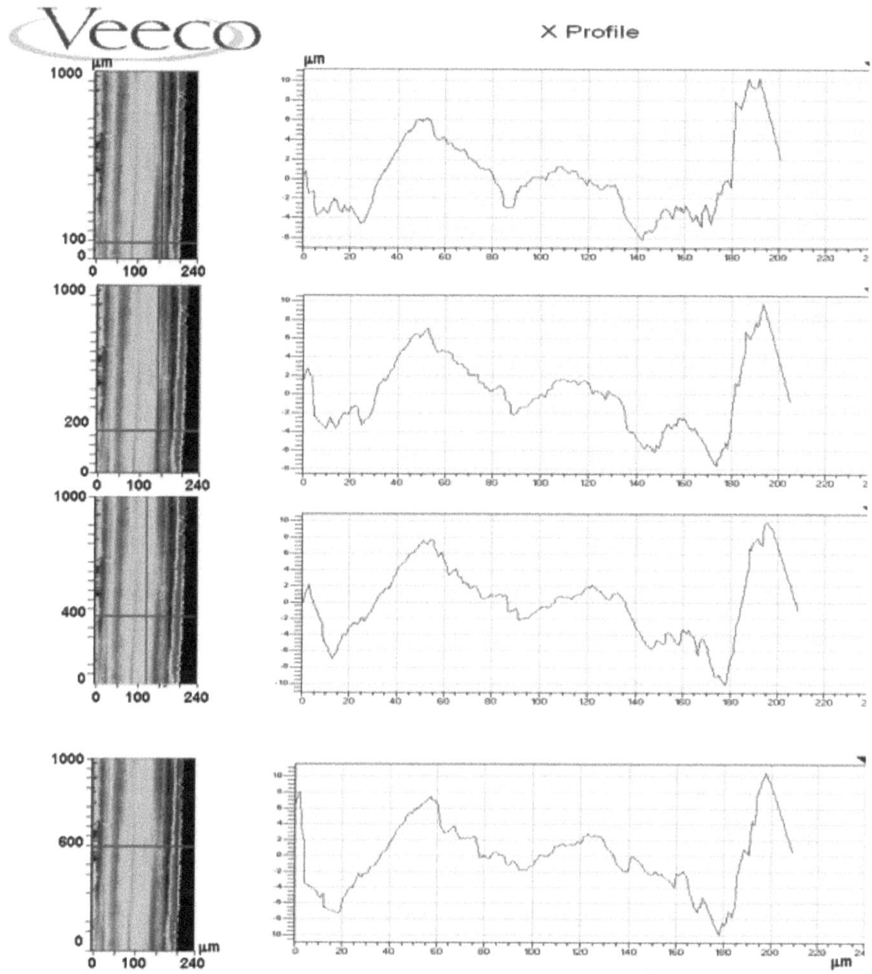

Figure 2.11 The roughness of the bottom channel [41].

characterized by a uniform cutting pattern all along the channel. The following data were deduced from the measurements: $R_t = 149$ µm, $R_q = 8.72$ µm, and $R_a = 4.01$ µm. These parameters were used to calculate channel cross-section, hydraulic diameter D_h, relative roughness $\varepsilon_a = R_a/D_h$, and $\varepsilon_q = R_q/D_h$. The parameters are: $D_h = 297$ µm, $\varepsilon_a = 1.1\%$, $\varepsilon_q = 2.9\%$.

Figure 2.12 depicts the average wall temperature versus the applied heat flux for both single- and two-phase flow. The experiments were conducted at various mass fluxes, which were calculated for the microchannel array cross-section. For single-phase flow, at given values of mass flux, the

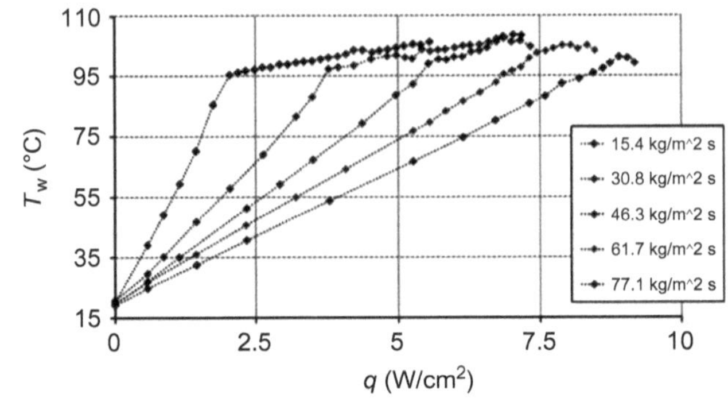

Figure 2.12 The dependence of the wall temperature on the heat flux [41].

temperature of the heater increased linearly with increasing heat flux. With the intention of obtaining the entire boiling curve, the power applied to the test module was increased by small increments, while the total fluid flow rate through the heat sink was maintained at a constant level. It can be seen that at a certain value of heat flux, the slope of the curves shown in Fig. 2.12 decreases. Such a phenomenon may be associated with the initiation of a boiling flow. The ONB can thus be specified as the point beyond which the gradients of the two curves become noticeably different.

The ONB phenomenon and associated instabilities can be analyzed based on the pressure drop–flow rate characteristics of heated channels. When boiling occurs, the pressure drop increases significantly with increasing of the heat flux. A typical characteristic curve is depicted in Fig. 2.13. The curve is obtained at different values of heat flux for a given constant value of mass flux. The ONB on a characteristic curve can be specified as the point beyond which the gradients of the two curves (single phase and two phase) become noticeably different [43]. The segment of the curve to the left of ONB is stable. Pressure drop and temperature fluctuations can develop once the heat flux increases above the ONB point where the slope of the curve changes. Further increases in heat flux leads to a sharp increase in the pressure drop. The liquid temperature variations in the experiments with heating are relatively large. They reduce the liquid viscosity, thereby decreasing the liquid single-phase friction factor and leading to smaller channel pressure drops. The average results on boiling incipience obtained from the two methods described earlier are presented in Table 2.4. The table lists the measured incipient heat flux for various fluid mass velocities. It indicates that at incipient ONB the heat flux increases with increased fluid mass flux. The results are consistent with those reported by

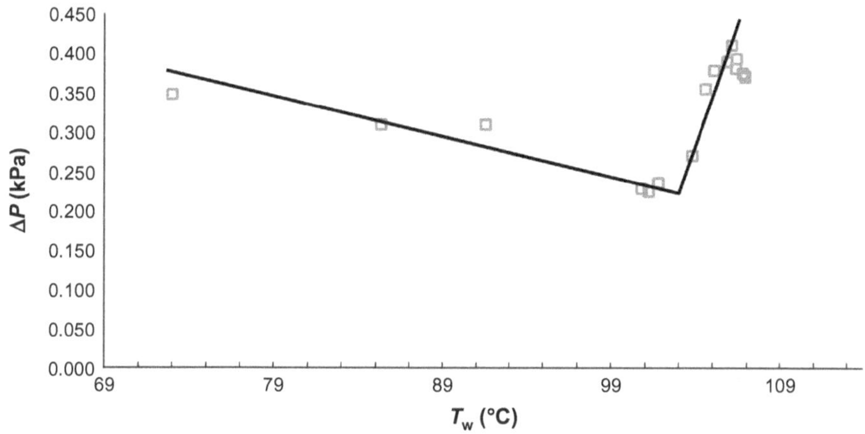

Figure 2.13 The dependence of the pressure drop on the temperature of the heated bottom wall. $m = 46.3$ kg/m²s.

Table 2.4 Incipience of nucleation boiling in parallel micro-channels [41]

Mass flux		**Heat flux on the heater**	**Onset of nucleate boiling**	
$\dot{m}$ **(kg/m²s)**	U **(m/s)**	$q^{\cdot} \times 10^4$ **(W/m²)**	$\Delta T_{(ONB)}$ **(K)**	***St***
15.4	0.0155	2.4	2.0	19.6
30.8	0.030	3.7	2.5	16.0
46.3	0.045	6.0	3.5	17.0
61.7	0.060	7.5	3.9	16.0
77.1	0.077	9.5	4.3	16.0

Qu and Mudawar [29], Liu *et al.* [32], and with the prediction by Yarin *et al.* [34].

Due to the importance of accurately accounting for the influence of subcooling on ONB, an analysis based on experimental data of Bergles and Rohsenow [35], Unal [37], Kennedy *et al.* [43], Liu *et al.* [32], as well as data obtained in Mosyak *et al.* [41] was performed. To estimate the value of the bulk liquid temperature at ONB, $T_{B,ONB}$, the energy and continuity equations should be considered. In terms of mean mass temperature, the thermal balance equation for a rectangular channel with three heated walls is [41]

$$\frac{q_{ONB} L_{ONB}(2a+b)}{\rho U_{in} C_p \left(T_{B,ONB} - T_{in}\right) ab} = 1 \tag{2.3}$$

If the value of $T_{B,ONB} \approx T_S$, Eq. (2.8) takes the form

$$\frac{q_{\mathrm{ONB}} L_{\mathrm{ONB}} (2a+b)}{\rho U_{\mathrm{in}} C_{\mathrm{p}} (T_{\mathrm{S}} - T_{\mathrm{in}}) ab} = 1 \tag{2.4}$$

where a is the channel height; b is the channel width; C_{p} is the heat capacity of the liquid at constant pressure; L_{ONB} is the distance from the channel inlet to the cross-section where ONB occurs; q_{ONB} is the heat flux at ONB; T_{in}, $T_{\mathrm{B,ONB}}$, T_{S} are the fluid temperatures at the inlet, bulk temperature at the cross-section where ONB occurs, and saturation temperature, respectively; U_{in} is the fluid velocity at the inlet; and ρ is the fluid density.

The bulk temperature $T_{\mathrm{B,ONB}}$ is close to the saturation temperature T_{S} when the value calculated using Eq. (2.4) does not differ significantly from unity. From the experimental results reported by Unal [37], Kennedy *et al.* [43], Liu *et al.* [32], it may be concluded that at ONB, the parameter $D = T_{\mathrm{W,ONB}} - T_{\mathrm{B,ONB}}/T_{\mathrm{S}} - T_{\mathrm{in}}$ is of the order of 0.01, where $T_{\mathrm{W,ONB}}$ is the wall temperature at ONB. When the value D is in the range 0.1–0.2, as in experiments by Bergles and Rohsenow [35], the ONB occurred at values of bulk temperature, $T_{\mathrm{B,ONB}}$ that are significantly less than saturation temperature.

Figure 2.14 illustrates the dependence of the fluid bulk temperature at the ONB cross-section on the inlet flow temperature. The dark area corresponds to the region where the local fluid temperature T exceeds the saturation temperature T_{S}. In this region, the maximum probability of bubble embryo formation takes place. If the temperature $T_{\mathrm{W,ONB}}$ does not differ significantly from the saturation temperature T_{S}, a bubble nucleates in a low subcooled region, as observed here.

Although the water temperature at the entry to the reservoir was 20 °C, the temperature at the inlet of the micro-channels was significantly higher due to pressure variations. Single-phase water flow in parallel channels is mostly stable, since the flow rates and pressure drop levels are high enough. As heating is applied at low mass flow rates, and vapor is generated, the oscillatory motion of the liquid and vapor could be self-sustaining as long as certain heating conditions are maintained. The system departs from the stable operating conditions and often morphs to flow regimes where hydrodynamic instability becomes a possibility. The vapor jet not only moves downstream but also upstream, and the space that it occupied increases. In this case, the temperature at the inlet of the micro-channels increases [44].

2.2.2 The Influence of Surface Roughness on Boiling Incipience

We compared our results [41] with macro-scale nucleation criteria of saturated boiling in terms of active nucleating cavity radius and wall excess

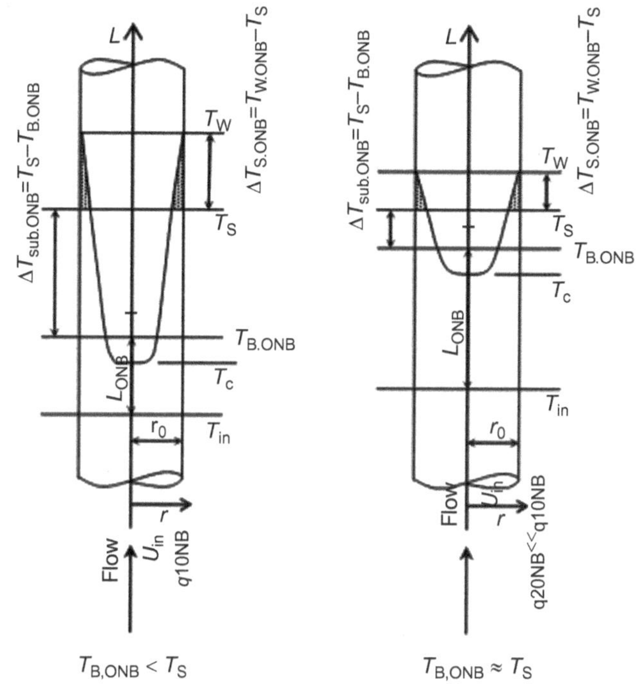

Figure 2.14 Dependence of fluid bulk temperature at ONB cross-section on inlet flow temperature [41].

temperature as reported in the literature. In the case where the upper limit of available cavity sizes was restricted to the radius r^*_{max}, the incipient boiling conditions according to Hino and Ueda [39] may be calculated using Eq (2.5).

$$q_{ONB} = \frac{k_L}{r^*_{max}} \Delta T_{ONB} - \frac{2\sigma k_L T_S}{h_{LG}\rho_G \left(r^*_{max}\right)^2} \tag{2.5}$$

Figure 2.15 shows the nucleation criterion calculated for our experimental results using Eq. (2.5). According to Collier and Thome [45], a bubble will nucleate from the cavity of critical size

$$r_{c,crit} = \left(\frac{2\sigma k_L T_S}{h_{LG}\rho_G q_{ONB}}\right)^{0.5} \tag{2.6}$$

Calculation by use of Eq. (2.6) is also shown in Fig. 2.15. Comparison between the values of the surface roughness in the study by Mosyak *et al.* [41] matches the theoretical effective cavity radius. Thus, the substantial agreement between experimental results and theory supports the idea that for

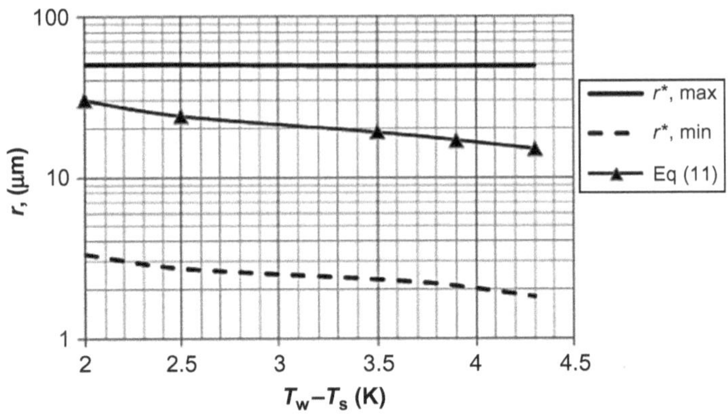

Figure 2.15 The dependence of the radius of the activated cavity on the wall excess temperature [41].

subcooled boiling in micro-channels, nucleation incipience conditions can be estimated by conventional theory. The experimental data obtained by Kennedy *et al.* [43] together with the main results that appeared in the open literature in the last years are used by Mosyak *et al.* [41] to highlight conditions for incipient boiling in conventional size and micro-channels. It is shown that the ONB in terms of heat flux, wall excess temperature, and the upper limit of available cavity size does not depend on scaling effects. For this reason, the trend to generate new correlations for the prediction of ONB in micro-channels does not stimulate any improvement in the knowledge of the physics of the problem.

2.2.3 Effect of Inlet Velocity on Incipient Boiling Heat Flux

Variation of the boiling Stanton number, $St_{ONB} = q_{ONB}/[\rho_L U_{in} c_p (T_S - T_{in})]$, versus the Reynolds number based on inlet flow velocity and hydraulic diameter is shown in Fig. 2.16. Figure 2.16A illustrates the dependence of the boiling Stanton number at the ONB point on the inlet flow velocity in the Reynolds number ($U_{in} = 0.015$–0.072 m/s) obtained in the present study. Experiments carried out by Liu *et al.* [32] at $P = 1$ bar for different values of the inlet flow temperature that varied in the range $T_{in} = 41.2$–92.0 °C are shown in Fig. 2.16B. One can see that the value of the St_{ONB} did not change significantly with the Reynolds number in these experiments.

Analysis of the data shown in these figures (Fig. 2.16A and B) makes it possible to illustrate the salient features of the dependence of q_{ONB} on U_{in} at various values of the inlet temperature. This dependence may be

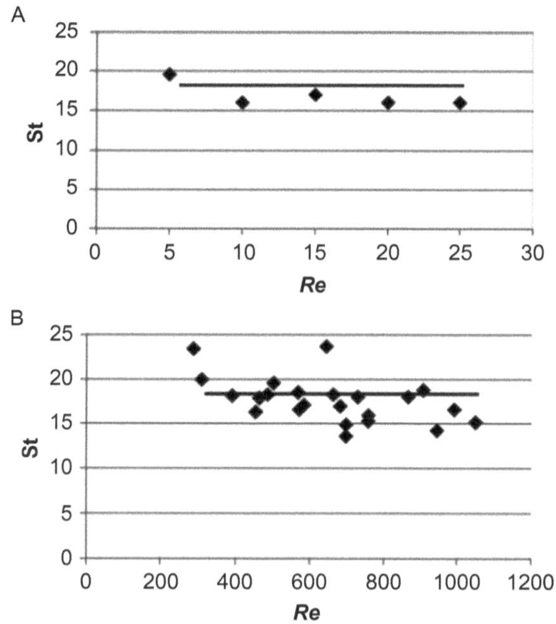

Figure 2.16 The variation of the boiling Stanton number, with the Reynolds number. (A) Experiments by Mosyak et al. [41]. (B) Experiments by Liu et al. [32].

approximated by the following lines in Fig. 2.17A: the solid line corresponds to the case of $T_{in} << T_S$ and the dotted line corresponds to the case of $T_{in} \approx T_S$. Experimental results reported by Liu *et al.* [32] and shown in Fig. 2.17B agree qualitatively with predictions shown in Fig. 2.17A.

The experimental results indicate that parameters which affect incipience of nucleation in micro-channels, such as cavity radius and wall excess temperature, are well predicted by the theoretical nucleation criteria which were developed for conventional size channels.

3. BOILING HEAT TRANSFER IN MICRO-CHANNELS

3.1. Heat transfer coefficient

The heat transfer coefficient for flow boiling through a horizontal rectangular channel with low aspect ratio (0.02–0.1) was studied by Lee and Lee [46]. The mass flux in these experiments ranged from 50 to 200 kg/m^{2}s, maximum heat flux of 15 kW/m^2, and quality range from 0.15 to 0.75 which

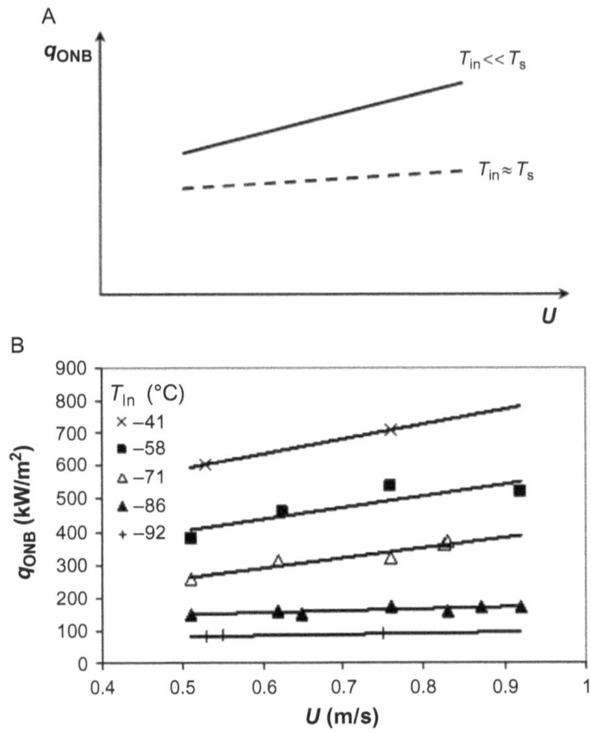

Figure 2.17 The variation of the incipient boiling heat flux with the inlet flow velocity: (A) Analytical predictions by Yarin et al. [34]. (B) Experiments by Liu et al. [32].

corresponds to annular flow. The experimental data showed that under the given experimental conditions, forced convection plays a dominant role.

The detail experimental study of flow boiling heat transfer in two-phase heat sinks was performed by Qu and Mudawar [47]. It was shown that the saturated flow boiling heat transfer coefficient in a micro-channel heat sink is a strong function of mass velocity and depends only weakly on the heat flux. This result, as well as those of Lee and Lee [46], indicates that the dominant mechanism for water micro-channel heat sinks is forced convective boiling but not nucleate boiling.

Heat transfer characteristics for saturated boiling were considered by Yen *et al.* [48]. From this study of convective boiling of HCFC123 and FC72 in micro-tubes with inner diameters of 190, 300, and 510 μm, it can be seen that in the saturated boiling regime, the heat transfer coefficient monotonically decreased with increasing vapor quality, but was independent of mass flux.

The convective and nucleate boiling heat transfer coefficient was the subject of experiments by Grohmann [49]. The measurements were performed in micro-tubes of 250 and 500 μm in diameter. Nucleate boiling metastable flow regimes were observed. Heat transfer characteristics for nucleate and convective boiling in micro-channels with different cross-sections were studied by Yen *et al.* [48]. Two types of micro-channels were tested: a circular micro-tube with a 210 μm diameter and a square micro-channel with a 214 μm hydraulic diameter. The heat transfer coefficient was higher for the square micro-channel because the corners acted as effective nucleation sites.

Several popular macro-channel correlations and recently recommended small-channel correlations were examined by Lee and Mudawar [50]. Predictions were adjusted for three-sided wall heating and rectangular geometry using the following relation:

$$h_{tp} = h_{tp,cor}\frac{Nu_3}{Nu_4} \tag{2.7}$$

where $h_{tp,cor}$ is the value predicted from a correlation for uniform circumferential heating, and Nu_3 and Nu_4 are the single-phase Nusselt numbers for laminar flow with three- and four-side wall heating, respectively [51].

$$Nu_3 = 8.235\left(1 - 1.883\beta + 3.767\beta^2 - 5.814\beta^3 + 5.361\beta^4 - 2.0\beta^5\right) \tag{2.8}$$

and

$$Nu_4 = 8.235\left(1 - 2.042\beta + 3.085\beta^2 - 2.477\beta^3 + 1.058\beta^4 - 0.186\beta^5\right) \tag{2.9}$$

where β is the ratio of the channel depth to its width.

Experiments by Lee and Mudawar [50] revealed the range of parameters at which heat transfer is controlled by nucleate boiling or annular film evaporation. The first of these processes occurs only at low qualities ($x < 0.05$) corresponding to very low heat fluxes; the second one is encountered at moderate ($0.05 < x < 0.55$) or high ($x > 0.55$) qualities that correspond to high heat fluxes. New correlations were suggested by Lee and Mudawar [50]. They are based on the Martinelli parameter X and account for micro-channel effects not represented in the prior correlations.

Table 2.5 summarizes the new correlations for the three quality regions. The low and high-quality regions are based solely on the Martinelli parameter while the mid-range includes the effects of Bo and We_{fo} as well. Overall,

Table 2.5 Two-phase flow boiling in micro-channels

x_e	Correlation	Data	MAE (%)
0–0.05	$h_{tp}=3.856X^{0.267}h_{sp,L}$ $X^2=\frac{(dp/dz)_L}{(dp/dz)_G},\ h_{sp,L}=\frac{Nu_3k_L}{d_h}$ $X_{vv}=\left(\frac{\mu_L}{\mu_G}\right)^{0.5}\left(\frac{1-x_e}{x_e}\right)^{0.5}\left(\frac{v_L}{v_G}\right)^{0.5}$ $X_{vt}=\left(\frac{f_L Re_G^{0.25}}{0.079}\right)^{0.5}\left(\frac{1-x_e}{x_e}\right)^{0.5}\left(\frac{v_L}{v_G}\right)^{0.5}$ $Re_G=\frac{Gx_e d_h}{\mu_G}$	50 Water data points	11.6
0.05–0.55	$h_{tp}=436.48Bo^{0.522}We_L^{0.351}X^{0.665}h_{sp,L}$ $Bo=\frac{q}{Gh_{LG}},\ We_{fo}=\frac{v_L G^2 d_h}{\sigma}$	83 R-134a data points 157 Water data points	11.9
0.55–1.0	$h_{tp}=\max\{(108.6X^{1.665}h_{sp,G}),h_{sp,G}\}$ $h_{sp,G}=\frac{Nu_3k_G}{d_h}$ for laminar gas flow $h_{sp,G}=0.023Re_G^{0.8}Pr_G^{0.4}$ for turbulent gas flow	28 R-134a data points	16.1

Heat transfer coefficient [50].

liquid convection is important for both the low- and mid-quality regions, while convection to vapor becomes important for the high-quality region. For the latter, the low viscosity of R-134a vapor yields vapor Reynolds numbers corresponding to turbulent flow at high-heat-flux conditions despite the small hydraulic diameter of the micro-channel. Thus, the single-phase vapor term in the high-quality correlation must allow for laminar or turbulent vapor flow. Table 2.5 shows that the effect of the Martinelli parameter is important for each of the three quality ranges. The correlations show that the heat transfer coefficient is proportional to the Martinelli parameter raised to a positive exponent.

In Table 2.5, the parameters are defined as follows: *Bo* is the boiling number, d_h is the hydraulic diameter, f is the friction factor, h is the local heat transfer coefficient, k is the thermal conductivity, *Nu* is the Nusselt number, *Pr* is the Prandtl number, q is the heat flux, v is the specific volume, X is the Martinelli parameter, X_{vt} is the Martinelli parameter for laminar liquid–turbulent vapor flow, X_{vv} is the Martinelli parameter for laminar liquid–laminar vapor flow, x_e is thermodynamic equilibrium quality, z is the streamwise coordinate, μ is the viscosity, ρ is the density, σ is the surface tension; the subscripts are L for saturated liquid, LG for property difference between saturated vapor and saturated liquid, G for saturated vapor, sp for single phase, and tp for two phase.

The predictive accuracy of the correlation was measured by the mean absolute error, defined as

$$\mathrm{MAE} = \frac{1}{N}\sum\left[\frac{\left|h_{\mathrm{tp,pred}} - h_{\mathrm{tp,exp}}\right|}{h_{\mathrm{tp,exp}}} \times 100\%\right] \tag{2.10}$$

It may be seen from Table 2.5 that MAE is in the range of 11–16.1%.

The work by Steinke and Kandlikar [52] focused on obtaining heat transfer data during flow boiling in micro-channels. An experimental investigation was performed using water in six parallel, horizontal micro-channels with a hydraulic diameter of 207 μm. The channels had a slightly trapezoidal cross-section, with the top and bottom widths differing by about 15 μm. The average channel dimensions were 214 μm wide by 200 μm deep and 57.15 mm long. The range of parameters were mass flux from 157 to 1782 $\mathrm{kg/m^2s}$, heat flux from 5 to 930 $\mathrm{kW/m^2}$, inlet temperature of 22 °C, quality from subcooled to 1.0, and atmospheric pressure at the exit. The results were correlated in the form

$$h_{\mathrm{tp}} = 0.6683\,Co^{-0.2}(1-x)^{0.8}f_2 h_{\mathrm{L}} + 1058\,Bo^{0.7}(1-x)^{0.8}F\cdot h_{\mathrm{L}} \tag{2.11}$$

$$Co = \left(\frac{\rho_{\mathrm{G}}}{\rho_{\mathrm{L}}}\right)^{0.5}\left(\frac{1-x}{x}\right)^{0.8} \tag{2.12}$$

$$Bo = \frac{q}{m h_{\mathrm{LG}}} \tag{2.13}$$

where Co is the convection number given in Eq. (2.12), Bo is the boiling number defined in Eq. (2.13), f_2 is a multiplier, h_{L} is the heat transfer coefficient for all-liquid flow, x is the quality, and h_{LG} is the latent heat. The F number for water is 1.0, and the multiplier f_2 is 1.0 for micro-channel flow.

The empirical correlation (3.5) predicts a heat transfer coefficient about twice as high as that measured by Qu and Mudawar [47] during flow boiling of water ($x < 0.15$) and during flow boiling of R-134a ($x = 0.4$–0.8). That correlation also overpredicts the experimental data obtained by Yen *et al.* [48] during convective boiling of HCFC123 and FC72 in $d = 190$ μm tubes in the range of $x = 0.4$–0.9.

The extent to which incoming liquid will be vaporized in micro-channels is a design parameter that depends on the intended application.

In micro-scale refrigeration systems, the change in vapor quality may be substantial, e.g., of the order of 0.0–0.8. In electronic cooling applications, the equilibrium vapor quality may remain at zero or be very small to capture the high heat transfer coefficients of subcooled flow boiling without the need to incorporate a condenser. The objective of the study by Hetsroni *et al.* [53] was to measure heat transfer coefficients by means of IR for boiling of water in parallel micro-channels at vapor quality $x=0$–8%.

Figure 2.18 shows the dependence of the saturated flow boiling heat transfer coefficient h on the thermodynamic equilibrium quality x at the outlet of the test section. The data were obtained in the range of mass flux from 95 to 340 kg/m^{2}s. As shown in Fig. 2.18, h decreases appreciably with increasing x. This trend agrees with results reported by Qu and Mudawar [47], Kandlikar [54], and Kandlikar and Balasubramanian [55]. The overall range of $h=10$–30 kW/m^{2}K is fairly similar to that measured by Qu and Mudawar [47], $h=20$–45 kW/m^{2}K. A decrease of heat transfer coefficient with increasing heat flux does not support the nucleate boiling mechanism observed in channels of $d_h > 3$ mm, which is normally associated with a significant increase in h with increasing heat flux [45].

To fully understand the heat transfer phenomenon, the effect of both heat flux and of mass flux on the heat transfer coefficient should be addressed. Using a high speed IR technique, Hetsroni *et al.* [56] demonstrated that both these mechanisms occurred during saturated flow boiling of Vertrel XF in micro-channels. This technique was also employed by Hetsroni *et al.* [57] to evaluate a mathematical expression for convective boiling in micro-channels at low values of mass flux 31.6–338 kg/m^{2}s and vapor quality

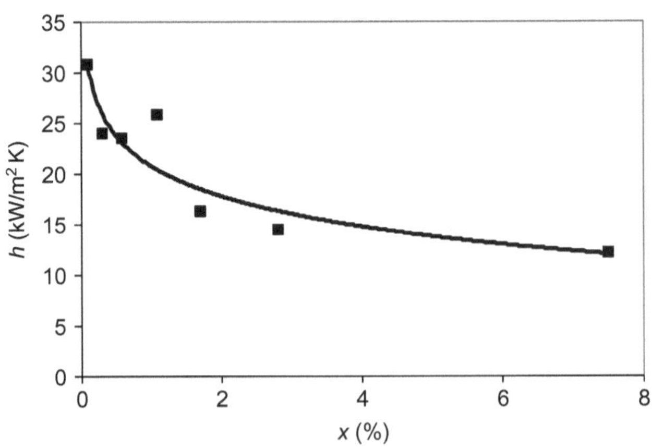

Figure 2.18 Dependence of heat transfer coefficient on vapor quality [53].

$x = 0.01$–0.08. For flow boiling heat transfer, the Nusselt number $Nu = hd/k$ was used, where d is the characteristic dimension, h is the heat transfer coefficient, k is the thermal conductivity of the liquid. The distinctive feature of the experiments is low values of mass flux and vapor quality. The Eotvos number $Eo = d(\rho_L - \rho_G)d^2/\sigma$ reflects the effect of confinement on boiling (where g is the acceleration due to gravity; ρ_L and ρ_G are the liquid and vapor density, respectively; d is the characteristic dimension; and σ is the surface tension). The combined effect of heat flux and mass flux was taken into account in the boiling number, $Bo = q/mh_{LG}$, where q is the heat flux, m is the mass flux, and h_{LG} is the latent heat of vaporization. These foregoing dimensionless groups were used to describe the heat transfer phenomenon.

The results are shown in Fig. 2.19. Comparison between dependence of the heat transfer coefficients on boiling number, which was obtained in micro-channels of $d_h = 220$ μm for boiling of water and ethanol, shows that the results are different, as expected. In reality, the effects will depend on the fluid properties. At a boiling number $Bo = 1.16 \times 10^{-3}$, the heat transfer coefficient is about $h = 15{,}000$ W/m^{2}K for water. At about the same boiling number ($Bo = 1.14 \times 10^{-3}$), the heat transfer coefficient is about $h = 10{,}000$ W/m^{2}K for ethanol. Figure 2.19 also shows that for the same value of the boiling number, the heat transfer coefficient increases with

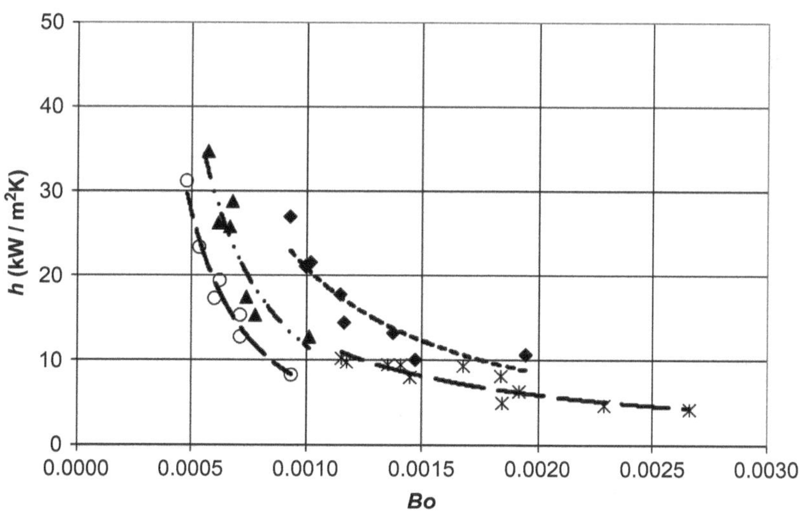

Figure 2.19 Dependence of average heat transfer coefficient on boiling number ○ $D_h = 100$ μm, water, ▲ $D_h = 130$ μm, water, ◆ $D_h = 220$ μm, water, Ж$D_h = 220$ μm, ethanol [57].

an increase in the hydraulic diameter. For example, for a boiling number of about $Bo = 0.8 \times 10^{-3}$, the heat transfer coefficients were about $h = 9000$, 12,000, and 22,000 W/m^2K for $d_h = 100$, 130, and 220 μm, respectively. Figure 2.20 shows the dependence of the Nu/Eo on the boiling number, Bo. All fluid properties are taken at the saturation temperature. This dependence can be approximated, with standard deviation of 18%, by the relation:

$$Nu/Eo = 0.030\, Bo^{-1.5} \tag{2.14}$$

3.2. Flow instabilities

Two-phase flow instabilities in micro-channels are more intense than in conventional channels due to the low flow velocities and the confined space available for the bubbles to grow. Instabilities in micro-channels have been reported by many authors, but in many cases with different oscillation amplitude and frequency, under similar mass and heat flux conditions. One possible reason for this was proposed by Wang *et al.* [58] who suggested that the geometric configuration of the inlet and outlet regions can significantly influence the type of instabilities observed. Qu and Mudawar [59] have investigated flow boiling of water in a heat sink containing 21 copper parallel micro-channels. They reported severe pressure drop oscillations caused by interaction between the vapor generated within the channels and in the compressible volume of the inlet manifold. The authors found that a

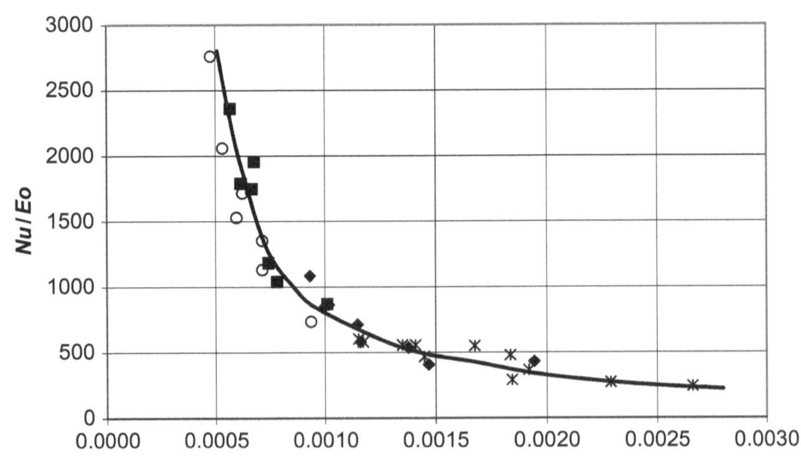

Figure 2.20 Dependence of Nu/Eo on Bo ○ $D_h = 100$ μm, water, ▲ $D_h = 130$ μm, water, ◆ $D_h = 220$ μm, water, Ж $D_h = 220$ μm, ethanol [57].

throttling valve installed immediately upstream of the heat sink significantly reduced these instabilities. Zhang *et al.* [60] have used doped silicon sensors to measure transient pressure fluctuation frequencies during boiling in a single micro-channel with water as the working fluid. The authors observed frequencies ranging from 3 to 40 Hz and recorded up to 138 kPa transient pressure fluctuations due to bubble nucleation.

Wu and Cheng [61–63] reported two oscillatory flow boiling modes in micro-channels depending on the heat to mass flux ratio. It was found that mass flux and pressure oscillated out of phase in these two boiling modes, and the large amplitude oscillations of pressure and temperature could be self-sustained. Wang *et al.* [64] experimentally investigated dynamic instabilities of flow boiling of water in parallel micro-channels as well as in a single micro-channel. They classified these instabilities into stable and unstable regimes depending on the heat to mass flux ratio. Two types of unstable oscillations were reported, one with long-period oscillations and another with short-period oscillations in temperature and pressure.

Kandlikar *et al.* [65] have assessed the effects of an inlet pressure restrictor and fabricated nucleation sites as a means of stabilizing the two-phase flow instabilities in micro-channels. The authors found that fabricated nucleation sites in conjunction with the pressure restrictor having an area of 4% of the channel cross-sectional area completely eliminated the instabilities associated with the reverse flow. Kosar *et al.* [66] have experimentally investigated the effects of inlet orifices with various geometries on the suppression of flow boiling instabilities in parallel micro-channels. They have shown that for sufficiently high inlet pressure restrictions imposed by the inlet orifices, parallel-channel and upstream compressible volume instabilities are eliminated. Kuo and Peles [67] experimentally studied the effects of pressure on flow boiling instabilities in micro-channels. They found that high system pressure moderates instabilities by reducing the void fraction, the superheat needed to activate bubble nucleation, and bubble departure diameter. Local transient temperature measurements showed lower magnitudes and higher frequencies of oscillations at high system pressure. Explosive generation of vapor and the temporal behavior of temperature and pressure in parallel triangular micro-channels have been observed by Hetsroni *et al.* [14,53,57].

3.2.1 Fluctuation of pressure drop, fluid, and heated wall temperatures

Hetsroni *et al.* [57] carried out experimental investigations on boiling instability in parallel micro-channels by simultaneous measurements of temporal variations of pressure drop, fluid, and heater temperatures. The channel-to-channel interactions may affect pressure drop between the inlet and the outlet manifold as well as associated temperatures of the fluid in the outlet manifold and heater temperature. Figure 2.21 illustrates this phenomenon for pressure drop in a heat sink that contains 13 micro-channels of $d_h = 220$ μm at a mass flux $m = 93.3$ kg/m^{2}s and a heat flux $q = 200$ kW/m^2. The temporal behavior of the pressure drop in the whole boiling system is shown in Fig. 2.21A. The considerable oscillations were caused by the flow pattern alternation, that is, by the liquid/two-phase alternating flow in the micro-channels. The pressure drop Fast Fourier Transform (FFT) is presented in Fig. 2.21B. Under the conditions of the given experiment, the period of pressure drop fluctuation is about $\tau = 0.36$ s. These results differ significantly from those reported by Wu and Cheng [63]. In the range of

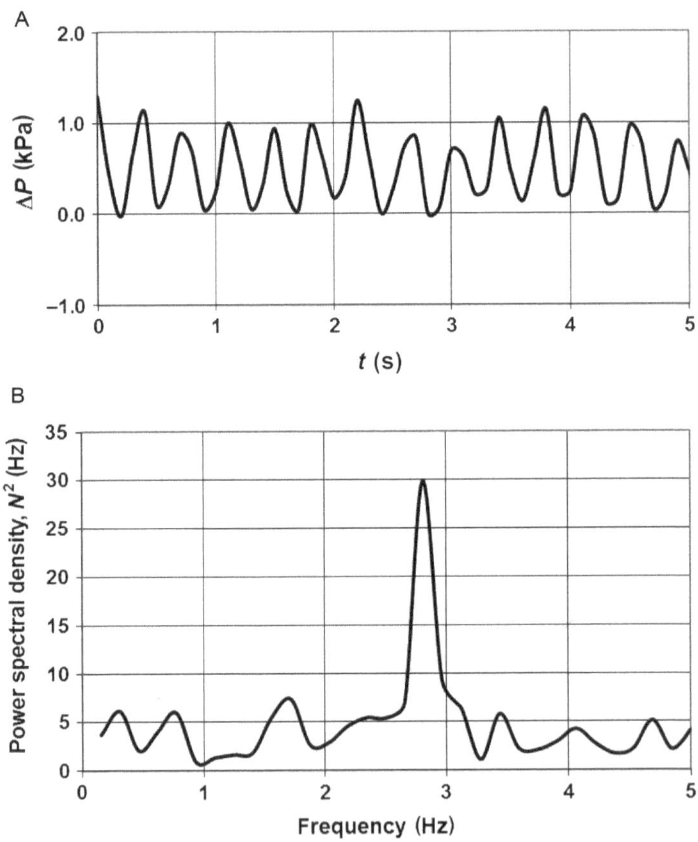

Figure 2.21 Time variation of pressure drop at $q = 200$ kW/m^2. (A) Pressure drop fluctuations, (B) pressure drop amplitude spectrum [57].

average values of mass flux $m = 112–146$ kg/m^{2}s and heat flux $q = 135–226$ kW/m^2 the authors observed a much longer oscillation period (from $\tau = 15.4$ to 202 s). In the experiments conducted by Wu and Cheng [63], the water in the pressure tank was moved by compressed nitrogen gas to the test section. According to the authors, when boiling occurred in the test section, the pressure drop across the test section was suddenly increased due to generation of vapor bubbles. This increase in pressure drop caused a decrease in mass flux. The long-period pressure drop fluctuations may be connected to the period of increasing or decreasing of the incoming mass flux. The former depends not only on boiling in the microchannels of heat sinks, but also on the nitrogen pressure in the water tank and on the total length of the pipe connecting the water tank to the test section. In our experiments [41], the mass flow rate was independent of pressure drop fluctuations, and the oscillation periods were very much different from those recorded by Wu and Cheng [63].

The pressure drop fluctuations provide insights into the thermal behavior of the fluid in the outlet manifold. The pressure drop fluctuation frequency is representative of the oscillations in the system. Figure 2.22A and B shows time variations and FFT of the fluctuation component of the fluid temperature, respectively. From Fig. 2.22A, it can be seen that the average fluid temperature at the outlet manifold is less than the saturation temperature. This results in the fact that only single-phase liquid comes to the outlet manifold through some of the parallel micro-channels.

The time variation of the mean and maximum heater temperature is presented in Fig. 2.23. The mean heater temperature (i.e., the average temperature of the whole heater) changed in the range of $\Delta T_{av} = 10$ K. The maximum heater temperature changed in the range of $\Delta T_{max} = 6$ K. Comparison between Figs. 2.21–2.23 shows that the time period (frequency) is the same for the pressure drop, the fluid temperature at the outlet manifold, and the mean and maximum heater temperature fluctuations. It also allows one to conclude that these fluctuations are in phase.

When the heat flux was further increased to $q = 260$ kW/m^2, the liquid-phase period did not disappear and continuous two-phase flow was not observed throughout the parallel micro-channels. The behavior of the heater temperatures is shown in Fig. 2.24. The average values of the whole and maximum heater temperatures are about $T_{mean} = 110$ °C and $T_{max} = 120$ °C. The maximum heater temperature varied in the range of $\Delta T_{max} = 20$ °C (i.e., from 110 to about 130 °C). This is somewhat unexpected. While numerous types of flow pattern observations in

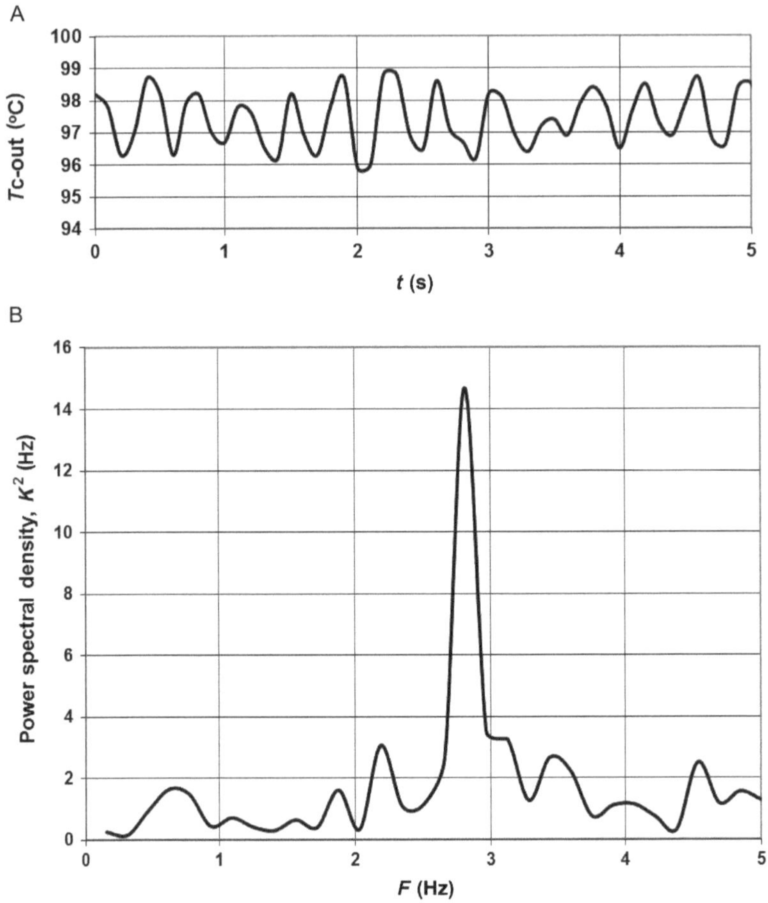

Figure 2.22 Time variation of fluid temperature at the outlet manifold, $q = 200$ kW/m^2. (A) Temperature fluctuations, (B) temperature amplitude spectrum [57].

micro-channel flow have been reported in the literature, this issue has received very limited attention, despite its importance for the design and safe operation of a heat sink. CHF generally refers to a sudden decrease in the heat transfer coefficient for a surface on which boiling is occurring. For a heat-flux-controlled system, exceeding CHF can lead to a sudden increase in the average surface temperature. Under conditions discussed earlier ($m = 93.3$ kg/m^{2}s and heat flux $q = 260$ kW/ m^2), the average temperature of the whole heater was about $T_{mean} = 110$ °C, whereas the maximum temperature periodically reaches the value of 130 °C, which for most heat sinks can lead to catastrophic system failure.

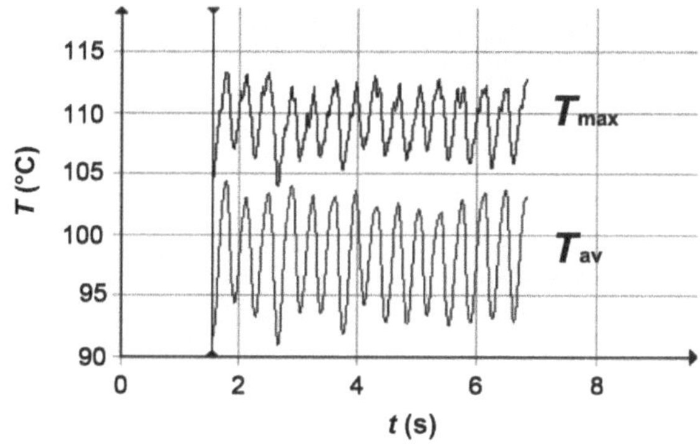

Figure 2.23 Time variation of average and maximum heater temperature at $q = 200$ kW/m^2 [57].

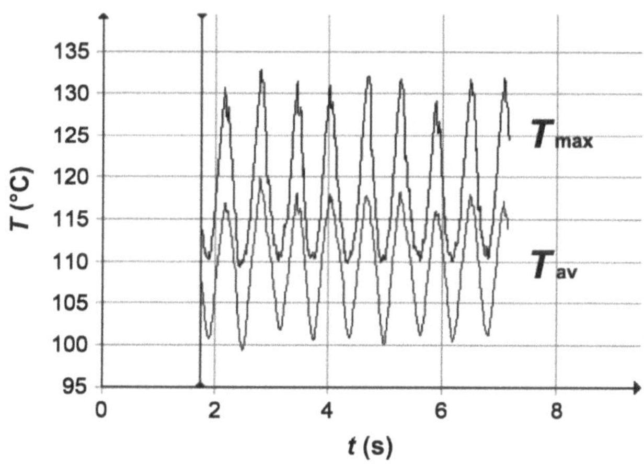

Figure 2.24 Time variation of average and maximum heater temperature at $q = 220$ kW/m^2 [57].

The paper by Diaz and Schmidt [68] presents an experimental investigation of flow boiling heat transfer in a single 0.3×12.7 mm^2 rectangular micro-channel. Water and ethanol were employed as test fluids. The test section, which was made of the nickel alloy Inconel 600, is electrically heated. IR images of the test section were recorded at a frequency of 150 Hz. In the range of heat flux $q = 67.7$–359.8 kW/m^2 and mass flux of 200, 400, and 500 kg/m^{2}s, the heat transfer coefficient was found to decrease with increasing quality for water. On the other hand, different trends could be observed for ethanol above the low quality region. The heat transfer coefficient increased with increasing quality for a mass flux of 100 kg/m^{2}s at low heat fluxes $q = 11.5\text{–}56.0\ \text{kW/m}^2$. For mass flux of 500 $\text{kg/m}^2\text{s}$, the heat transfer coefficient decreased with quality with increasing heat flux in the range of $q = 66.8\text{–}188.3\ \text{kW/m}^2$.

Barber *et al.* [69] carried out simultaneous data acquisition of pressure, temperature, wall heat transfer parameters, and flow visualization. This has been made possible due to a transparent metallic deposit of Tantalum on the exterior wall of rectangular micro-channel of hydraulic diameter 727 μm and the use of an IR camera. This tantalum deposit is both conductive and transparent at the thickness sputtered, hence enabling simultaneous uniform heating and visualization of the micro-channel flow. A uniform heat flux of 4.26 kW/m^2 was applied to the micro-channel and an inlet liquid mass flow rate held constant at 1.13×10^{-5} kg/s. In conjunction with obtaining high-speed images, a sensitive IR camera was used to record the temperature profiles on the exterior wall of the micro-channel, and a data acquisition system was used to record the pressure fluctuations over time. The working fluid chosen was *n*-pentane, for its low saturation temperature and hence its low power requirements to induce boiling. The IR camera allowed the measurement of the local temperature on the micro-channel wall, and hence enabled the deduction of the heat transfer coefficient along the micro-channel while simultaneously visualizing the fluid flow inside. The frame rate of the IR system was up to 50 Hz.

In Fig. 2.25, the data over a time period of 80 s are presented. The various peaks in the temperature, average heat transfer coefficient, and the pressure data are well correlated. There are periodic fluctuations in both the temperature, heat transfer, and pressure data of approximate frequency 0.067 Hz, with the magnitude of the temperature fluctuations being approximately 16 °C and the magnitude of the pressure fluctuations being approximately 25 mbar. The pressure fluctuations occurred just after the peak of the temperature fluctuation has been reached. The pressure continued to fluctuate during the negative gradient region of the temperature peak. The pressure has a more constant increasing tendency (from 5 to 10 mbar) during the positive gradient region of the temperature peak. The maximum fluctuation in the average heat transfer coefficient was approximately 1000 $\text{W/m}^2\text{K}$ at time $t = 33$ s, with average fluctuations of around 500–700 $\text{W/m}^2\text{K}$. Vapor instabilities are believed to be responsible for triggering flow reversals and high fluctuations in both temperature and pressure.

Spatial and temporal variations of channel wall temperature during flow boiling in micro-channel flows using IR thermography are presented and analyzed by Krebs *et al.* [70]. In particular, the top channel wall temperature

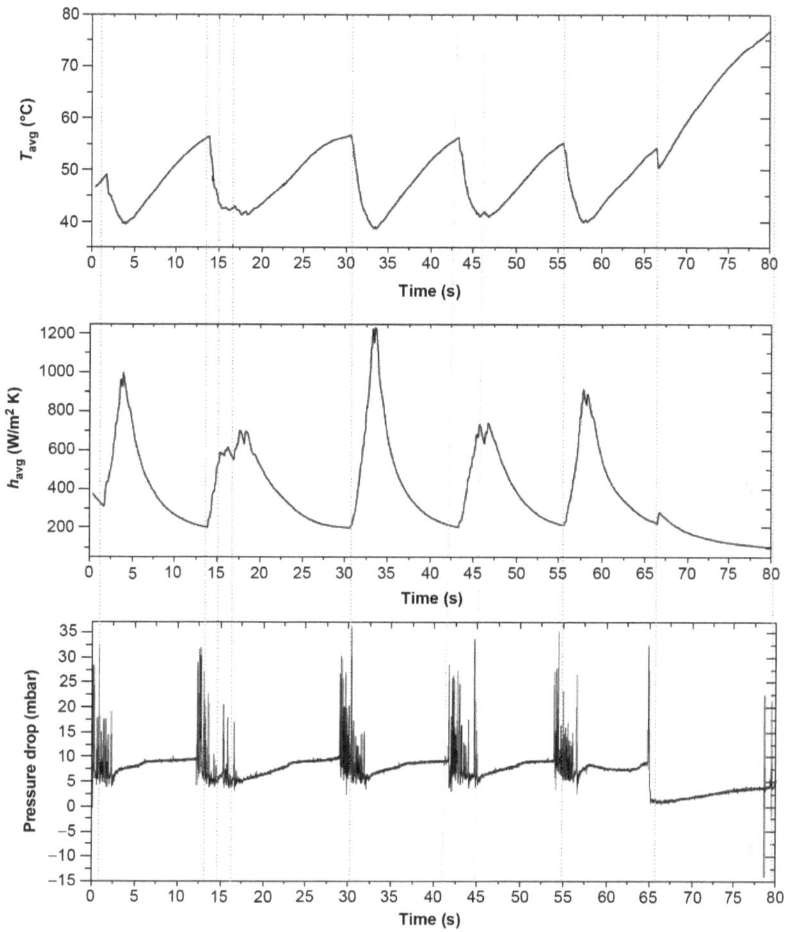

Figure 2.25 Simultaneous data measurements of average temperature profile and pressure drop across the micro-channel. Experimental conditions: uniform heat flux applied to the micro-channel $q=4.26$ kW/m^2, constant inlet liquid mass flow rate $m=1.13\times10^{-5}$ kg/s [69]. (For color version of this figure, the reader is referred to the online version of this chapter.)

in a branching micro-channel silicon heat sink was measured nonintrusively. By the use of this technique, time-averaged temperature measurements, with a spatial resolution of 10 μm, are presented over an 18 mm × 18 mm area of the heat sink. For a mass flux of 93 kg/m^{2}s and a heat flux of 200 kW/m^2, they found that the average surface temperature, exit fluid temperature, as well as pressure drop showed a periodic cycling at a frequency between 2 and 3 Hz. The amplitude of the heater surface temperatures was about 10 °C, while that of the fluid at the exit was about 2 °C.

4. NUCLEATION CHARACTERISTICS OF HEATERS

4.1. Nucleation site density (NSD)

In this section, the discussion will be directed to the hot and dry areas appearing on the whole surface as well as hot and dry spots appearing within the bubble bases. Relevant to these issues is the number of bubbles that are found (at any time) per unit area of the heater surface, NSD. For boiling heat transfer, due to the high frequency of physical processes of interest, it is essential to have high speed imaging and storing capability. A first attempt in using this method in boiling research was made by Myers's group at UCSB in the early 1970s. They used thin heater plates covered with liquid crystals. Raad and Myers [71] and separately by means of a high-speed IR camera, Sgheiza and Myers [72] studied nucleation in pool boiling. Technological limitations at the time, however, allowed only qualitative results. Later, Kenning [73] developed further the liquid crystal thermographic technique and obtained interesting results for low heat flux nucleate boiling [74]. Hetsroni *et al.* [75] carried out a study of pool boiling from a horizontal heating surface facing upward. Boiling occurred over a constantan foil, 50 μm thick, attached to the window at the bottom of the vessel. The foil was heated by DC current and the heat flux on the wall could be regulated and measured. The IR image of the heater was recorded from below by an IR radiometer. The roughness parameters measured by a profile measuring instrument are given in Table 2.6. The experiments showed the unique capacity of the IR technique to measure wall temperature distributions with high spatial resolution over an area encompassing many nucleation sites and over long periods. For example, the temperature distributions on the heater at constant heat flux $q = 90$ kW/m^2 show variations of wall temperature of about 17 K as can be seen in Fig. 2.26.

Detailed experimental study of nucleate pool boiling under highly controlled conditions using electrically heated, vapor-deposited submicron metallic films (see Section 1) was performed by Theofanous *et al.* [19,21].

Table 2.6 Roughness parameters [75]

Type of boiling surface	Maximum peak-to-valley height R_{max} (μm)	Average Roughness R_{av} (μm)
Before boiling	1.32	0.100
After boiling	5.40	0.670

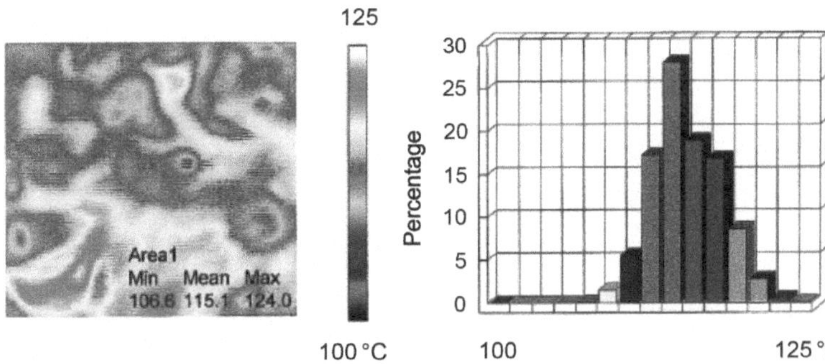

Figure 2.26 Thermal patterns and histograms of the temperature at pool boiling of water on the flat plate ($q = 90$ kW/m^2) [75]. (For color version of this figure, the reader is referred to the online version of this chapter.)

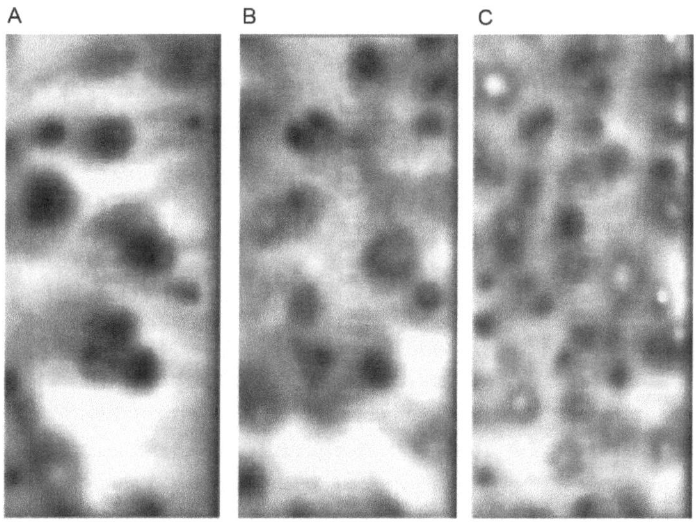

Figure 2.27 IR thermometry images of a fresh heater (F1) at three different heat fluxes, $q = 406$, 536, and 807 kW/m^2 [19].

A sampling of the IR records obtained in those tests is shown in Figs. 2.27 and 2.28. These are complete original data in that the reading from every pixel has been converted to temperature using the calibration scale and in that the whole heater surface is imaged. The temperatures in the figures are represented by a gray scale that varies in each case so that the "light" end corresponds to the highest temperature in the image and the "dark" end to the lowest one. Generally, the light end is at ~150 °C and the dark end at ~100 °C. The dark circular spots (or discs) are a visualization of the

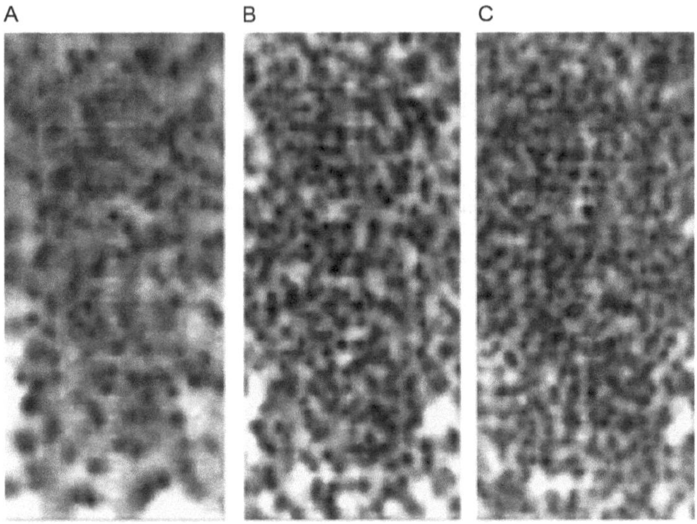

Figure 2.28 IR thermometry images of an aged heater (A1) at three different heat fluxes, $q = 348$, 1051, and 1517 kW/m^2 [19].

cooled areas under bubbles growing on the heater. NSD increases with heat flux and also the much higher density of nucleation sites found on aged heaters, as compared to fresh heaters. Quantitatively, these results are depicted in Fig. 2.29 for three fresh heaters (F1, F4, F9) and in Fig. 2.30 for three aged heaters (A1, A3, A4). The difference in NSD between fresh and aged heaters is by about one order of magnitude. Also can be seen is a linear dependence on heat flux and the effect of water quality (Fig. 2.29) and the effect of heavy aging (Fig. 2.30). Other noteworthy features of Figs. 2.27 and 2.28 are that uniformity of nucleation increases with heat flux. Activation of nucleation sites can be either regular or irregular. Correspondingly, they form what may be called regular and irregular bubbles. At regular sites, the nucleation is more or less periodic, while the irregular activation has a "silent" period between aperiodic bubbling cycles. With the increase of heat flux, irregular bubbles become more regular. Also can be seen is the discrepancy between results obtained by Theofanous *et al.* [19] and previous works published in the literature. Gaertner and Westwater [76] and Wang and Dhir [77] reported hundreds of nuclei per cm^2, even on aged heaters. Theofanous *et al.* [19] found nucleation densities that are well under 100 per cm^2. On the other hand, while the Benjamin and Balakrishnan [78] data at first appear to be in line with what was found by the present author, on a closer examination they appear to be too low by a factor of 4. For example, it can

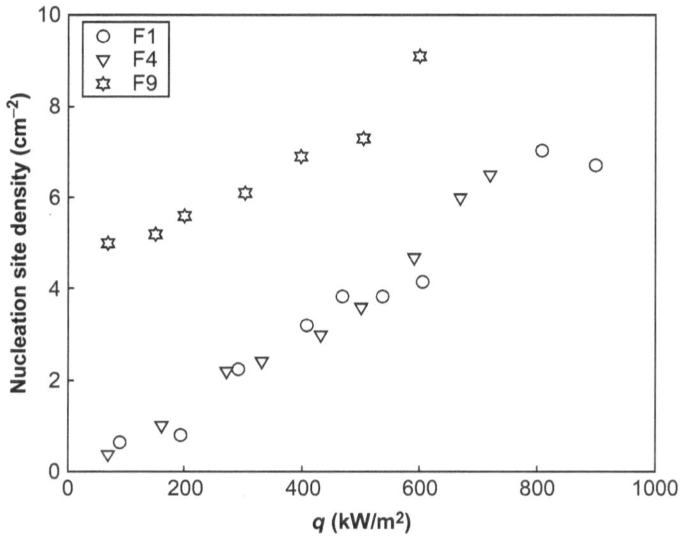

Figure 2.29 NSD (n = A) as a function of the heater's heat flux, q, on fresh heaters. Test F9 was run with clean distilled water. Tests F1 and F4, with HPLC class water [19].

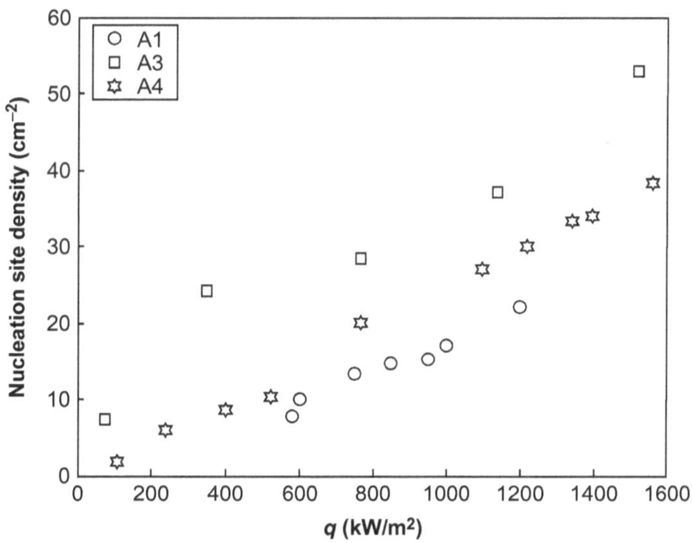

Figure 2.30 NSD (n = A) as a function of the heater's heat flux, q, on aged heaters. Heater A1 was aged by pulse heating in air. Heaters A3 and A4 were heavily aged by repeated pulse heating and boiling in water [19].

be expected that aluminum to be well aged (it oxidizes rapidly even in air), so their $N \sim 8$ per cm^2 at ~ 1 MW/m^2 is to be compared with the $N \sim 30$ per cm^2 for the heater A3 (Fig. 2.30). Finally, the previously deduced trends [19] with superheat and heat flux seem to be off. For example, compare Gaertner and Westwater's [76] $N \sim q^2$ with results $N \sim q$ shown in Figs. 2.29 and 2.30. Also, the Wang and Dhir [77] and Kocamustafaogullari and Ishii [79] dependencies $N \sim \Delta T_s^m$ with $m = 6$ and 4.4, respectively, are quite steep. The data indicate that parameters such as cavity size, or wetting angle, are significant in heterogeneous nucleation, and that the principal cause/mechanism remains to be found.

4.2. Dryout

4.2.1 Flat surface with nano-scale roughness

The dark circular spots shown in Figs. 2.27 and 2.28 were called by Theofanous *et al.* [19] as cold spots. The spatial variation of heater temperature across a cold spot at different moments during a bubbling cycle is shown in Fig. 2.31. At 200 kW/m^2 the bubble base is large, about 3 mm in diameter. During the bubbling cycle the heater surface under the bubble and a liquid sub-layer attached to it remain cold. The wall superheat varies from 6 K at the cold spot center to about 20 K in the peripheral ring. This variation corresponds to a decrease in heat transfer associated with the

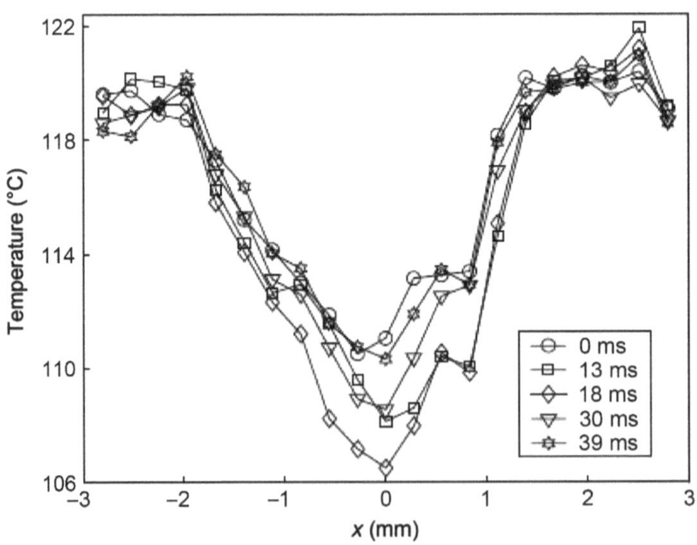

Figure 2.31 Temperature profiles across a cold spot of regular bubble on a fresh heater at five time moments. Heater F1 at $q = 200$ kW/m^2 [19].

gradual thickening of the liquid meniscus beneath the bubble. The IR thermometry data show the local temperature within the cold spots being affected by the dynamics of the surrounding liquid, e.g., due to nucleation and growth of vapor bubbles in the proximity.

As the surface heat flux increases, the IR images show bright spots appearing within the bubble bases. These bright spots, typically 1–2 mm in diameter, represent overheating of the heater surface. At low heat fluxes, life duration of hot spots is controlled by regular bubbling cycles and is generally of short duration, 10–20 ms. The hot spot temperature is peaked only 5–10 K higher than the surrounding fluid, and they are periodically replaced by cool areas under a newly growing bubble. As heat flux increases, the hot spot maximum superheat may reach 50–60 K. Such a hot spot can be seen in Fig. 2.32.

4.2.2 Channel surface with micro-scale roughness

The first results of heat transfer characteristics during flow boiling in micro channels with rectangular cross-section using a thermo-graphic measuring method were reported by Hapke *et al.* [80]. IR technique was used also in experiments conducted by Hapke *et al.* [80]. The channels were designed as Joule heating pipes. Thus, the evaporation was achieved under conditions

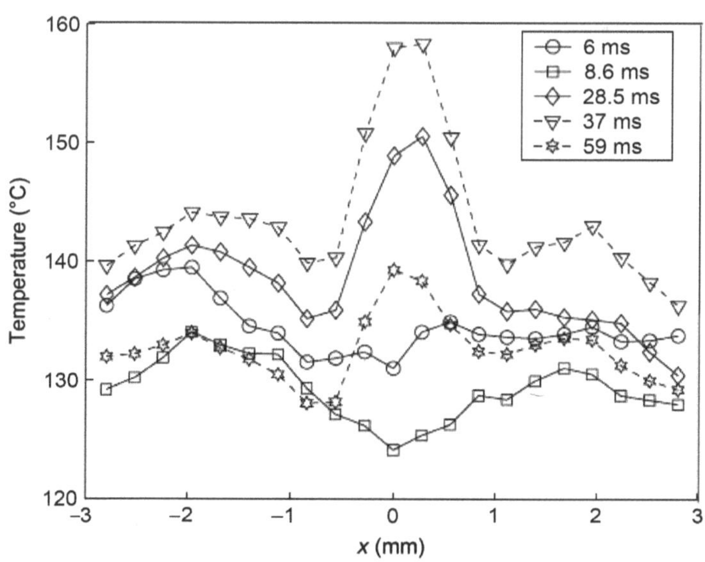

Figure 2.32 Temperature profiles across of a hot spot (900 kW/m^2) at five different time moments [19].

of nearly constant heat fluxes at the wall of the channel. The channels investigated had a width b of 10 mm and heights d of 700 and 300 μm. The length of each channel was 200 mm. Water and n-heptane were used as test fluids. The mass flux was varied from 25 to 350 kg/m^{2}s and the heat flux from 20 to 400 kW/m^2. In order to determine the outer wall temperature, the axial distribution was measured using the thermography system. Images of different boiling stages can be seen in Fig. 2.33. The high spatial and temporal resolution of this thermographic measuring method makes it possible to detect the axial position of the different boiling regions. Furthermore, it allows conclusions to be made on which flow conditions occur in the different sections of the channel.

4.2.3 Explosive boiling

Flow and thermal visualization of water boiling in parallel micro-channels showed that behavior of vapor bubbles, occurring in micro-channels at low Reynolds numbers, was not similar to annular flow with interposed intermitted slugs of liquid between two long vapor trains. This process may be regarded as explosive boiling with periodic wetting and dryout [53]. Tests were performed in the range of inlet Reynolds number 25–60, mass flux 95–340 kg/m^{2}s, and heat flux 80–330 kW/m^2. The test module was fabricated of a square-shape silicon substrate 15 × 15 mm, 530 μm thick, and utilized a Pyrex cover, 500 μm thick, which served as both an insulator and a transparent cover through which flow in the micro-channels could be observed. The Pyrex cover was anodically bonded to the silicon chip, in order to seal the channels. In the silicon substrate, 21 parallel micro-channels were etched, the cross-section of each channel was an isosceles triangle with base $a = 250$ μm. The angles at the base were 55°.

In two-phase flow, the flow began as subcooled single-phase and changed under conditions of the present experiments into explosive two-phase

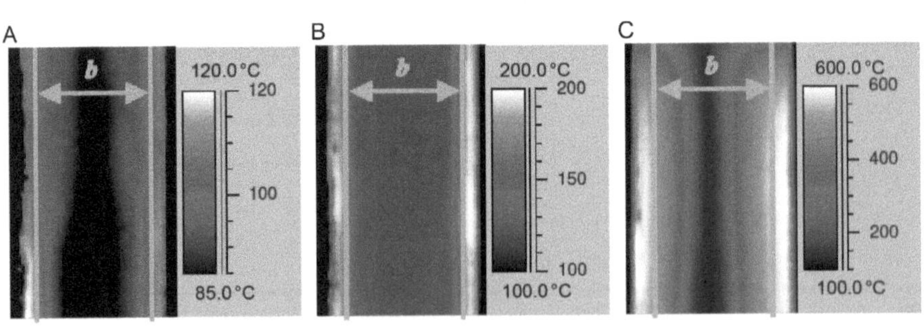

Figure 2.33 (A) Initial boiling, (B) flow boiling, (C) dryout [80].

flow. The vapor may be generated only in part of the parallel micro-channels. The location of the saturation temperature is unknown. The simultaneous images obtained from flow and thermal visualization were analyzed to distinguish the part of the parallel micro-channels where boiling occurred. Then the temperature on the heater along each micro-channel was measured and the area available for boiling heat transfer was determined. Figure 2.34 illustrates a typical IR image of the heater obtained at $m = 95$ kg/m^{2}s and $q = 160$ kW/m^2. The flow moves from the bottom to the top, the area of the whole heater is 1 × 1 cm (Area 1 is the whole large square). Area 1 shows the temperature distribution in the streamwise and spanwise directions. The mean temperature on the whole heater was less than the saturation temperature, which indicates that boiling did not occur in the part of this area. There are a hot spots of Area 1 where the boiling occurred, since $T_{max} = 113.0$ °C. If the mean temperature, was less than saturation temperature it indicates that boiling did not occur in part of the test module. From Fig. 2.34 one can see that the mean temperature of Area 1 is $T_{mean} = 96.1$ °C. There are the following two mechanisms: (i) heating fluid from the inlet temperature to saturation temperature, (ii) single phase fluid

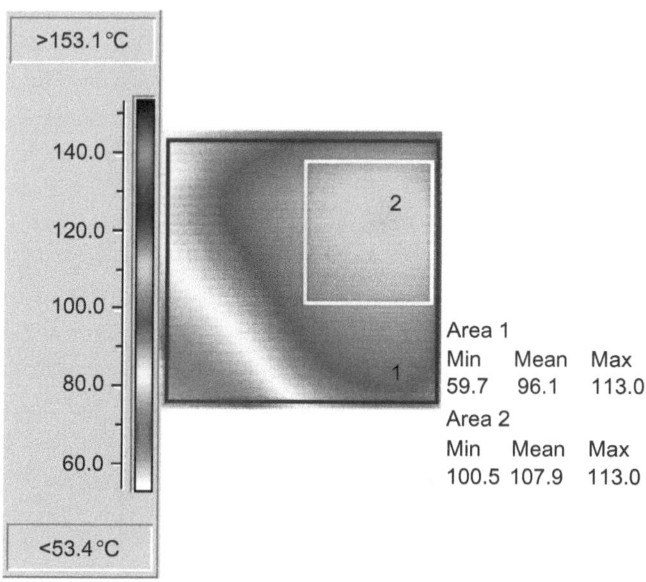

Figure 2.34 Thermal field on the heater, $F = 1 \times 1$ cm^2. $\dot{m} = 95$ kg/m^{2}s, $q = 160$ kW/m^2. 1. Area of the whole heater. 2. Area of the part of the heater, where saturated flow boiling occurs [53]. (For color version of this figure, the reader is referred to the online version of this chapter.)

flow into certain micro-channels. In the present study we consider part of the whole heater, along which the minimum temperature on the channel wall exceeds the saturation temperature. Area 2 (marked as the small square) shows the temperature distribution in the streamwise and spanwise directions at such part of the heater. The mean temperature of this area is $T_{mean} = 107.9\ °C$, the minimum and maximum temperatures are $T_{min} = 100.5\ °C$ and $T_{max} = 113.0\ °C$. The mean temperature of Area 2 was used for calculating the mean temperature at the heated channel wall and the boiling heat transfer coefficient.

One can see that the temperature field of the heater is nonuniform. The inlet fluid temperature was less than saturation temperature. That leads to significant increase in the temperature on the heater in the streamwise direction. Although the channels were made in such a way that the cross-section of all channels would be uniform, the flow rate through each channel may not be equal. The distribution of flow rate through the channels depends on the connection of the test section to the inlet and outlet manifolds. This unevenness leads to nonuniformity of the heater temperature in the spanwise direction.

Figure 2.35A–O illustrates a typical example of alternate two-phase flow pattern at a distance of 1000–1500 μm downstream from the inlet of the test section. In Fig. 2.35A–O a single triangular channel and the walls between the adjacent channels (white strips) are depicted in the central part and in the peripheral parts of each image, respectively. The field of view (including the channel and the walls) is 0.6 mm × 0.6 mm, the flow moves from the bottom to the top, the mass flux is $m = 95\ kg/m^2s$, the heat flux is $q = 160\ kW/m^2$. Figure 2.35A shows the water flow that moved through the micro-channel. When the water moves along the channel it is heated up to saturation temperature and then the elongated bubble vents in a short time. Figure 2.35B displays the onset of two-phase reversal flow, which evolved due to venting of the elongated bubble downstream from the observation point. From Fig. 2.35A and B one can conclude that at the specified conditions of this experiment, the life time of the elongated bubble did not exceed 0.001 s. Annular flow was powered by venting of the elongated bubble and had a short duration (about 0.004 s).

Figure 2.35G and H displays the onset of dryout, where liquid droplets are accumulated at the bottom of the triangular channel. Figure 2.35I–K shows that the amount of liquid phase decreases, and only few droplets of water remain on the channel bottom. Figure 2.35L shows the beginning

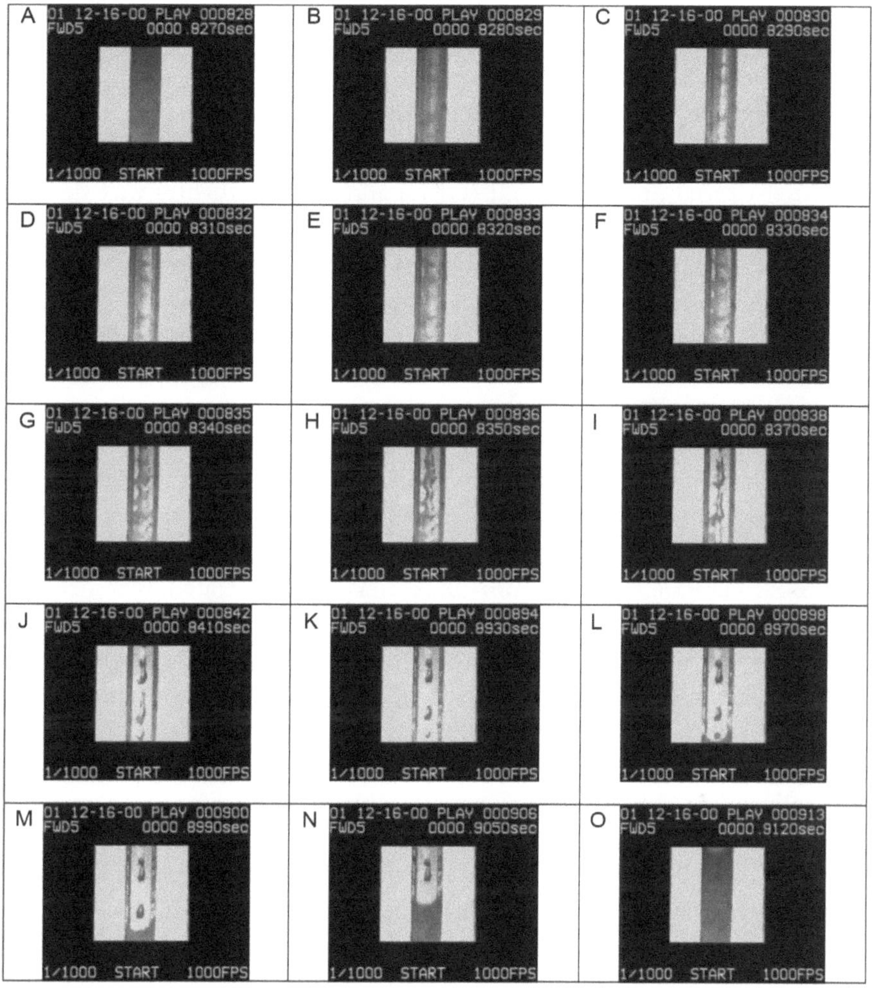

Figure 2.35 Flow pattern near inlet manifold $\dot{m} = 95$ kg/m^{2}s, $q = 160$ kW/m^2 [53].

of wetting and the channel is filled with liquid (Fig. 2.35M–O). Flow pattern depicted in Fig. 2.35A–O were also observed by Serizawa *et al.* [81] in steam–water flow in a 50-μm silicon tube. The authors reported that at low liquid flow rates, partially continuous liquid film flow changed to rivulets or even to discrete liquid lumps or large liquid droplets. The periodic phenomenon described earlier reveals that the entire channel acts like the area beneath a growing bubble, going through periodic drying and rewetting. The cycle was repetitive with venting of the elongated bubble.

5. THE BOILING CRISIS PHENOMENON

5.1. CHF measurements in micro-channels

As boiling in micro-channel heat sinks is an attractive method for cooling computer central processing unit (CPU) and other high-heat flux devices (such as laser diodes), it is of crucial importance to accurately predict the CHF in the small-diameter channels. CHF or burnout is a limiting value for safe operation of heat dissipation applications and refers to replacement of liquid in contact with the heated surface with a vapor blanket. The thermal conductivity of the vapor is very low compared to the liquid and the surface heat transfer coefficient drops dramatically, resulting in sudden increase of the surface temperature and possible failure of the cooled device.

The first experimental investigation of the CHF in multi-micro-channels was carried out by Bowers and Mudawar [82], who tested a cooling element with an array of 17 circular channels, 0.51 mm in diameter, 28.6 mm in length, in a 1.59-mm-thick nickel block, heated over a central 10 mm square section. Boiling curves were generated that terminated in well-defined CHF measurements, which were found to be independent of the liquid subcooling and almost directly proportional to the mass flux, for which a dimensionless correlation was proposed. Jiang *et al.* [83] developed two multi-micro-channel heat sinks integrated with a heater and an array of implanted temperature sensors. There were 58 or 34 channels of rhombic shape, having a hydraulic diameter of 0.040 or 0.080 mm, respectively, in their 10 mm × 20 mm test section. CHF data were taken for once-through water flow entering at $\Delta T=20$ K subcooling. The CHF condition was characterized by a rapid rise in the temperature sensors. The critical power limit was found to be proportional to the total volumetric flow rate. In a recent paper, Qu and Mudawar [84] reported a comprehensive study of the CHF in rectangular micro-channels. Their heated block contained 21 0.215 × 0.821 mm channels. The heat flux was reported based on the area of the three active sides of the channel. De-ionized, de-aerated water was supplied over a range of $G=86$ to 368 kg/m^{2}s, with an inlet temperature of 30–60 °C and outlet pressure of 1.13 bar. The dependence of the heat flux on the mass flux is shown in Fig. 2.36 ($q_{p,m}$ is the CHF based on channel heated inside area, $q_{eff,m}$ is the CHF based on the heat sink's top platform area). One can see that the CHF increases monotonically with increasing G for both inlet temperatures. What is quite surprising is that the inlet temperature T_{in} has an insignificant effect on the CHF. Interestingly,

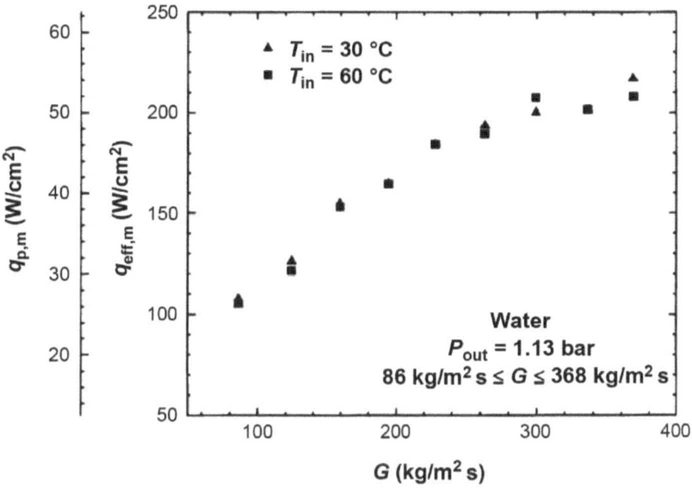

Figure 2.36 Variation of CHF with mass velocity. *Reprinted from Qu and Mudawar [84].*

these CHF trends relative to the mass flux G and inlet temperature T_{in} mirror those of Bowers and Mudawar [82] for refrigerant R-113 in circular micro-channel heat sinks. While the trend of increasing CHF with increasing G is quite common, the lack of inlet temperature effect on the CHF seems to be unique to two-phase micro-channel heat sinks, not to single micro-channels. A key difference between these is the aforementioned amplification of parallel channel instability prior to the CHF. As discussed earlier, this amplification causes back flow of vapor into the upstream plenum, which results in strong mixing of the vapor with the incoming liquid. Regardless of how subcooled the incoming liquid, the mixing action appears to increase the liquid temperature close to the local saturation level as it approaches the channel inlet.

The CHF correlation (2.15) was developed by Qu and Mudawar [84] for water in a rectangular micro-channel heat sink, as well as Bowers and Mudawar's [82] CHF data for R-113 in a circular micro-channel heat sink.

Their CHF correlation is:

$$\frac{q_{p,m}}{G h_{LG}} = 33.43 \left(\frac{\rho_G}{\rho_L}\right)^{1.11} We_L^{-0.21} \left(\frac{L}{d_h}\right)^{-0.36} \tag{2.15}$$

Since the CHF for micro-channel heat sink databases shows no dependence on inlet subcooling, these databases were correlated without the subcooling multiplier.

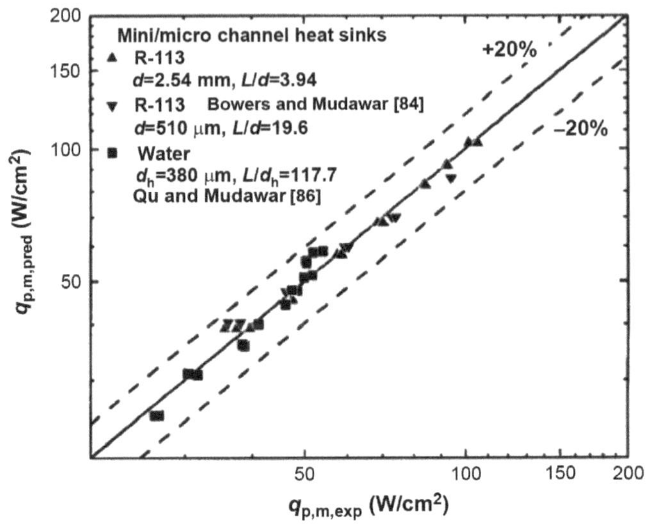

Figure 2.37 Comparison of CHF data for water and R-113 in mini/micro-channel heat sinks correlation (2.15) [84].

Figure 2.37 compares the predictions of this correlation with the flow boiling CHF data for water both in the rectangular micro-channel heat sink [84] and in the circular micro-channel heat sinks [82]. The overall mean absolute error of 4% demonstrates its predictive capability for different fluids, circumferential heating conditions, channel geometries, channel sizes, and length-to-diameter ratios.

A series of tests were performed by Wojtan *et al.* [85] to determine the saturated CHF in 0.5 and 0.8 mm inner diameter micro-channel tubes as a function of refrigerant mass velocity, heated length, saturation temperature, and inlet liquid subcooling. The tested refrigerants were R-134a and R-245fa and the heated length of the micro-channel varied between 20 and 70 mm. Figure 2.38 shows the evolution of the CHF as a function of mass velocity. All conditions (heated length and temperature of subcooling of the refrigerant) were the same for both diameters. As can be seen, the CHF increased with increasing mass velocity. The CHF for the 0.8 mm micro-channel is higher than that for the 0.5 mm one (Fig. 2.38) and the difference (30–50%) becomes greater as the mass velocity increases. The influence of the heated length on the CHF in both micro-tube diameters at constant mass velocity is depicted in Fig. 2.39. The highest CHF was measured for the shortest heated length. With the heated length for the 0.8 mm diameter tube enlarged from 20 to 70 mm, the drop in the CHF value is

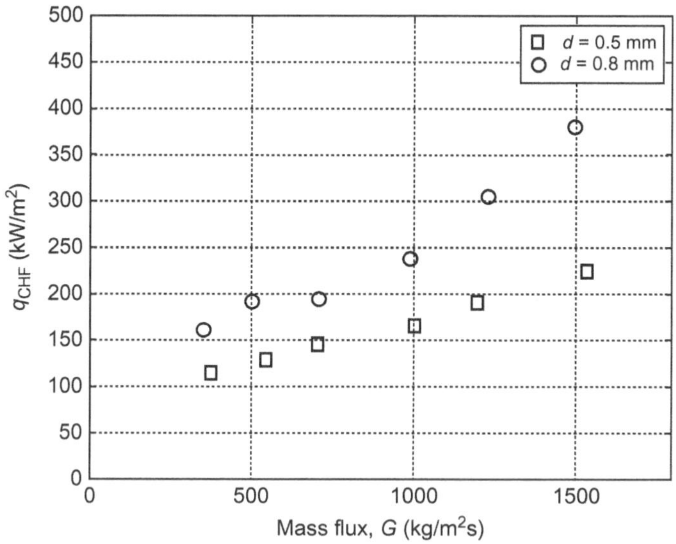

Figure 2.38 Variation of CHF as a function of the mass velocity in 0.5 and 0.8 mm tubes. R-134a, $\Delta T_{sub}=8$ K, $T_{sat}=35$ °C, $L=70$ mm [85].

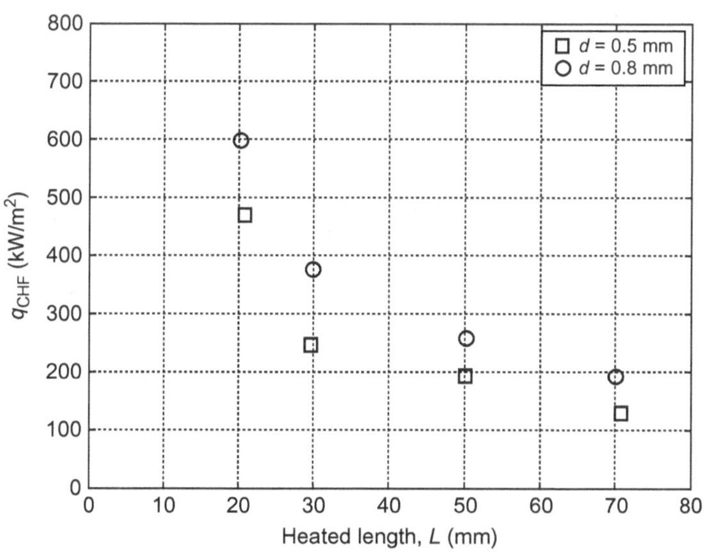

Figure 2.39 Variation of CHF as a function of the heated length in 0.5 and 0.8 mm tubes. R-134a, $G=500$ kg/m^{2}s, $\Delta T_{sub}=8$ K, $T_s=35$ °C [85].

$q_{CHF}=400$ kW/m^2. It confirms that the heated length, besides the mass velocity, is one of the most important parameters in the design of heat sinks.

Figure 2.40 shows the influence of liquid subcooling on the CHF. Only three experimental points are available as measurements at stronger subcoolings proved impossible. On the basis of the limited data, it can be concluded that CHF does not change significantly for the subcooling range $\Delta T=4.5$ to 12 K. This observation is in agreement with the results of Qu and Mudawar [84] for micro-channel heat sinks.

Comparison of the results obtained by Wojtan *et al.* [85] and the correlation presented by Qu and Mudawar [84] are shown in Fig. 2.41. As can be seen there is significant scatter between the respective results. One can conclude that there is a lack of consistency in the reported data. Available data sets for flow boiling CHF of water in small-diameter tubes are given in Table 2.7, as presented by Zhang *et al.* [86]. There are 13 collected data sets in all. Considering only data for tube diameters less than 6.22 mm, and then eliminating duplicate data and those not meeting the heat balance calculation, the collected database included a total of 3837 data points (2539 points for saturated CHF and 1298 points for subcooled CHF), covering a wide range of parameters, such as outlet pressures from 0.101 to 19.0 MPa, mass fluxes from 5.33 to 1.34×10^5 kg/m^{2}s, CHFs from 0.094 to 276 MW/m^2, hydraulic diameters of channels from 0.330 to 6.22 mm, length-to-diameter ratios from 1.00 to 975, inlet qualities from –2.35 to 0, and outlet thermal

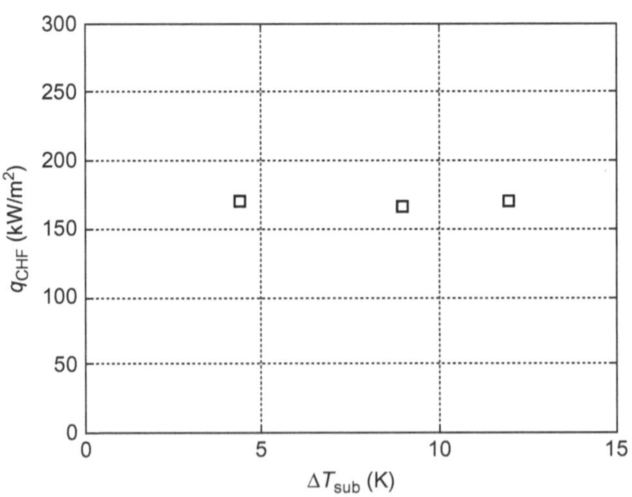

Figure 2.40 Influence of refrigerant subcooling on CHF. R-134a, $G=1000$ kg/m^{2}s, $T_s=35$ °C, $L=70$ mm, $d=0.5$ mm [85].

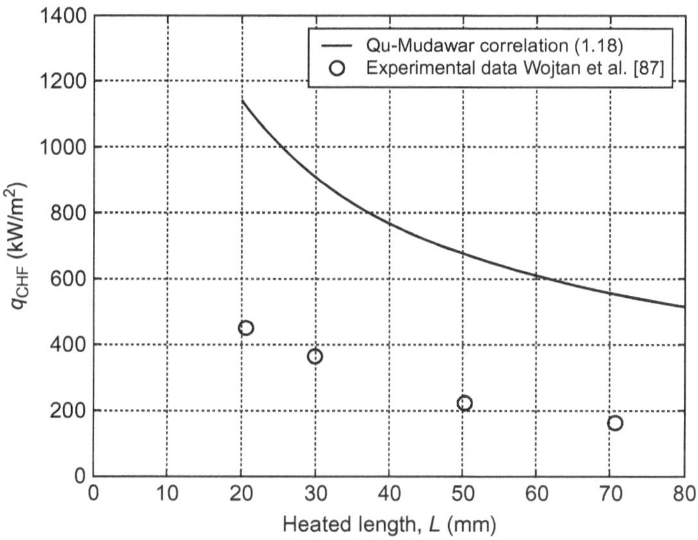

Figure 2.41 Comparison of the experimental results by Wojtan et al. (2006) to the correlation (2.15) of Qu and Mudawar (2004). R-134a, $G = 500$ kg/m^{2}s, $\Delta T_{sub} = 8$ K, $T_s = 35$ °C, $d = 0.5$ mm [85].

equilibrium qualities from -1.75 to 1.00. Zhang *et al.* [86] compared some correlation with each database. The comparison results are tabulated in Table 2.8. Mean deviation is defined as $(1/N)\sum|(q_{e.exp} - q_{c.cal})/q_{e.exp}| \times 100\%$, and a bold font in Table 2.8 denoting the smallest of the mean deviations predicted by four correlations including a new correlation, and an underlined font being the smallest except for the correlation by Zhang *et al.* [86]. The best agreement with experimental data on saturated CHF is the correlation by Zhang *et al.* [86]:

$$Bo = 0.0352\left[We_D + 0.0119\left(\frac{L}{D_h}\right)^{2.31}\left(\frac{\rho_G}{\rho_L}\right)^{0.361}\right]^{-0.295} \times \left[\left(\frac{L}{D_h}\right)^{-0.311}\left(2.05\left(\frac{\rho_G}{\rho_L}\right)^{0.170} - x_{eq.in}\right)\right] \quad (2.16)$$

where $Bo = q_c/(h_{LG}m)$ is the boiling number, $We_D = (m^2 D_h)/(\sigma\rho_L)$ is the Weber number, q_c is the CHF, h_{LG} is the latent heat of evaporation, G is the mass flux, $x_{eq,in}$ is the thermodynamic equilibrium quality at the inlet, ρ_G and ρ_L are the density of saturated vapor and liquid, respectively, D_h is the hydraulic diameter, L is the heated length, and σ is the surface tension.

Table 2.7 Collected database for flow boiling CHF of water in small-diameter tubes [87]

Symb.	Reference	d_h (mm)	l/d_h	p_o(MPa)	G(kg/m^{2}s)	$x_{eq,in}$ (%)	$x_{eq,0}$ (%)	$q_{c,exp}$ (MW/m^2)
○	Thompson and Macbeth [126]	1.02–5.74	11.7–792	0.103–19.0	$13.0–1.57\times10^4$	−235 to −0.031	−44.8 to 99.9	0.113–21.4
△	Lowdermilk *et al.* [127]	1.30–4.78	25.0–250	0.101–0.690	$27.1–3.42\times10^4$	−29.1 to −0.032	−3.02 to 99.1	0.167–41.6
▽	Becker *et al.* [128]	3.93–6.07	164.8–382	1.13–6.97	$470–5.45\times10^3$	−67.5 to −23.4	0.086–9607	1.59–5.66
◫	Griffel [129]	6.22	147	6.89–10.3	1.85×10^3–1.39×10^4	−91.4 to −6.38	−7.07 to 30.4	2.60–7.81
⬡	Nariai *et al.* [130]	1.00–3.00	3.33–50.0	0.101–1.05	4.30×10^3–2.99×10^4	−31.5 to −3.07	−18.3 to 5.55	4.50–66.1
⏀	Inasaka and Nariai [131]	3.00	33.3	0.290–1.05	4.30×10^3–2.99×10^4	−31.5 to −12.5	−18.3 to −3.99	7.30–44.5
◁	Inasaka [132]	1.00–3.00	3.17–50.4	0.101	6.71×10^3–2.09×10^4	−15.6 to −6.66	−13.2 to 1.10	4.64–67.0
□	Celata *et al.* [133]	2.5	40	0.585–2.61	1.12×10^4–4.00×10^4	−46.1 to −18.7	−35.6 to 0.6	1.21–60.6
⬠	Vandervort *et al.* [134]	0.330–2.67	1.66–26.2	0.131–2.28	5.03×10^3–4.18×10^4	−28.2 to −1.91	−22.6 to 28.4	4.60–124
Φ	Lezzi *et al.* [135]	1.00	239–975	1.90–7.20	7.76×10^2–2.74×10^4	−64.4 to 0	64.3–0.976	0.285–2.36

◊	Kureta [136]	1.00–6.00	1.00–113	0.101	$5.33–1.91\times10^4$	−17.2 to −0.032	−14.6 to 99.5	0.0935–158
◊	Roach *et al.* [137]	1.13–1.45	110–141	0.336–1.04	$256–1.04\times10^4$	-27.9 to -13.1	36.2–97.4	0.860–3.70
▷	Mudawar and Bowers [138]	0.406–2.54	2.36–34.2	0.250–17.2	$5.00\times10^3–1.34\times10^5$	−189 to −11.7	−175 to −6.23	9.40–276
Total (13 databases)		0.330–6.22	1.00–975	0.101–19.0	$5.33–1.34\times10^5$	−235 to 0	−175 to 99.9	0.0935–276

Table 2.8 Assessment of CHF correlations for flow boiling of water [87]

	Mean Deviation (%)						
	Correlations for saturated CHF				Correlations for subcooled CHF		
Reference	Bowring [101]	Katto and Ohno [106]	Shah [103]	Zhang *et al.* [85]	Inasaka and Nariai [93]	Celata *et al.* [95]	Hall and mudawar[86]
Thompson and Macbeth [126]	**12.3**	15.2	12.6	17.8	18.7	21.8	7.68
Lowdermilk *et al.* [127]	32.4	21.9	15.7	**11.2**	30.3	13.2	32.7
Becker *et al.* [128]	**5.51**	7.14	10.0	11.5			
Griffel [129]	5.61	5.47	**4.43**	17.2	4.48	21.3	5.97
Nariai *et al.* [130]	73.8	35.7	**21.8**	29.0	16.4	31.4	17.8
Inasaka and Nariai [131]	81.2	37.1			9.25	17.1	7.48
Inasaka [132]			**14.3**	38.3	18.9	28.0	21.7
Celata *et al.* [133]					23.2	28.7	14.8
Vandervort *et al.* [134]	82.5	16.9	42.9	**16.7**	17.3	24.6	29.6
Lezzi *et al.* [135]	32.3	**7.84**	15.3	9.40			
Kureta [136]	52.5	51.9	37.9	**20.2**	42.7	47.0	35.6
Roach *et al.* [137]	18.6	33.5	**16.4**	18.0			
Mudawar and Bowers [138]					83.5	38.7	18.8
Total (13 databases)	29.3	26.4	20.6	**16.8**	30.5	30.1	19.2

Hall and Mudawar [87] provided a comprehensive review of the current state of the knowledge of subcooled CHF for flow of boiling water in channels and designed a statistical correlation with five parameters based on almost all available subcooled CHF databases in the literature:

$$Bo = \frac{C_1 We_D^{C_2} (\rho_L/\rho_G)^{C_3} \left[1 - C_4 (\rho_L/\rho_G)^{C_5} x_{eq,in}\right]}{1 + 4 C_1 C_4 We_D^{C_2} (\rho_L/\rho_G)^{C_3+C_5} (L/D_h)} \tag{2.17}$$

where the Weber number $We_D = m^2 D_h/\rho_f \sigma$, $C_1 = 0.0722$, $C_2 = -0.312$, $C_3 = -0.644$, $C_4 = 0.900$, and $C_5 = 0.724$. The correlation was developed using a total of 4860 data points and predicted CHF with a rms error of 14.3% in the following parametric ranges: $0.1 < p_o < 20$ MPa, $0.25 < D_h < 15.0$ mm, $2 < L/D_h < 200$, $300 < m < 30{,}000$ kg/m^{2}s, $-2.00 < x_{eq,in} < 0.00$, and $-1.00 < x_{eq,o} < 1.00$.

A theoretical model for the prediction of the CHF of refrigerants flowing in heated, round micro-channels has been developed by Revellin and Thome [88]. The model is based on the two-phase conservation equations and includes the effect of the height of the interfacial waves of the annular film. Validation has been carried out by comparing the model with experimental results presented by Wojtan *et al.* [85], Qu and Mudawar [84], and Bowers and Mudawar [82]. More than 96% of the data for water and R-113, R-134a, R245fa were predicted within ± 20%.

5.2. Physical approach based on IR measurements

Theofanous *et al.* [21] examined IR records to study the CHF problem and the flow regimes that govern micro-hydrodynamics in pool boiling. The temperature of every pixel on the IR record can be converted to a micro-layer thickness (δ) by $\delta = k_l(T_w - T_s)/q$, where T_w and T_s are the heater and saturation temperatures, respectively. Sample results for a fresh and a heavily aged heater are shown in Figs. 2.42 and 2.43, respectively. Sequential arrangements of frames allowed one to visualize the full dynamic pattern of the micro-hydrodynamics. They observed that the average thickness of the micro-layer in an aged heater is in the range from 10 to 15 μm, while for fresh heaters this range changes to 20–30 μm near-CHF conditions.

The experimental data indicate that the dry spot growth is constrained and guided by neighboring active nucleation sites. Furthermore, the data obtained in burnout experiments show a direct correlation between CHF and NSD. The IR images showed an increasing order and regularity of

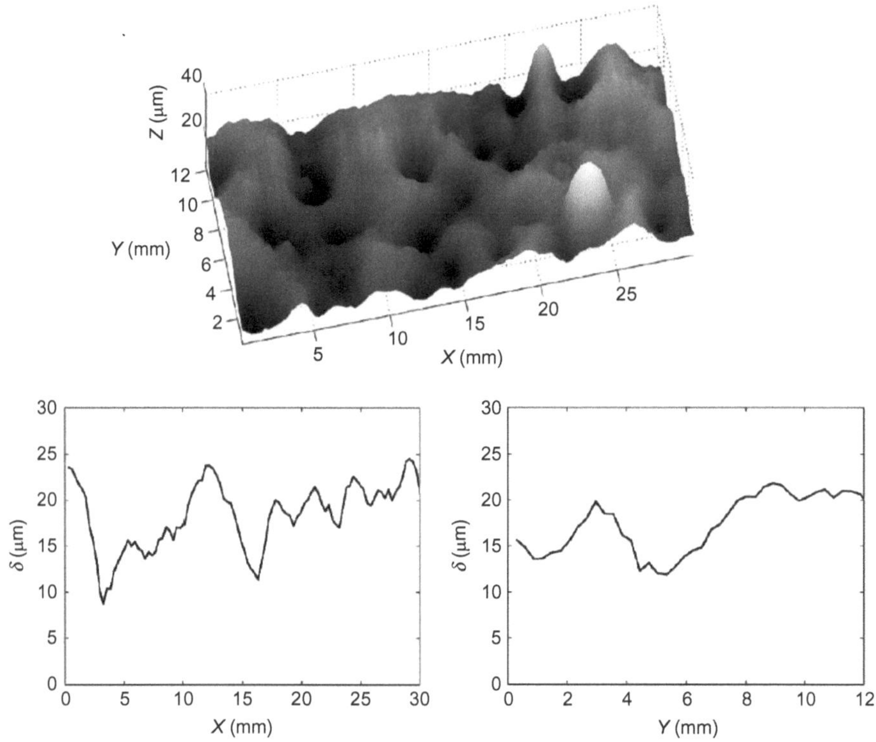

Figure 2.42 Map of microlayer thickness on a fresh heater (test F1) near burnout at $q=850$ kW/m^2. The thickness map presented is pixel (250 μm × 250 μm)—averaged and with an accuracy of ±5% (relative) due to time response of the glass heater [21].

the thermal pattern as the heat flux increases. This contrasts with an increasingly chaotic behavior of the two-phase flow dynamics above the heater as evidenced by the X-ray images of the boiling zone. Thus, the authors conclude that boiling heat transfer is independent of the complex two-phase flow hydrodynamics above the heater. This separation of scales creates a focus of inquiry for the dynamics of micro-layer sitting and vaporizing on the heater surface as an autonomous system. This in turn means that such an extended micro-layer, and its rupture, can be studied on its own, by direct observation, both from above, as well as by high-speed IR thermometry from below.

By analogy with pool boiling it may be assumed that the minimum initial thickness of the micro-layer is also the dominant parameter of CHF in micro-channels. The initial thickness of the liquid film is a key parameter of the explosive boiling. The term initial liquid film thickness is defined by Hetsroni *et al.* [57] as the average thickness of fluid, evenly distributed during period t, over the surface of the circular micro-channel, after venting

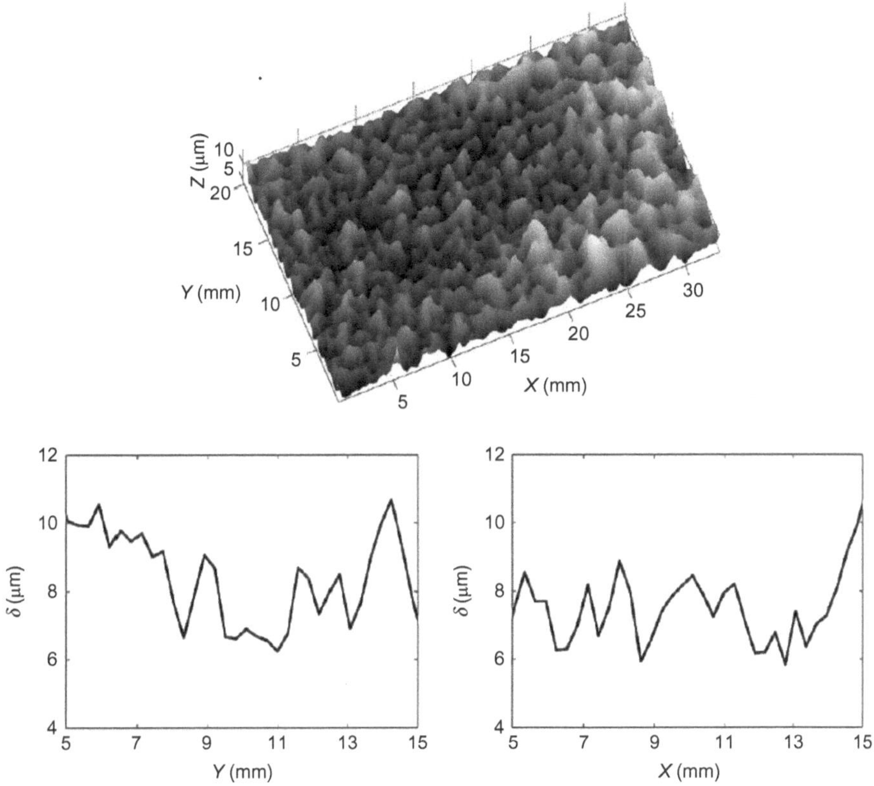

Figure 2.43 Map of microlayer thickness on a heavily aged heater (test A4) near burn-out at $q = 1500$ kW/m^2. The thickness map presented is pixel (250 μm × 250 μm)—averaged and with an accuracy of ±5% (relative) due to time response of the glass heater [21].

of the elongated bubble. This surface is located downstream of the ONB and may be characterized by the heated length, L, and hydraulic diameter, d_h. Detailed study of this phenomenon was conducted by Hetsroni *et al.* [57] using both IR and high speed measurements under condition of periodic flow boiling of water and ethanol in parallel triangular micro-channels. Tests were performed in the range of hydraulic diameter 100–220 μm, mass flux 32–200 kg/m^{2}s, heat flux 120–270 kW/m^2, and vapor quality $x = 0.01$–0.08.

5.2.1 Period between successive events

Figure 2.44 shows the dependence of the dimensionless period of phase transformations (i.e., the time between bubble venting), τ^*, on the boiling

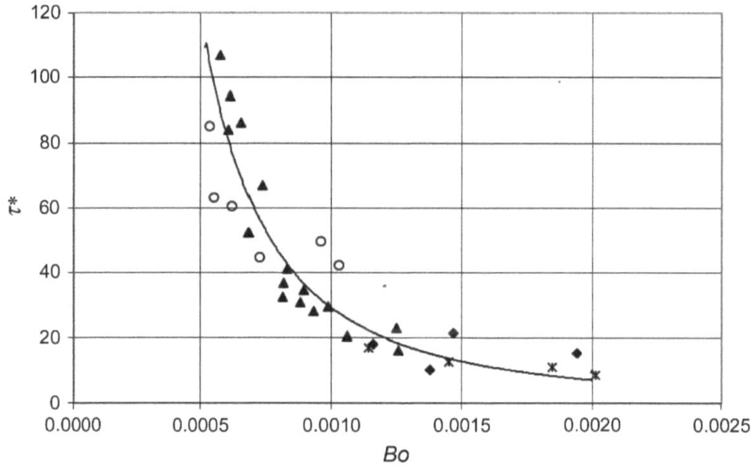

Figure 2.44 Dependence of dimensionless time interval between cycles on boiling number. ○ $D_h = 100$ μm, water, ▲ $D_h = 130$ μm, water, ЖК $D_h = 220$ μm, water, ЖК$D_h = 220$ μm, ethanol [57].

number, *Bo* ($\tau^* = \tau / U d_h$, $Bo = q / m h_{LG}$, τ is the period between successive events, U is the mean velocity of single phase flow in the micro-channel, d_h is the hydraulic diameter of the channel, q is heat flux, m is mass flux, h_{LG} is the latent heat of vaporization). The dependence τ^* on the Bo can be approximated, with standard deviation of 16%, by

$$\tau^* = 0.000030 Bo^{-2} \tag{2.18}$$

5.2.2 The initial thickness of the liquid film

It was assumed that during the period, *t*, the liquid film has disappeared due to evaporation. The heat removed from the wall surface is the same as required for the liquid film evaporation during the period, *t*.

The heat balance is:

$$\pi \delta d_h L \rho_L = q \pi d_h L t / h_{LG} \tag{2.19}$$

where ρ_L is the liquid density.

The average liquid thickness, δ, can be calculated as:

$$\delta = q t / \rho_L h_{LG} \tag{2.20}$$

Variation of initial film thickness for water versus heat flux is presented in Fig. 2.45. Figure 2.46 shows the dependence of the dimensionless initial liquid thickness of water and ethanol, δ^*, on the boiling number, *Bo*, where

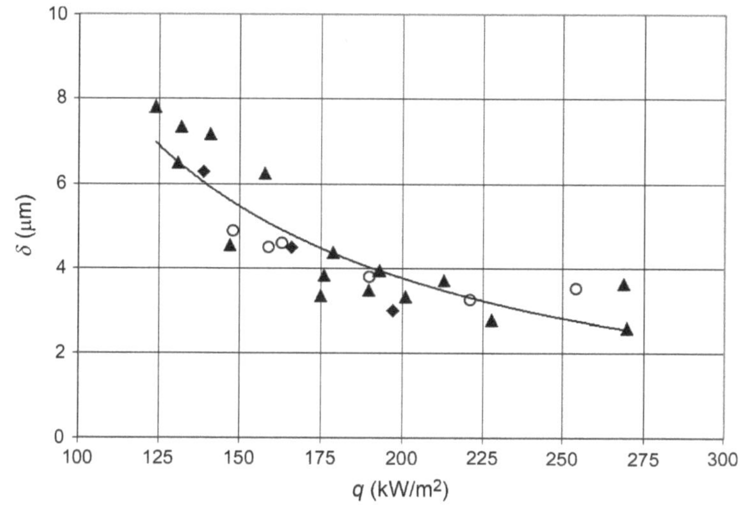

Figure 2.45 Variation of initial film thickness for water versus heat flux ○ $D_h = 100$ μm, water, ▲ $D_h = 130$ μm, water, $D_h = 220$ μm, water [57].

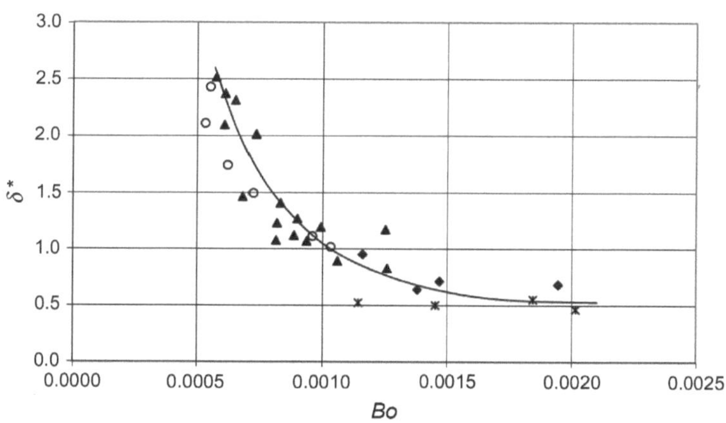

Figure 2.46 Dependence of dimensionless initial film thickness on boiling number ○ $D_h = 100$ μm, water, ▲ $D_h = 130$ μm, water, ♦ $D_h = 220$ μm, water, Ж $D_h = 220$ μm, ethanol [57].

$\delta^* = \delta U/\nu$, U is the mean velocity of single phase flow in the micro-channel, ν is the kinematic viscosity of the liquid at saturation temperature. The dependence of δ^* on Bo can be approximated, with standard deviation of 18%, by

$$\delta^* = 0.00015 Bo^{-1.3} \tag{2.21}$$

For explosive boiling the film thickness decreases with increasing heat flux from 125 to 270 kW/ m^2 from about 8 to 3 μm. This range of values is of the same order of magnitude as those given by Moriyama and Inoue [89] and by Thome *et al.* [90] for R-113 in small spaces (100–400 μm). Decreasing liquid film thickness with increasing heat flux is a distinct feature of dryout during explosive boiling. Under these conditions at which the instantaneous temperature of the heater surface exceeds 125 °C, the value of δ was in the range of 3 ± 0.6 μm. This value may be considered as the minimum initial film thickness. When the liquid film reached the minimum initial film thickness, δ_{min}, CHF regime occurred. According to Thome *et al.* [90] δ_{min} is assumed to be on the same order of magnitude as the surface roughness. The values of the minimum initial film thickness calculated by Thome *et al.* [90] for R-113 at saturation temperature 47.2 °C were in the range of 1.5–3.5 μm. The evaporation of the liquid layer on the wall was employed by Qu and Mudawar [47] and Thome *et al.* [90] to predict the heat transfer coefficient.

6. EFFECT OF SURFACE ACTIVE AGENTS (SURFACTANTS) ON BOILING CHARACTERISTICS

6.1. Properties of surfactants

In boiling heat transfer, it is usually desirable to transfer the largest possible heat flux with the smallest possible temperature difference between the heating surface and the boiling liquid, and to maximize the CHF. Various means have been developed with this aim in mind, including the use of additives to modify the liquid properties. The process of nucleate boiling is the total sum of the processes of bubble initiation, growth, and departure. Though these individual processes have been studied much, it is difficult to predict the effect of the physical properties of surface active agents (surfactants) on the main boiling characteristics, such as the relationship between the heat flux and the temperature difference.

In contrast to the momentum and scalar transfer in turbulent pipe flow with surfactants, which shows a reduction in the friction factor and the heat transfer coefficient, the study of surfactant solutions in the pool boiling shows a significant enhancement of the boiling mechanism. The role of surface-active solutes was explored for 0.1–1.0% aqueous solutions of a commercial surfactant. They found that the boiling curves (q vs. Δt) were shifted laterally in varying degrees, such that heat transfer was higher than that for pure water (q is the heat flux, $\Delta t = t_s - t_{sat}$ is the superheat, t_s is the average surface temperature of the heater, t_{sat} is the saturation temperature of the solution). This is an important fact because, if proved to be applicable under industrial boiling conditions, it may lead to a considerable increase in the power level of all boilers without any increase in size or operating temperature. One interesting field of application of boiling and evaporation is in desalination of seawater, which is becoming essential in some arid regions. It was already shown by Sephton [91] that addition of small amounts of surfactants to seawater can substantially enhance the boiling process and reduce the price of the desalinated water to an acceptable level. At that time, the research was discontinued because the environmental impact of surfactants was not known.

Since the concentrations are usually low, addition of the surfactant to water causes no significant change in the saturation temperature and the majority of other physical properties, except for the surface tension and, in some cases, the viscosity. There have been a large number of studies to determine the boiling enhancement mechanism caused by addition of surfactants to water. Frost and Kippenhan [92] investigated boiling of water with varying concentrations of surfactant "Ultra Wet 60 L." They found an increase in heat transfer and concluded that it resulted from the reduced surface tension. Heat transfer in nucleate pool boiling of dilute aqueous polymer solutions was measured by Kotchaphakdee and Williams [93] and compared with results for pure water. Photographs showed distinct differences in bubble size and dynamics between polymeric and nonpolymeric liquids. Gannett and Williams [94] concluded that surface tension was not relevant in explaining the enhancement effect and reported that viscosity could be a generally successful correlating parameter. Nucleate boiling curves for aqueous solutions of drag-reducing polymers have been measured experimentally by Shah and Darby [95] and by Paul and Abdel-Khalik [96]. The explanation of observed changes in the boiling curves was based only on how the polymer additives changed the solution viscosity. Polymer type, concentration, and molecular weight were important only insofar as they affect the solution viscosity. Yang and Maa [97] studied pool boiling of dilute surfactant solutions. The surfactants used in this study were sodium lauryl benzene sulfonate and sodium dodecyl sulfate (SDS). Since all experiments were carried out under very low concentrations, it was concluded that these additives had no notable influence over the physical properties of the boiling liquid, except surface tension, which was significantly reduced. This study showed that the surface tension of the boiling liquid had significant influence on the boiling heat transfer coefficient.

Pool boiling experiments were carried out by Tzan and Yang [98], for relatively wide ranges of surfactant concentration and heat fluxes. The results verify again that a small amount of surface-active additive makes the nucleate boiling heat transfer coefficient of water considerably higher. It was also found that there is an optimum additive concentration for the highest heat flux. Beyond this optimum point, further increase in the concentration of the additive lowers the boiling heat transfer coefficient. Wu *et al.* [99–101] reported experimental data on the effect of surfactants on nucleate boiling heat transfer in water with nine additives. Anionic, cationic, and nonionic surfactants were studied at concentration up to 400 ppm (parts per million). The enhancement of heat transfer was related to the depression of static surface tension. Boiling heat transfer coefficients were measured by Ammerman and You [102] for an electrically heated platinum wire immersed in saturated water, and in water mixed with three different concentrations of SDS (an anionic surfactant). Their results showed that addition of an anionic surfactant to water caused an increase in the convection component and a corresponding reduction in the latent heat component of the heat flux in the fully developed boiling region. The enhancement of heat transfer at boiling of water, which is caused by the addition of an anionic surfactant, appears to be influenced by this relative change in these heat flux components. The comprehensive reviews on the heat transfer in nucleate pool boiling of aqueous surfactants and polymeric solutions have been published by Kandlikar and Alves [103] and by Wasekar and Manglik [104]. It is shown that surfactant additives at low concentrations can enhance the nucleate boiling heat transfer significantly.

6.2. Pool boiling heat transfer

Saturated pool boiling on a heated surface and on a heated tube was studied and the effect of the surface tension and viscosity on the heat transfer coefficient has been examined by Hetsroni *et al.* [75].

In this study, they used the cationic surfactant Habon G of molecular weight 500 for the boiling enhancement. The cation of the surfactant is hexadecyldimethyl hydroxyethyl ammonium and the counter-ion is 3-hydroxy-2-naphthoate. The surfactant molecules form large rod-shaped micelles. It was shown by Zakin *et al.* [105] that although microstructure of Habon G was mechanically degraded under high shear conditions, it recovered quickly—no matter how many times it was broken up by shear. The objective of this study was to determine how the nucleate boiling is affected by the addition of Habon G to water and to generalize the data on the heat transfer enhancement for a wide range of concentration of surfactants.

6.2.1 Physical properties of solutions

The measurements of physical properties were carried out over a wide range of temperatures and for various concentrations. All the solutions used were prepared by dissolving the powdered surfactant in de-ionized water, with gentle stirring, over a period of several days. Concentrations investigated here were 65–1060 ppm. Typical results are presented and discussed below.

The shear viscosity of all surfactant solutions was determined in the temperature range 25–60 °C with Rheometrics Fluids Spectrometer using a Couette system. The standard deviation was 4%. Figure 2.47 shows the effect of shear rate ω on shear viscosity η for a 530 ppm Habon mixture, at different temperatures. The curves come closer to one another for higher shear rate values. The magnitude of the shear viscosity as a function of the shear rate decreases, when the temperature of the solution increases. Moreover, the shear viscosity η does not change significantly in the range of the shear rate $\omega = 1$–1000 per second at $t=60$ °C. Based on this result, we studied the viscosity behavior of Habon G solution after boiling, using a Cannon–Fenske capillary viscometer, at high shear rates for temperatures of 55 °C and above.

Figure 2.48 shows the effect of temperature on the kinematic viscosity of the surfactant solution at various Habon G concentrations. One can see here the similar tendency of viscosity curve to approach that of pure water near the saturation temperature. Such behavior is more pronounced as the

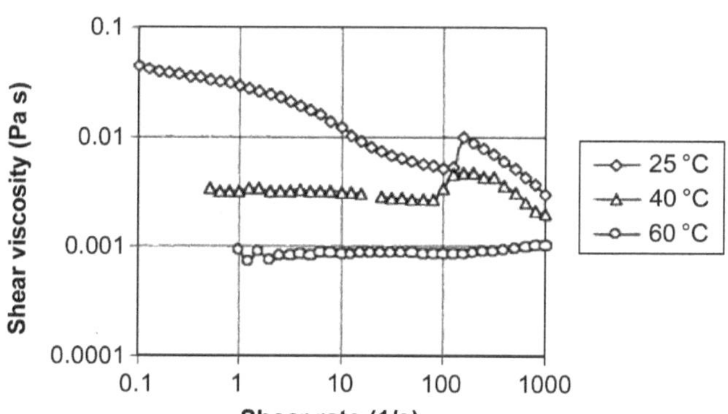

Figure 2.47 The shear viscosity of 530 ppm Habon G solution versus shear rate [75].

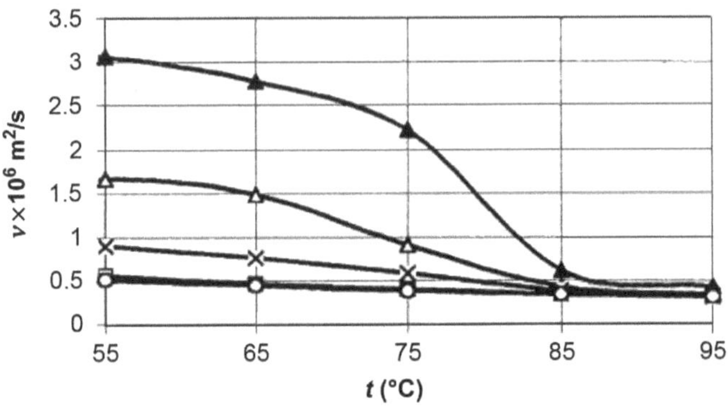

Figure 2.48 Kinematic viscosity of solution versus temperature at various Habon G concentrations. *Circles* (O) indicate water; Habon G, *boxes* (□) represent 130 ppm, *crosses* (×) represent 260 ppm, *empty triangles* (Δ) represent 530 ppm, *filled triangles* (▲) represent 1060 ppm [75].

concentration of the solution decreases. The authors also measured the thermal conductivity of Habon G solution. The apparatus was of the steady-state type. Both the clear water and the tested surfactant solution were enclosed in two identical cells. The top of each cell was made of 50 mm stainless steel foil, heated by DC current and the bottom was cooled by water. The difference between inlet and outlet water temperature did not exceed 0.2 °C. The thermal field of the cell top was measured by an IR radiometer. During these measurements, the thermal pattern associated with free convection was not observed. Figure 2.49 shows the dependence of thermal conductivity λ on the temperature t for the 530 ppm Habon G solution. The value of the thermal conductivity agrees well with that for pure water within the uncertainty of the measurements. The standard deviation of the thermal conductivity measurements was 2%.

The surface tension was measured in the temperature range of 25–70 °C with standard deviation of 2%. The data were obtained using a Surface Tensiometer System, which measures surface tension within the body of a test fluid by blowing a bubble of gas through two probes of different diameters inside the body. Figure 2.50 shows the magnitude of the surface tension as a function of the temperature for various concentrations of Habon G. As seen in this figure, the surface tension decreases with increase in both concentration and temperature. The temperature effect on the surface tension is much stronger at temperatures near the saturation temperature, whereas the opposite trend is observed for the viscosity.

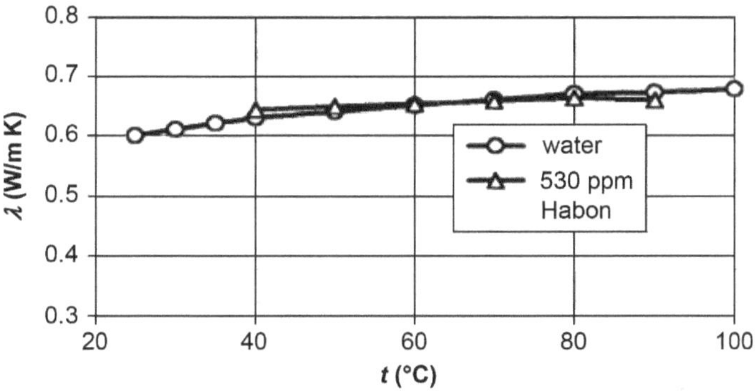

Figure 2.49 Thermal conductivity versus temperature [75].

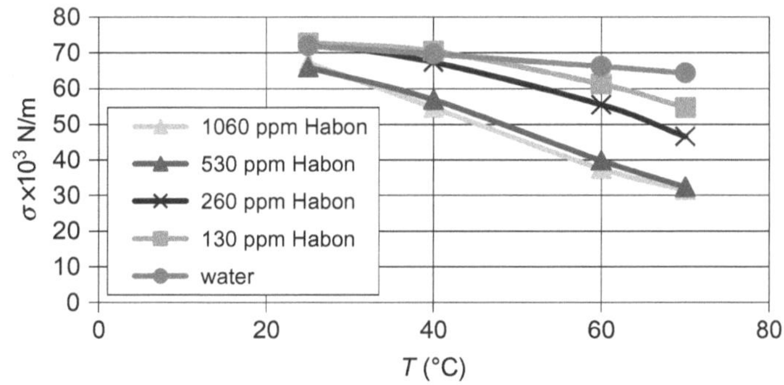

Figure 2.50 The surface tension as a function of temperature at various Habon G concentrations [75]. (For color version of this figure, the reader is referred to the online version of this chapter.)

6.2.2 Instrumentation

The growth of bubbles and the bubble motion near the heated surface were recorded by a high-speed video camera with recording rate up to 10,000 frames/s. The playback speed can be varied from a single frame to 250 frames/s. An IR radiometer was used to investigate the thermal patterns. The radiometer has a typical minimum detectable temperature difference of 0.1 K. Time response of this instrument is limited by the video system format (25 frames/s). The image has horizontal resolution of 256 pixels per line and 256 intensity levels. Since the foil was very thin, the temperature difference between the two sides of the foil did not exceed 0.2 K at a heat flux of 100 kW/m^2. Therefore, the time-averaged temperature was almost the same on both sides of the foil. The radiometer allows to obtain a quantitative thermal profile in the line mode, the average temperature in the area mode, and the temperature of a given point in the point mode. It was shown by Hetsroni and Rozenblit [13] that temperature distortions and phase shift in temperature fluctuations on the heated wall begin at $f=15$–20 per second. In the study, the highest frequency of the bubble departure was higher, so measurements of average temperatures and qualitative observation of the thermal structure on the heated bottom were limited. The surface temperature and the surfactant mixture temperature were measured with an accuracy of 0.1 °C. Electrical power was determined by means of a digital wattmeter with an accuracy of $\pm 0.5\%$.

6.2.3 Visualization of thermal pattern on the heated wall

Although IR thermography applied to boiling has relatively low frequency response, it is still more accurate than surface temperature measurement by micro-thermocouples or resistance thermometers. Its advantages are the extensive nature of the measurements and the absence of disturbance to the micro-geometry of the boiling surface. The examples given in this paper are just a small sample of the information contained in the recordings. They illustrate the advantages and limitations of IR thermography combined with video recording for the study of boiling heat transfer. The technique is necessarily limited to boiling on very thin walls, conditions that maximize the local variations in wall temperature and minimize lateral conduction.

The spatial distributions (Fig. 2.51) show variations of wall temperature of about 17 K for water (Fig. 2.51A) and 25 K for surfactant (Fig. 2.51B). With such wide ranges, it is clear that models for the bubble nucleation and growth that assume uniformity of wall superheat cannot be realistic. The IR thermography samples and the histograms of the thermal fields show that instantaneous values of the surface temperature, t_s, are lower for the Habon G solutions. This means that the average heat transfer coefficient in the surfactant solution increases as compared to boiling of water. Moreover, the larger width of the histogram for Habon G solution means that higher values of local heat transfer caused more intensive vaporization, which could happen more often than with boiling of water.

6.2.4 Boiling curves and heat transfer coefficients

In Fig. 2.52, experimental boiling heat transfer data are presented as a function of heat flux versus heater excess temperature (space–time average values at the fluid–solid interface). As the heat flux increases, the boiling curve shifts toward left as the concentration of Habon G increases.

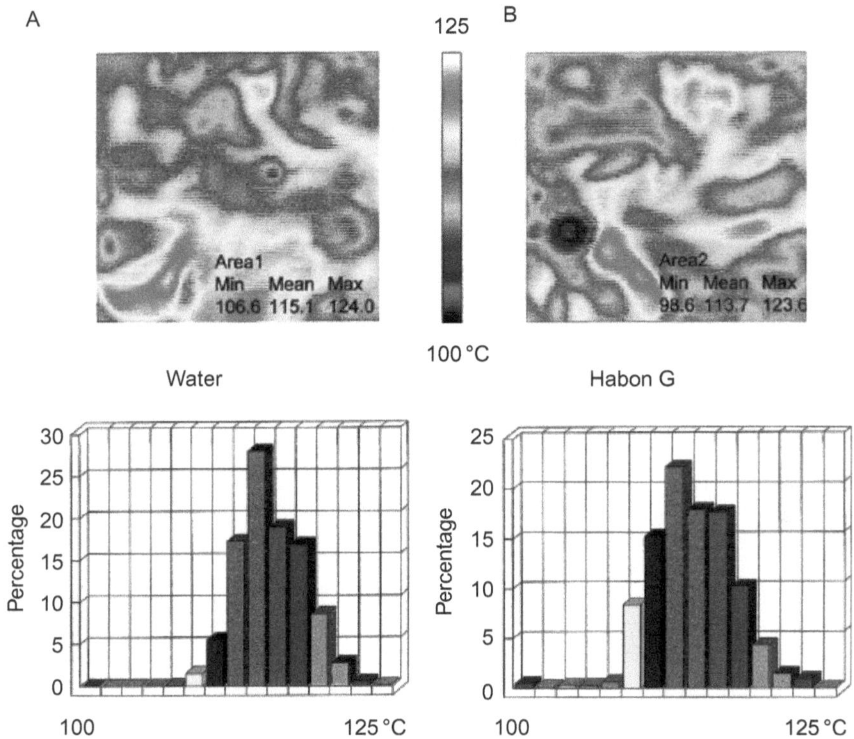

Figure 2.51 Thermal patterns and histograms of the temperature at pool boiling of water (A) and Habon G solution (B) on the flat plate ($q = 90$ kW/m^2) [75]. (For color version of this figure, the reader is referred to the online version of this chapter.)

It can be seen that the boiling curve at concentration of 1060 ppm is close to the curve for 530 ppm Habon G solution, and at high values of heat flux shifts toward right. Thus, it is evident that the influence of the surfactant on the boiling curve behavior has a maximum, depending on the concentration. For each concentration, 12 runs were performed: six runs for increasing heat flux and the remaining six for decreasing heat flux. Each point in Figs. 2.52 and 2.53 represents an average value obtained from these measurements. We did not observe any signs of hysteresis. The ONB point (in terms of a mean boiling excess temperature) is not affected by the surfactant concentration.

The effect of heat flux and additive concentration on the nucleate boiling heat transfer coefficient of Habon G solutions is more evident if the experimental data are expressed as a plot of heat transfer coefficient versus heat flux, as shown in Fig. 2.53. The heat transfer coefficient increases as the heat

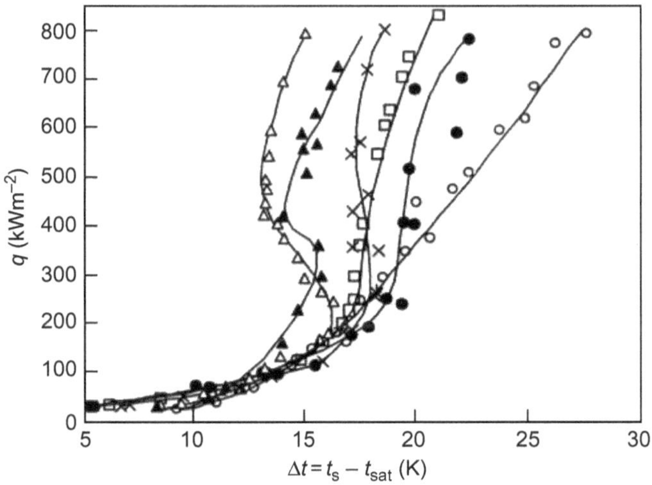

Figure 2.52 Boiling curves of water and aqueous Habon G solutions: ○—water; Habon G: •—65 ppm, □—130 ppm, × — 260 ppm, Δ—530 ppm, ▲—1060 ppm [75].

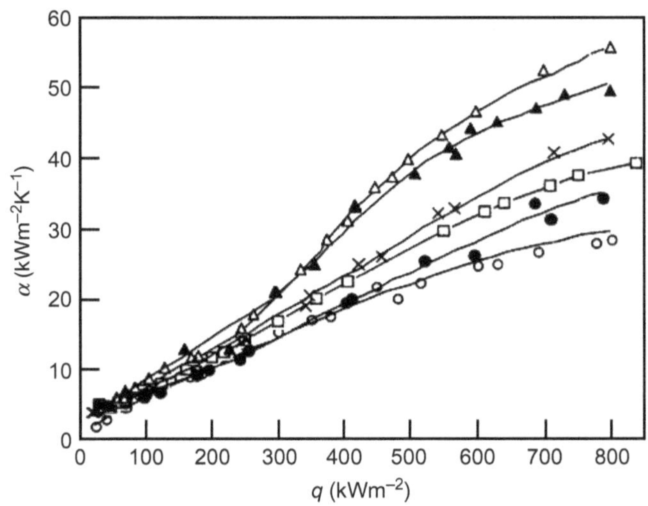

Figure 2.53 Boiling heat transfer coefficient (symbols as in Fig. 6.6) [75].

flux and concentration are increased, except when the heat flux level is higher than about $q = 300\ \mathrm{kW/m^2}$ and the concentration is higher than 530 ppm. Of these two trends, the former is consistent with results observed previously by Shah and Darby [95], Yang and Maa [97], and Tzanand and Yang [98]. The maximum in heat transfer coefficient at a certain concentration has seldom been reported. This may be attributed to the fact that the solutions tested were usually too dilute and/or the heat flux level was too low in most of the past experiments reported in the literature. Our data of heat transfer for 1060 ppm Habon G solution agree qualitatively with boiling heat transfer results reported by Tzanand and Yang [98]. They demonstrated that the effect of surfactant additives on nucleate boiling heat transfer decreases, when the concentration of the anionic surfactant SDS solution was higher than 700 ppm.

6.2.5 The Effect of Physical Properties of Surfactant Solution on Heat Transfer

A detailed investigation of the physical characteristics of the surfactant can demonstrate the effect of different properties of the mixture on the heat transfer. Figure 2.54 shows the dependencies of the relative surface tension σ/σ_w and the relative viscosity ν/ν_w as a function of Habon G concentration, where σ and σ_w are the surface tensions, ν and ν_w are the kinematic viscosities for the Habon G solution and pure water, respectively. One can see that the magnitude of relative surface tension decreases gradually from 1.0 for pure water to about 0.5 at 530 ppm surfactant solution. Further increase in the surfactant concentration does not significantly affect the value of the surface tension. On the other hand, the value of the kinematic viscosity (the right axis in Fig. 2.54) is practically equal to that of pure water at low surfactant concentrations (<300 ppm of Habon G), while further increase in additive concentration leads to significant increase in viscosity. Such complicated behavior of physical properties inevitably affects the complex behavior of the heat transfer coefficient in boiling.

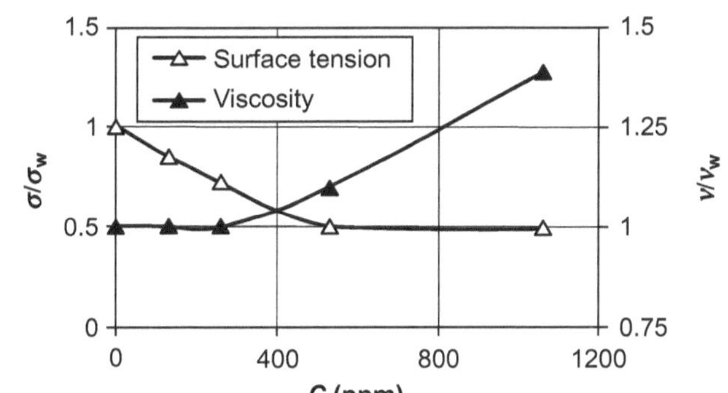

Figure 2.54 The relative surface tension s (at t=70 °C) and kinematic viscosity ν (at t=95 °C) as a function of Habon G concentration [75].

To clarify the effect of concentration on heat transfer, we compared our results with the data of investigations on heat transfer enhancement, where maximum heat transfer enhancement has also been reached at a certain concentration of surfactant solution. The comparison is shown in Fig. 2.55 as the enhancement of the heat transfer coefficient h/h_w-1 versus concentration of different surfactant solutions (where h and h_w are the heat transfer coefficients in the surfactant solution and pure water, respectively).

These curves have different values of heat transfer enhancement depending on the kind of surfactant and the heat flux. However, they have similar trends with increasing surfactant concentration. Heat transfer increases at low surfactant concentration, reaches a maximum, and further increase in the amount of additive leads to a decrease in the heat transfer coefficient. The value of this maximum depends on the heat flux (e.g., points A1 and A2 for Habon G solution) and on the kind of surfactant (points B1, B2 for SDS, [100]).

The curves for a given kind of surfactant reach their maximum at some definite value of surfactant concentration. Such a behavior may be explained by the effect of changing surface tension with surfactant concentration. In Fig. 2.56, the dependencies of the surface tension for the various surfactants discussed are shown. We can see that beginning from some particular value of surfactant concentration (which depends on the kind of surfactant), the value of the relative surface tension almost does not change with further increase in the surfactant concentration. Such a behavior agrees with results presented by Wasekar and Manglik [104] and may be referred to the c.m.c.

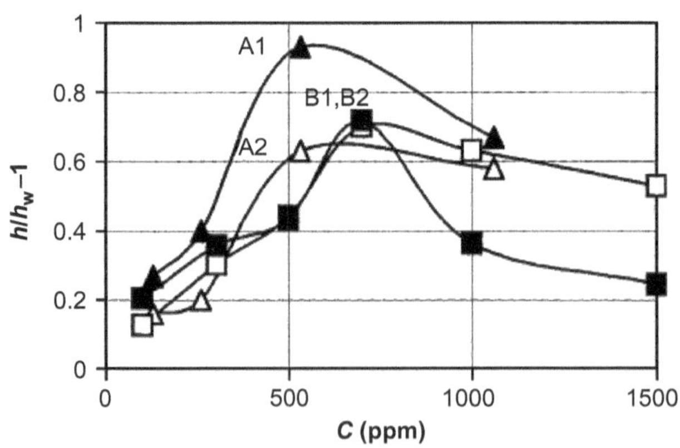

Figure 2.55 The heat transfer enhancement for various surfactant solutions at different heat fluxes as a function of surfactant concentration. Habon G: Δ—$q=400$ kW/m^2, ▲—$q=800$ kW/m^2, SDS [98]: □—$q=350$ kW/m^2, ■—$q=400$ kW/m^2 [75].

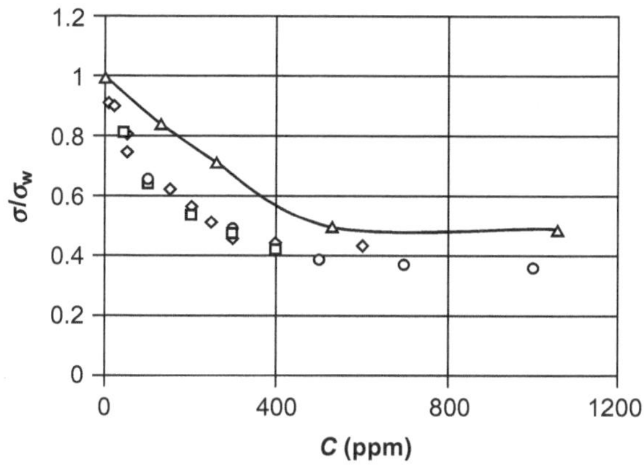

Figure 2.56 The nondimensional surface tension of various surfactants versus solution concentration (◊—SDS [100]), $t=25$ °C; ○—SDS [98], $t=25$ °C; □—SDS [99], $t=100$ °C; Δ—Habon G, $t=70$ °C [75].

of the surfactant. In Fig. 2.56, the data for Habon G are presented at the temperature $t=70$ °C, the data for SDS are reported at the temperature 25 °C [98,100], and at $t=100$ °C [101]. Unfortunately, the data of Wu *et al.* [101] are for concentration up to $C=400$ ppm only. However, it should be emphasized that the variation of the nondimensional surface tension as a function of the surfactant concentration shows the same behavior for various temperatures. The normalized nucleate boiling heat transfer coefficient may be related to normalized surface tension of the surfactant solution. We used the surfactant concentration where the change of relative surface tension reaches 90% of the complete change to normalize the concentration scale. The values $C_0=530$ and 700 ppm were chosen for Habon G and SDS solutions, respectively.

For normalization of the value of the heat transfer enhancement, we used its magnitude at the maximum for each curve. The result of such normalization is shown in Fig. 2.57. In this figure, C is the solution concentration, C_0 is the characteristic concentration, h is the heat transfer coefficient at given values of the solution concentration and the heat flux q, h_{max} is the maximum value of the heat transfer coefficient at the same heat flux, h_w is the heat transfer coefficient for pure water at the same heat flux q. Data from all the sources discussed reach the same value of 1.0 at the magnitude of relative surfactant concentration equal to 1.0.

Thus, the enhancement of heat transfer may be connected to the decrease in the surface tension value at low surfactant concentration. In such

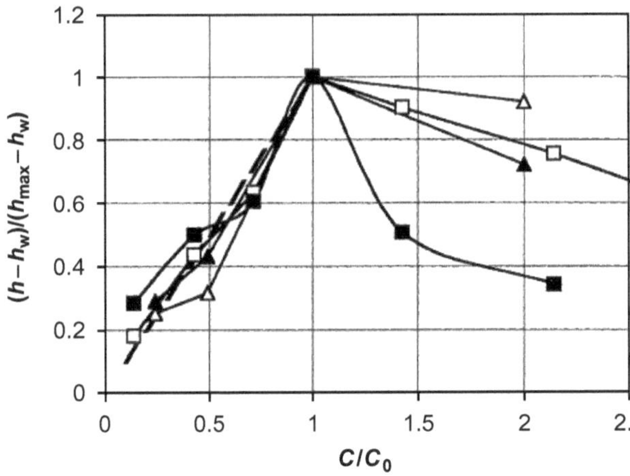

Figure 2.57 The excess heat transfer coefficient versus the surfactant concentration (Habon G: Δ—$q=400$ kW/m^2, ▲—$q=800$ kW/m^2, SDS [98]: □—$q=350$ kW/m^2, ■—$q=400$ kW/m^2) [75].

system of coordinates, the effect of the surface tension on excess heat transfer $(h-h_{\max})/(h_{\max}-h_{\mathrm{w}})$ may be presented as linear fit of the value C/C_0. On the other hand, the decrease in heat transfer at higher surfactant concentration may be related to the increased viscosity. Unfortunately, we did not find surfactant viscosity data in the other studies. However, we can assume that the effect of viscosity on heat transfer at surfactant boiling becomes negligible at low concentration of surfactant only.

The surface tension of a rapidly extending interface in surfactant solution may be different from the static value, because the surfactant component cannot diffuse to the absorber layer promptly. This may result in an interfacial flow driven by the surface tension gradient $\Delta\sigma$ (known as Marangoni flow).

We consider a fluid zone of thickness d across which the surface tension difference is $\Delta\sigma$: The Marangoni number $Ma=\Delta\sigma\cdot d/\rho vk$ is the controlling parameter of this type of flow that affects the heat transfer coefficient. As it is seen in Fig. 2.54, the surface tension decreases significantly, whereas the kinematic viscosity almost does not change with concentration increase at low solution concentration. The Marangoni number can be expressed as $Ma=Re\cdot Pr$, where $Re=\Delta\sigma\cdot d/\rho v^2$ and $\mathrm{Pr}=v/k=vcp/\lambda$ are the Reynolds number and Prandtl number, respectively. The density ρ, specific heat c, and thermal conductivity λ (Fig. 2.49) of Habon G solution are the same as for water. Thus, the Reynolds number increases whereas the Prandtl number almost does not change, at low surfactant concentration. Such a behavior of dimensionless parameters explains the increase in the heat transfer coefficient at low concentration of surfactant solution. Figure 2.54 shows that the surface tension almost does not change. In this case, the Marangoni effect acts in the opposite direction and suppresses the boiling heat transfer.

6.3. Boiling in confined narrow space

Boiling in confined narrow spaces becomes quite different from boiling process observed in a pool. This topic was studied experimentally by a number of researchers. Previous investigations [106–112] showed that confinement of a space for boiling led to enhanced heat transfer coefficient compared to unconfined boiling, but led to a decrease in the CHF.

Yao and Chang [108] assumed that effect of confinement on boiling depends on the ratio of the channel gap size to the capillary length scale, the latter being proportional to the departure diameter of isolated bubbles. The latter one is proportional to the square root of the liquid surface tension and may be connected to the Bond number $Bn = s\,[g\,(\rho_L - \rho_G)/\sigma]^{0.5}$, where s is the channel gap size, σ is the liquid surface tension, g is acceleration due to the gravity, ρ_L and ρ_G are density of the liquid and the vapor, respectively.

The first study on simultaneous effect of the space confinement and surfactant additive was carried out by Hetsroni *et al.* [113]. Natural convection boiling of water and surfactants at atmospheric pressure in narrow horizontal annular channels was studied experimentally in the range of Bond numbers $Bn = 0.185$–1.52. The flow pattern was visualized by high-speed video recording to identify the different regimes of boiling of water and surfactants. The authors reported that the additive of surfactant led to enhancement of heat transfer compared to water boiling at the same gap size; however, this effect decreased with decreasing gap size. CHF in surfactant solutions was significantly lower than that in water at the same gap size.

An experimental study has been carried out to investigate the heat transfer processes at natural convective boiling of water and surfactant solutions in narrow vertical channel at atmospheric pressure [114]. The gap size of the vertical channel was 1.0, 2.0, 3.0, and 80 mm. The latter position of the glass plate was considered as unconfined space. The heat flux was in the range $q = 19$–170 kW/m^2, the concentration of surfactant solutions was in the range $C = 0$–600 ppm. Alkyl (8–16) glycoside nonionic surfactant solution of molecular weight 390 g/mole was used. Before experimental runs the surface roughness of the heater, created during the boiling process, was examined by an atomic force microscope using such tools as the section analysis and line profiles. The results on boiling of water and surfactant solution of

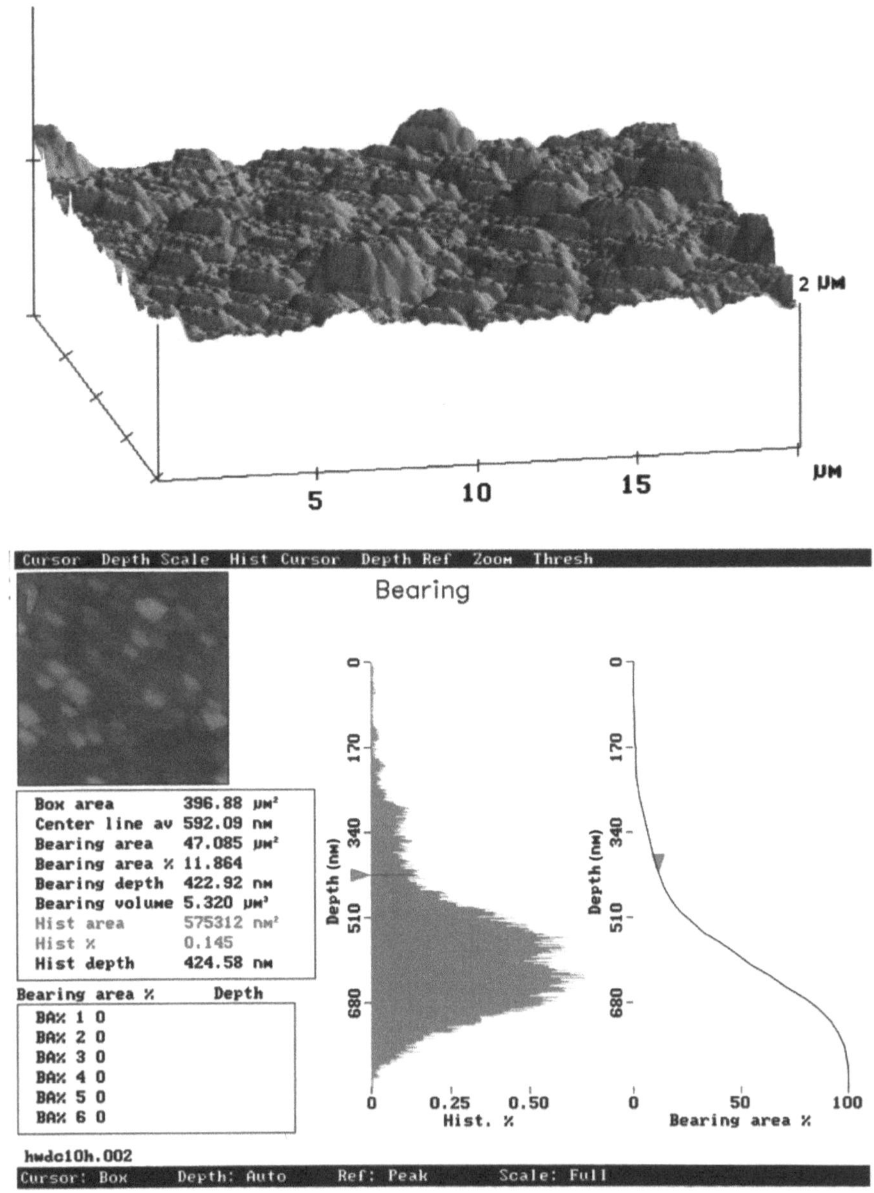

Figure 2.58 Surface roughness after boiling of water during 10 h [114]. (For color version of this figure, the reader is referred to the online version of this chapter.)

C=200 ppm after 10 h are presented in Figs. 2.58 and 2.59, respectively. As shown in these figures, the rms roughness of the stainless steel heater was 422.92 nm in the case of water boiling whereas it was 617.67 nm in the case of surfactant boiling.

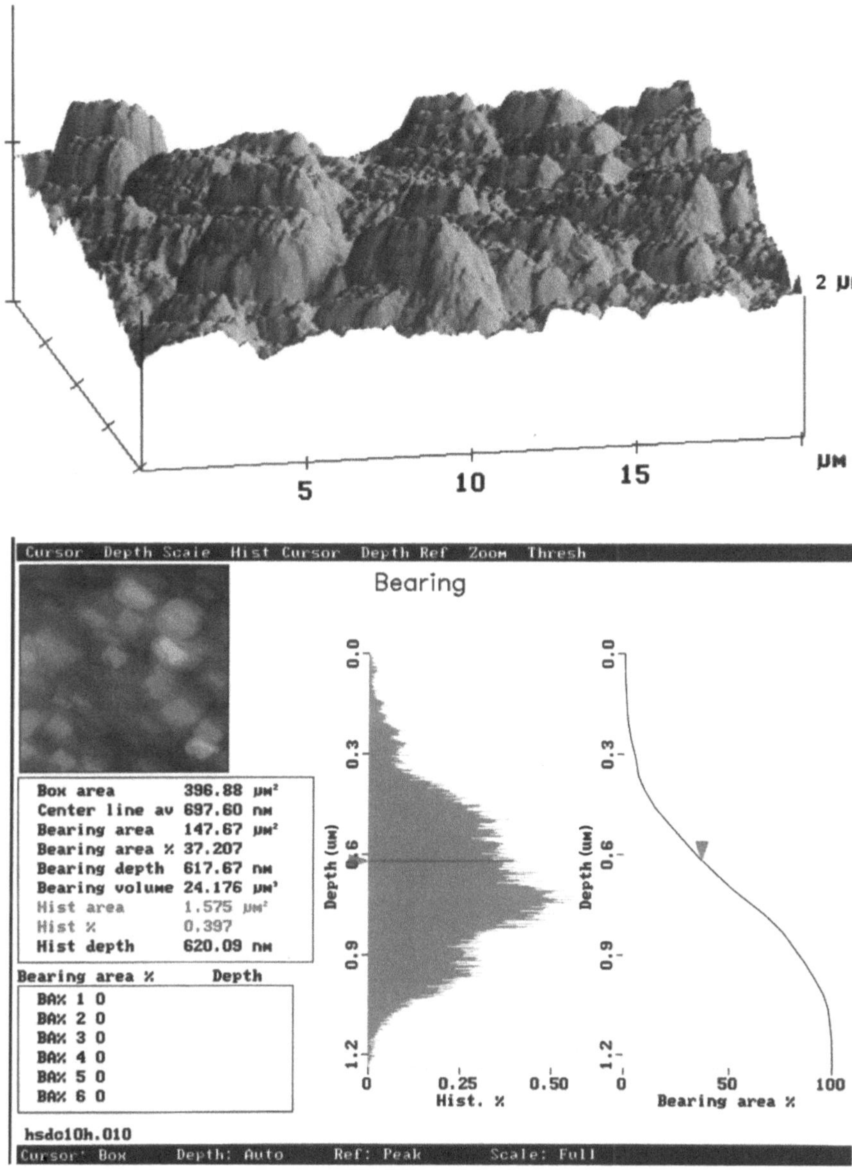

Figure 2.59 Surface roughness of heater after boiling of surfactant solution of $C=200$ ppm during 10 h [114]. (For color version of this figure, the reader is referred to the online version of this chapter.)

6.3.1 Boiling Curves and Average Heat Transfer

Data were taken for both increasing and decreasing heat fluxes. The total mass of the liquid in the test facility remained constant; no fresh liquid was introduced to "top off" the system. The investigation was carried out

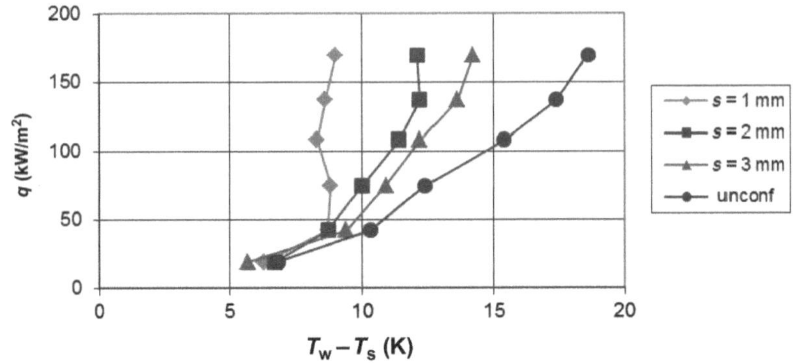

Figure 2.60 Boiling curves obtained in vertical confined open channel in clear water at various values of gap size [114]. (For color version of this figure, the reader is referred to the online version of this chapter.)

in the range of the Bond numbers of $0.4 \leq Bo \leq 47$. Figures 2.60 and 2.61 show boiling curves obtained in vertical confined open channel at various values of the gap size for clear water and surfactant solutions of various concentrations. The heat flux, q, is plotted versus the wall excess temperature $\Delta T_s = T_W - T_S$. Each point in Figs. 2.60 and 2.61 represents an average value obtained from the measurements. In these experiments we did not observe any signs of a hysteresis. It can be seen from Fig. 2.60 that the wall superheat in clear water was reduced at heat flux higher than 50 kW/m^2 as the gap size decreased.

Similar effect was observed in the solution of surfactants at concentrations of 200, 300, and 600 ppm (Fig. 2.61). The boiling curves for the boiling in surfactant solutions of the all concentrations in unconfined space shifted left relative to the boiling curve at the same conditions in the water. It should be noted that for the gap size of 3.0 mm the boiling curve was only slightly different from the one in unconfined space. This might have some validity since the bubbles in surfactant solutions were smaller in diameter than those in water. For low Bond numbers (of the order of unity or less), the squeezing effect is important since bubbles cannot grow naturally because the channel is narrower than the bubble diameter. For high Bond numbers, boiling can almost be considered as unconfined.

The decrease in the wall excess temperature may be considered as an enhancement of the heat transfer. According to Yang and Maa [115], boiling heat transfer in surfactant solutions is enhanced by the depression of the equilibrium surface tension but suppressed by the depression of the equilibrium contact angle.

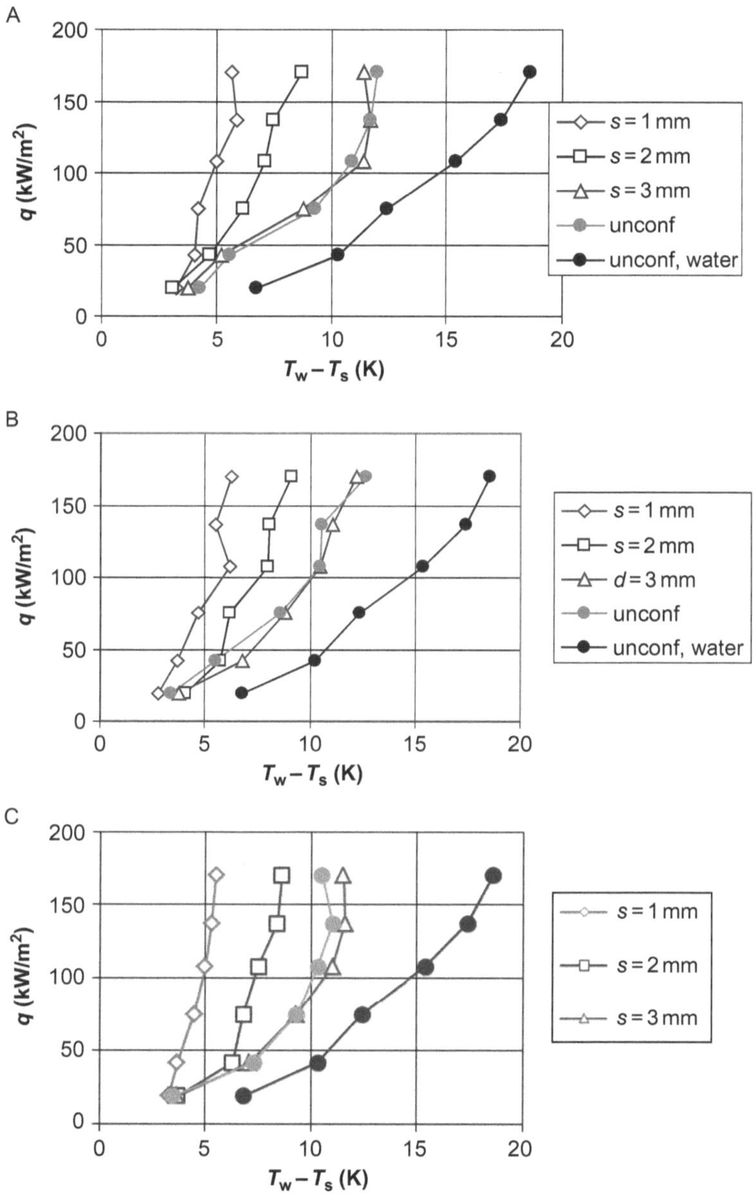

Figure 2.61 Boiling curves obtained in vertical confined open channel in the surfactant solution of various concentrations at various values of gap size: (A) – $C = 200$ ppm, (B) –$C = 300$ ppm, (C) – $C = 600$ ppm [114]. (For color version of this figure, the reader is referred to the online version of this chapter.)

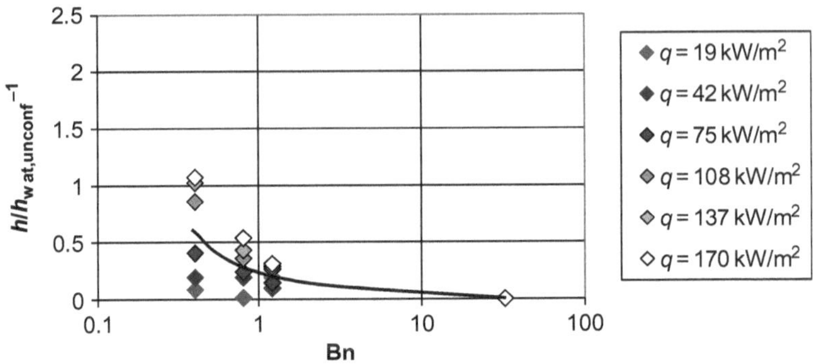

Figure 2.62 Effect of confinement on water boiling [114]. (For color version of this figure, the reader is referred to the online version of this chapter.)

Effect of the confinement on water boiling is presented in Fig. 2.62 as the dependence of the relative heat transfer coefficient ($h/h_{\mathrm{wat,unconf}} - 1$) on the Bond number, where h is the heat transfer coefficient in confined space and $h_{\mathrm{wat,unconf}}$ is the heat transfer coefficient in unconfined space. Figure 2.62 shows that under conditions of water boiling the decrease in the gap size leads to the enhancement of the heat transfer. This effect is more pronounced for $Bn < 1$ (small gap sizes).

In Fig. 2.63 the dependence of the relative heat transfer coefficient on the Bond number at different surfactant concentrations is presented. Figure 2.64 shows the dependence of the relative heat transfer coefficient on the concentration of surfactant solutions. The gap size is $s = 2$ mm. The dependence reaches the maximum at the concentration $C = 200$ ppm.

6.4. ONB in parallel micro-channels

6.4.1 *Effect of dissolved gases on ONB during flow boiling of water and surfactant solutions in micro-channels*

Desorption of the dissolved gases formed bubbles of gas and a limited amount of bubbles containing gas–water vapor mixture. As a result, boiling incipience occurred at a channel wall temperature below the saturation temperature. Steinke and Kandlikar [116] studied flow boiling in six parallel micro-channels, each having hydraulic diameter of 0.207 mm. During the flow boiling studies with water in these micro-channels, nucleation was observed at a surface temperature of $T_W = 90.5$ °C for the dissolved oxygen content of 8.0 ppm at a pressure of $P = 1$ bar. Comparison between water flow and surfactant solution was investigated by Klein *et al.* [117]. The experimental facility was designed and constructed as illustrated

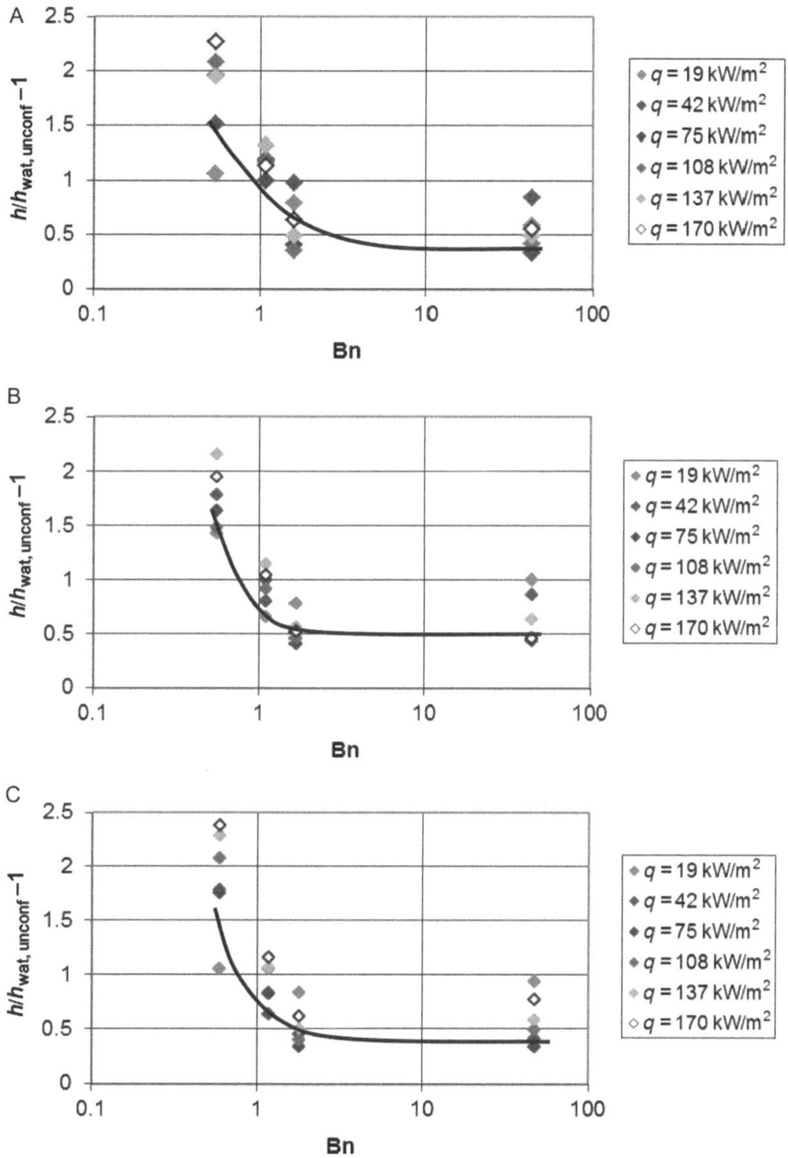

Figure 2.63 Dependence of the relative heat transfer coefficient on Bond number, (A) – $C=200$ ppm, (B) – $C=300$ ppm, (C) – $C=600$ ppm [114]. (For color version of this figure, the reader is referred to the online version of this chapter.)

schematically in Fig. 2.65. The test module consisted of inlet and outlet manifolds that were attached to the test chip (Fig. 2.66). The tested chip with the heater is shown in Fig. 2.67. It was made of a square shape 15 mm × 15 mm and 0.5-mm-thick silicon oxide wafer, which was later

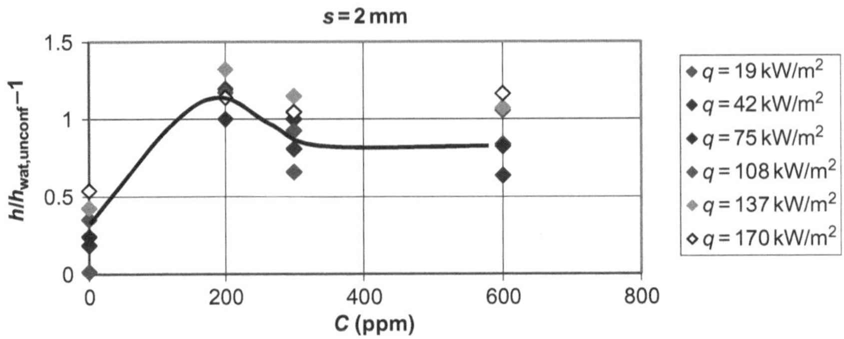

Figure 2.64 Dependence of the relative heat transfer coefficient on concentration of the surfactant solutions. The gap is $s = 2$ mm [114]. (For color version of this figure, the reader is referred to the online version of this chapter.)

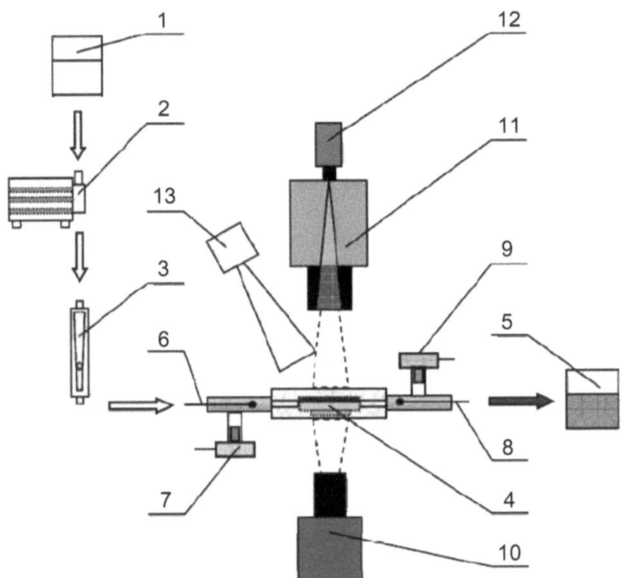

Figure 2.65 Schematic view of the experimental facility. *1* Inlet tank, *2* mini-gear pump, *3* rotameter, *4* test module, *5* exit tank, *6* inlet thermocouple, *7* inlet pressure gauge, *8* outlet thermocouple, *9* outlet pressure gauge, *10* high-speed IR camera, *11* microscope, *12* high-speed CCD camera, *13* external light source [117].

bonded to a 0.53-mm-thick Pyrex cover. On one side of the silicon wafer 26 micro-channels were etched, with triangular shaped cross-sections, with a base of 0.21 mm and a base angle of 54.7°. Using a microscopic lens, IR measurements can be performed up to 800 Hz with a 30-μm spatial resolution. The surfactant used was of the alkyl polyglucosides (APG) type.

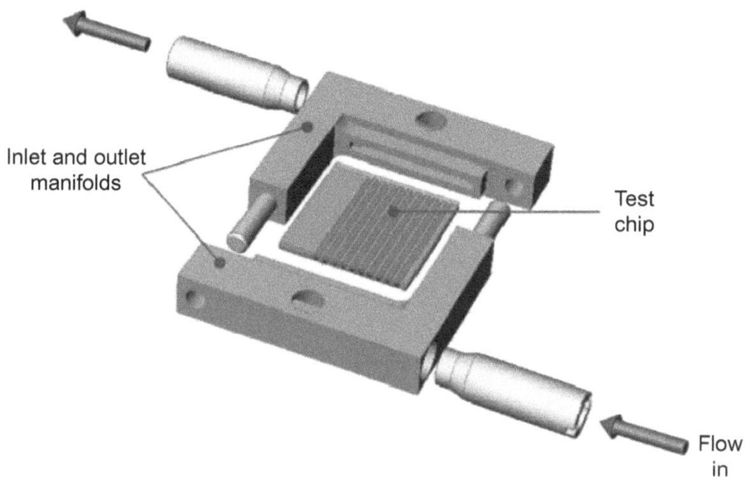

Figure 2.66 Test module [117]. (For color version of this figure, the reader is referred to the online version of this chapter.)

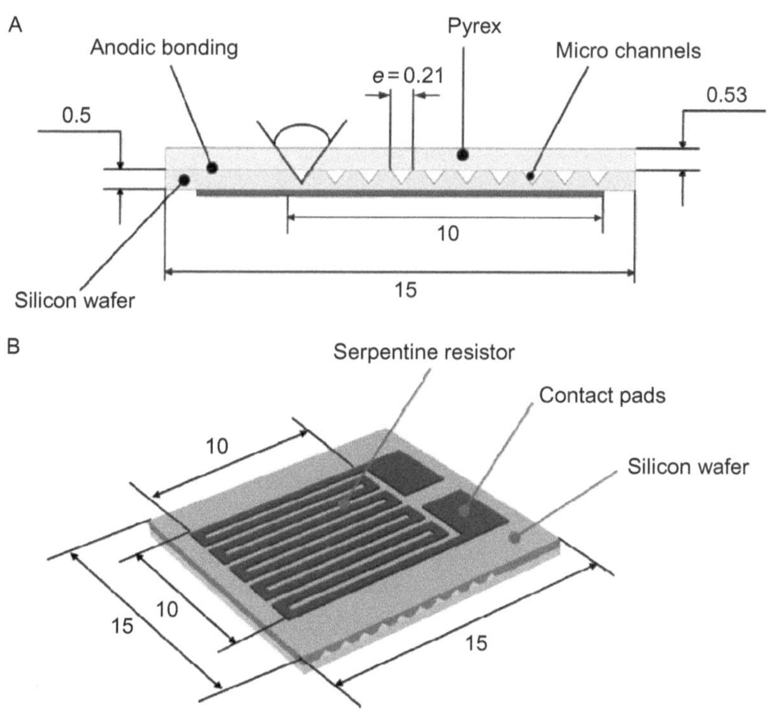

Figure 2.67 Test chip with heater: (A) cross-section, (B) heater [117].

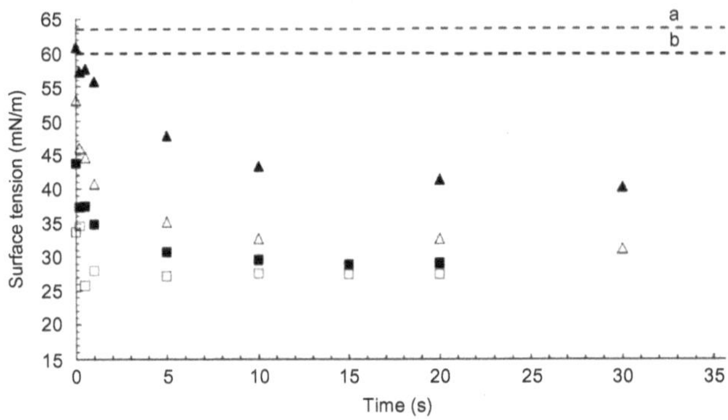

Figure 2.68 Surface tension of the APG solutions. Concentration and solution temperature (°C): $C=100$ ppm, *filled triangles* ( ) represent 75 °C, *empty triangles* (Δ) represent 95 °C; $C=300$ ppm, *filled squares* (■) represent 75 °C, *empty squares* (□) represent 95 °C [117].

Figure 2.68 shows the effect of APG additives on the dynamic and the static surface tension for different mass concentrations, measured at 75 and 95 °C. The dashed lines a,b represent the surface tension value for pure water at 75 and 95 °C. Solid points represent the APG data at 75 °C and the hollow points represent the APG data at 95 °C. Note that an increase in concentration decreases surface tension down to a value of 31 mN/m, compared to 59.9 mN/m for pure water. The temperatures on the heater $T_{W,ONB}$ and heat fluxes q_{ONB} corresponding to ONB in water and surfactant solution that contain dissolved gases are presented in Table 2.9. As can be seen in Table 2.9, ONB in APG solution of concentration $C=100$ ppm took place at significantly higher surface temperatures. It should be noted that the ONB in surfactant solutions may not be solely associated with static surface tension [118]. Other parameters such as heat flux, mass flux, kind of surfactant, surface materials, surface treatments, surface roughness, dynamic surface tension, and contact angle need to be considered as well.

6.4.2 Boiling incipience in degraded surfactant solutions

Under some conditions boiling incipience in surfactant solutions may be quite different from that in Newtonian fluids. Hetsroni *et al.* [113] presented results for natural convection boiling in narrow horizontal annular channels of a gap size 0.45–2.2 mm for alkyl (8–16) degraded solutions, i.e., solutions that were used previously for 6–10 experimental runs.

For degraded alkyl (8–16) solutions boiling occurred at wall superheat higher than that observed in fresh solutions and water. Incipience of boiling

Table 2.9 Onset of nucleate boiling in fluids that contain dissolved gases [117]

De-ionized Water			**APG-100 ppm, surfactant solution**		
Mass flux (kg/m^2s)	**q_{ONB} (W/cm^2)**	**$T_{W,ONB}$ (°C)**	**Mass Flux (kg/m^2s)**	**q_{ONB} (W/cm^2)**	**$T_{W,ONB}$ (°C)**
37.9	5.2	81.6	39.4	7.7	107.3
57.7	8.2	91.6	57.2	9.3	101.4
84.9	9.9	81.6	83.3	15.9	116.6
116.2	16.2	96.6	117.6	23.2	120.3
172.3	21.0	91.6	171.2	32.3	121.1

in both water and fresh surfactant solutions was accompanied by formation of small bubbles on the heated surface. However, a significant difference in the behavior of boiling patterns was observed. The formation of big vapor clusters took place before boiling incipience in degraded alkyl (8–16) solutions in the range of concentrations $C = 10–600$ ppm (weight part per million). This process is shown in Fig. 2.69A–C. The burst of such a cluster is shown in Fig. 2.69D. The cluster formation was accompanied by high wall superheat ($T_W - T_S$) in heat flux controlled experiments, where T_W is the temperature measured on the heated wall and T_S is the saturation temperature measured in the vessel. It should be stressed that these clusters were not gas (air) bubbles. The desorption of the dissolved gases formed bubbles of gas and a limited amount of bubbles containing gas–water vapor mixture. As a result, boiling incipience occurred at the heated wall at temperature below that of saturation temperature. In the present study such a phenomenon was not observed. We also measured, with a thermocouple, the fluid temperature T_f in the annular space between the heated tube and the inner wall of the glass tube. This temperature exceeded by 4–12 K the saturation temperature, depending on the concentration of the solution. Finally, the collapse of the cluster led to a reduction in wall superheat and saturated boiling regime occurred. For water boiling we did not observe the bubble coalescence at very small scales. For pool boiling of surfactant solutions, bubble coalescence was observed. There were clusters of small bubbles, which rose from the cavity. These bubbles were adjacent to each other and a cluster neck was not observed.

The bursting of vapor clusters before boiling incipience of degraded cationic surfactant Habon G solution was also observed by Hetsroni *et al.* [119]. Data were taken for both increasing and decreasing heat fluxes. The total

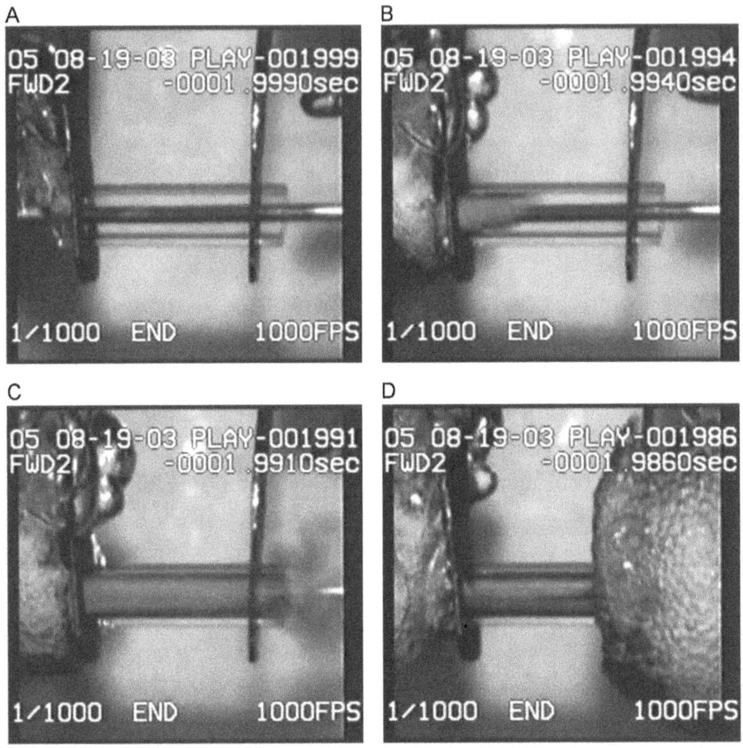

Figure 2.69 Boiling incipience in degraded solutions [113].

mass of the liquid in the test facility remained constant, thus no fresh liquid was introduced to "top off" the system. For water boiling in the gap sizes of 0.45, 1.2, 2.2, and 3.7 mm, the Bond numbers, $Bn = \delta(\sigma/g(\rho_L - \rho_G))^{-0.5}$, were 0.185, 0.493, 0.9, and 1.52, respectively, where δ is the gap size, σ is the surface tension, g is the acceleration due to gravity, ρ_L and ρ_G are the liquid and the vapor densities, respectively. Boiling of surfactant solutions was investigated in a gap size of 0.45 and 2.2 mm in the range of Bond numbers $Bn = 0.26$–1.26.

The results obtained at $Bn = 1.26$ are presented in Fig. 2.70, for different concentrations of surfactant solutions. The onset of boiling corresponds to the curve ABCD for the runs with increasing heat flux. It follows the curve DCA for decreasing heat flux. The measurements were repeated several times and the same phenomena were observed. Point B stands for the condition at which the fluid starts to boil when the heat flux is increasing (the typical process is shown in Fig. 2.69A–D). Zhang and Manglik [120] concluded that hysteresis occurred due to high wettability, which takes place at

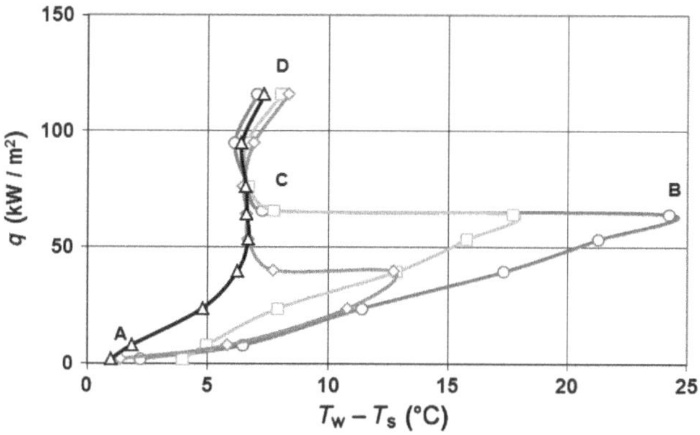

Figure 2.70 Boiling hysteresis in degraded Alkyl (8–16) solutions. ○ $C=300$ ppm, □ $C=100$ ppm, Ж $C=25$ ppm, △ $C=10$ ppm [113].

very high concentrations, $C > C_{CMC}$, (C_{CMC} is the critical micelle concentration). It should be stressed that in the present study hysteresis was observed in restricted boiling of degraded solutions as for pre-*CMC* solutions ($C < 300$ ppm) as for post-*CMC* solutions. It is speculated that molecules of degraded surfactant are more amenable to formation of a surfactant monolayer, which renders the interface less flexible and results in the dampening of interfacial motion. For alkyl (8–16) hysteresis occurs only in degraded solutions.

7. EXPERIMENTAL STUDY OF INTEGRATED MICRO-CHANNEL COOLING FOR 3D ELECTRONIC CIRCUIT ARCHITECTURES

Past work on two-phase micro-channel cooling was focused on cooling of 2D circuits and on demonstration of a single- or multichannel system ignoring the effects of flow distributions in a channel network. Three-dimensional circuit cooling faces a conjugate heat transfer with 3D thermal conduction and boiling convection in micro-channels. In a two-phase micro-channel network, each channel experiences flow instability due to the random formation and growth of void. The instability problem induced by the flow instability is more critical in micro-channel cooling of 3D circuits, since more micro-channel layers are coupled. These should be addressed to demonstrate an integrated micro-channel cooling network for 3D circuits.

For example, experiments may be performed to study 3D IC (Three-dimensional integrated circuit) cooling performance with micro-channels fabricated between two silicon layers using deep reactive ion etching and wafer bonding techniques [121]. Figure 2.71 illustrates four different 3D stack schemes for a given flow direction. To simulate nonuniform power distributions in practical 3D ICs, the device is divided into logic circuitry

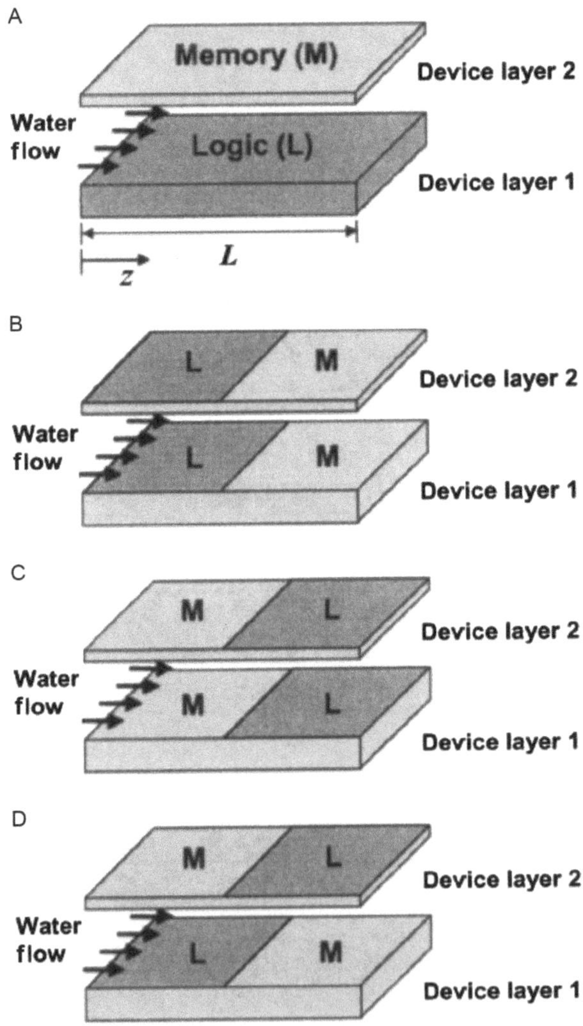

Figure 2.71 Two-layer 3D circuit layouts for evaluating the performance of micro-channel cooling. The areas occupied by memory and logic are the same and the logic dissipates 90% of the total power consumption [121].

and memory, where 90% of the total power is dissipated from the logic and 10% from the memory. This experiment assumes that the heat generation represents the power dissipation from the junctions and also from interconnect Joule heating. For case (a), the logic circuit occupies the whole device layer 1, while the memory is on the device layer 2. In the other cases, each layer is equally divided into memory and logic circuitry. For case (b), a high heat generation area is located near the inlet of the channels, while it is near the exit of channels for case (c). Case (d) has a combined thermal condition in which layer 1 has high heat flux and layer 2 has low heat dissipation near the inlet. IR measurements are needed to confirm numerical predictions, in particular for the case of a strong spatial variation in the heat flux between regions on the chip.

Further experimental and theoretical work is required to find a relation for two-phase convective heat transfer coefficient. A 3D conjugate conduction/convection simulation is required to calculate the wall temperature under conditions of 3D nonuniform heat generation. Another challenge to be addressed in future work will be the optimization of the micro-channel geometries and operating conditions with restriction from the circuit.

8. UNCERTAINTY

In general, the result of a measurement is only an approximation or estimate of the value of the specific quantity subject to measurement, and thus the result is complete only when accompanied by a quantitative statement of its uncertainty. The uncertainty of the result of a measurement generally consists of several components, which may be grouped into two categories according to the method used to estimate their numerical values: those which are evaluated by statistical methods and those which are evaluated by other means [18]. The summary of the standard uncertainty components for the value of the heat transfer coefficient under condition of single-phase flow in a micro-channel measured by IR technique may be calculated according to the Standard [122] in Table 2.10.

The paper by Patil and Narayanan [123] discusses the theory and procedure of the measurement of the liquid temperature of an opaque fluid near the channel wall for transparent channel walls. Uncertainties in measured and estimated variables are presented in Table 2.11.

A technique for thermal visualization and determination of quantitative, spatially resolved time series of wall temperature during flow boiling in a micro-channel heat sink was presented by Krebs *et al.* [70]. Spatial and temporal variations of channel wall temperature during flow boiling

Table 2.10 Summary of standard uncertainty components of heat transfer coefficient

Standard uncertainty component $u(x_i)$	Source of uncertainty	Value of standard uncertainty $u(x_i)$	$c_i \equiv \left\|\frac{\partial f}{\partial x_i}\right\|$	$\frac{u_i(h)}{h} = \frac{\|c_i\| u_{xi}}{h}$
$u(\dot{m})$	Flow rate	1.7×10^{-6} kg/s	$C\frac{T_{f,out}-T_{f,in}}{dl\left(\bar{T}_{W,IR}-\bar{T}\right)_f}$	1.02×10^{-2}
$u(T_{f,out} - T_{f,in})$	Measured difference between inlet and outlet liquid temperatures	0.14 K	$C\frac{\dot{m}}{dl\left(\bar{T}_{W,IR}-\bar{T}_f\right)}$	7.04×10^{-3}
$u(T_{W,IR} - T_f)$	Measured difference between the wall and the liquid temperatures	0.17 K	$C\frac{\dot{m}\left(T_{f,out}-T_{f,in}\right)}{dl\left(\bar{T}_{W,IR}-\bar{T}_f\right)^2}$	3.2×10^{-2}
$u(d)$	Uncertainty in the estimation of the capillar diameter	5×10^{-6}m	$C\frac{\dot{m}\left(T_{f,out}-T_{f,in}\right)}{d^2l\left(\bar{T}_{W,IR}-\bar{T}_f\right)}$	5.02×10^{-3}
$u(l)$	Uncertainty in the estimation of the test section length	5×10^{-5} m	$C\frac{\dot{m}\left(T_{f,out}-T_{f,in}\right)}{dl^2\left(\bar{T}_{W,IR}-\bar{T}_f\right)}$	3.58×10^{-4}

$C = 1.33 \cdot \times 10^3$; degrees of freedom – 8; $v_{eff}(h) = 15$; $k = 2.13$; $^u(h)/_h = 0.0347$
$U_{95} = ku(h)\mathrm{W/m^2K}$; $\frac{U_{95}(h)}{h} \cdot 100\% = 7.38\%$.

Table 2.11 Uncertainties in measured and estimated Variables [123]

Measured Variable	Total Error	Comments
Wetted perimeter	0.0148 μm (0.004%)	Based on deviation from a 50 · 135 μm rectangular geometry
Channel cross-section	2.0%	Based on deviation from a 50 · 135 μm rectangular geometry
D_h (μm)	135 μm^2 (2.0%)	Based on wetted perimeter and channel cross-section
Camera spatial resolution, x (μm)	10	
Mass flow rate, $\dot{m}$(g/s)	2.3% ($Re=297$) to 4.0% ($Re=204$)	Calibrated using a set flow rate from a syringe pump; data averaged over 20 min
Water radiation flux leaving the heat sink, $C\varepsilon_f \bar{T}_{Si}$	0.26%	Includes bias and precision errors in intensity and calibration error in thermocouple used for fluid temperature measurement Calibration error of thermocouple
Surface temperature, T_{sur} (°C)	0.15	Calibration error of thermocouple
Estimated Variable		
T_f (°C)	0.91 ($Re=204$)	Based on measurements
	1.33 ($Re=251$)	
	1.04 ($Re=285$)	
	0.60 ($Re=297$)	
Re	3.3% ($Re=297$) to 4.7% ($Re=204$)	Includes uncertainty in geometry and flow rate

micro-channel flows using IR thermography are presented and analyzed. In particular, the top channel wall temperature in a branching micro-channel silicon heat sink is measured nonintrusively. Using this technique, time-averaged temperature measurements, with a spatial resolution of 10 μm, are presented over an 18 mm × 18 mm area of the heat sink.

Sources of error for spatial uncertainty includes: (a) horizontal micro-traverse spatial uncertainty that includes both resolution and repeatability, and (b) repeatability in the x- and y-direction pixel shifts determined by the spatial cross-correlation program. The combined spatial uncertainty from these two sources was estimated to be 10 μm. The procedures for quantitative measurement, such as use of an antireflective coating and a detailed calibration were discussed. Results indicate that temperatures can be obtained with a spatial resolution of 10 μm and a temperature uncertainty varying from 0.9 °C at 25 °C to 1.0 °C at 125 °C.

9. CONCLUSIONS

Reliable measurement and control of temperature in the micro-scale are essential for further developing various micro-devices. Many temperature measurement methods traditionally applied to macro-devices are evolving into more advanced techniques applicable to micro-devices taking into consideration enhanced spatial, temporal, and temperature resolutions. The thermo-chromic liquid crystal may be employed for full-field mapping of temperature fields. The good results obtained by the widespread use of IR thermography in experimental studies of convective heat transfer and boiling in micro-channels have proved to be an effective tool in overcoming several limitations of the standard sensors originating both from the measurement and the visualization techniques. Recently, IR has been developed to measure the temperature of the fluid and wall in a micro-channel, using a transparent cover. Measurement of the temperature field of a micro-object by an IR camera has a number of .difficulties: the small size of the object causes a substantial amount of background IR radiation. The problem of background influence on the object temperature measurement should be taken into account; and, of course the IR radiometer has to be carefully calibrated in the temperature range where it is to be used.

In general, the result of measurement is only an approximation or estimate of the value of the specific quantity subject to measurement, and thus the result is complete only when accompanied by a quantitative statement of its uncertainty. Because the reliability of evaluations of components of uncertainty depends on the quality of the information available, it is recommended that all parameters upon which the measured parameter depends be varied to the fullest extent practicable so that the evaluations are based as much as possible on observed data.

Two-phase flow maps and heat transfer prediction methods exist for vaporization in macro-channels and are inapplicable in micro-channels.

Due to the predominance of surface tension over the gravity forces, the orientation of micro-channel has a negligible influence on the flow pattern. The models of convection boiling should correlate the frequencies, length and velocities of the bubbles, and the coalescence processes, which control the flow pattern transitions, with the heat flux and the mass flux. The vapor bubble size distribution must be taken into account.

The flow pattern in parallel micro-channels is quite different from that found in a single micro-channel. At the same values of heat and mass flux, different flow regimes exist in a given micro-channel depending on the time interval. Moreover, at the same time interval different flow regimes may exist in each parallel micro-channel. At low vapor quality heat flux causes a sudden release of energy into the vapor bubble, which grows rapidly and occupies the entire channel. The rapid bubble growth pushes the liquid–vapor interface on both caps of the vapor bubble, at the upstream and the downstream ends, and leads to a reverse flow. The existence of periodic dry zone involves the periodic appearance of hot spots leading to rewetting or wetting of the surface. The instabilities cause fluctuations in the pressure drop and decrease in the heat transfer coefficient. It was found that the temporal behavior of temperature fluctuations corresponds to that of pressure fluctuations. The frequencies of the pressure drop oscillations typical for high-amplitude/low frequency instabilities were in the range of 1–5 Hz. This phenomenon may be regarded as explosive boiling. In the case of uniform heat flux, the hydraulic instabilities cause irregularity of temperature distribution on the heated surface. In the case of nonuniform heat flux, the irregularity increases drastically. Two-phase micro-channel heat sinks do not maintain both streamwise and spanwise uniformity of heat sink temperatures, when hydraulic instabilities occur.

The large heated wall temperature fluctuations are associated with the CHF. The CHF phenomenon is different from that observed in a single channel of conventional size. A key difference between micro-channel heat sink and single conventional channel is the amplification of the parallel-channel instability prior to CHF. As the heat flux approached CHF, the instability, which was moderate over a wide range of heat fluxes, became quite intense and should be associated with maximum temperature fluctuation of the heated surface. The dimensionless experimental values of the heat transfer coefficient may be correlated using the Eotvos number and the boiling number.

It is noteworthy that several studies presented very different results for both the heat transfer at flow boiling and CHF in micro-channels. This is generally due to differences in many parameters that characterize these studies such as the geometry, the hydraulic diameter, the shape and surface roughness of the channels, the fluid nature, the boundary conditions, the flow regimes, and the measuring technique. Such a large variety of experimental conditions often makes it difficult to apply the results of a given study to other investigations.

A new experimental approach was developed and employed to study the physics of nucleate boiling heat transfer and pool boiling crisis. It allows direct visualization of the heat transfer patterns on the heated wall and thus the quantitative characterization of the key processes that underlie the boiling phenomenon all the way to the occurrence of crisis. This is achieved by means of high speed, high resolution IR thermometry of even nano-scale heaters. Significant new insights were gained from direct observations and of the origin and dynamics of hot spots. The hot spots formed within bubble base were identified as dry spots, which serve as precursors of burnout at high heat fluxes. The data obtained in burnout experiments show a direct correlation between CHF and NSD. NSD was found to increase with the degree of heater aging.

The addition of small amount of surfactants makes the boiling behavior quite different from that for pure water. For water, bubble action is seen to be extremely chaotic, with coalescence during the rise. Bubbles formed in surfactant solutions were much smaller than those in water and the surface became covered with them faster. The boiling excess temperature becomes smaller and vapor bubbles are formed easily. Boiling of surfactant solutions, when compared with that in pure water, was observed to be more vigorous. Surfactant solutions promote activation of nucleation sites in a clustered mode, especially at lower heat fluxes. The boiling curves of surfactants differ significantly from the boiling curve of pure water. Experimental results demonstrate that the heat transfer of the boiling process can be enhanced considerably by the addition of small amount of surfactant. The heat transfer increases monotonously at an increase in the concentration.

Confined boiling of water and surfactant solutions under condition of natural convection also causes a heat transfer enhancement. Additive of surfactant leads to enhancement of heat transfer compared to water boiling in the same gap size; however, this effect decreased with decreasing gap size. For the same gap size, CHF decreases with increase in channel length. CHF in surfactant solutions is significantly lower than in water.

IR technique may be used to address the fundamental thermal management problems faced by designers of 3D circuits, specifically the limited surface area available for cooling and the large vertical thermal resistance between the bottom layer of the device and the cooling technology.

REFERENCES

[1] B. Palm, Heat transfer in microchannels, Microscale Therm. Eng. 5 (2001) 155–175.

[2] S.M. Ghiaasiaan, S.I. Abdel-Khalik, Two-phase flow in microchannels, Adv. Heat Transfer 34 (2001) 145–254.

[3] C.B. Sobhan, S.V. Garimella, A comparative analysis of studies on heat transfer and fluid flow in microchannel, in: Proceedings of the International Conference on Heat Transfer, 2000, pp. 80–92.

[4] S. Kandlikar, Fundamental issues related to flow boiling in minichannels and microchannels, Exp. Therm. Fluid Sci. 26 (2002) 389–407.

[5] I. Hassan, P. Phutthavong, M. Abdelgawad, Microchannel heat sinks: An overview of the state-of-the-art, Microscale Therm. Eng. 8 (3) (2004) 183–205.

[6] J.M. Buchlin, Convective heat transfer and infrared thermography, J. Appl. Fluid Mech. 3 (1) (2010) 55–62.

[7] G. Hetsroni, A. Mosyak, E. Pogrebnyak, R. Rozenblit, Infrared temperature measurements in micro-channels and micro-fluid systems, Int. J. Therm. Sci. 50 (2011) 853–868.

[8] T. Astarita, G. Cardone, G.M. Carlomagno, C. Meola, A survey of infrared thermography for convective heat transfer measurements, Opt. Laser Technol. 32 (2000) 593–610.

[9] G.M. Carlomagno, L. de Luca, Infrared thermography in heat transfer, in: W.I. Yang (Ed.), Handbook of Flow Visualization, Hemisphere Publishing Corporation, Washington, DC, 1989, pp. 531–553.

[10] T. Sikanen, T. Zwinger, S. Tuomikoski, S. Franssila, R. Lehtiniemi, C.M. Fager, T. Kotiaho, A. Pursula, Temperature modeling and measurement of an electrokinetic separation chip, Microfluid. Nanofluidics 5 (2008) 479–491.

[11] G. Hetsroni, R. Rozenblit, L.P. Yarin, A hot foil infrared technique for studying the temperature field on a wall, Meas. Sci. Technol. 7 (1996) 1418–1427.

[12a] G.M. Carlomagno, G. Cardone, Infrared thermography for convective heat transfer measurements, Exp. Fluids 49 (2010) 1187–1218.

[12b] T. Astarita, G.M. Carlomagno, Infrared Thermography for Thermo-Fluid-Dynamics, Springer-Verilog, Berlin-Heidelberg, ISBN 978-3-642-29507-2 (2013).

[13] G. Hetsroni, R. Rozenblit, Heat transfer to a liquid–solid mixture in a flume, Int. J. Multiphase Flow 20 (1994) 671–689.

[14] G. Hetsroni, A. Mosyak, Z. Segal, G. Ziskind, A uniform temperature heat sink for cooling of electronic devices, Int. J. Heat Mass Transfer 45 (2002) 3275–3286.

[15] I. Tiselj, G. Hetsroni, B. Mavko, A. Mosyak, E. Pogrebnyak, Z. Segal, Effect of axial conduction on the heat transfer in micro-channels, Int. J. Heat Mass Transfer 47 (2004) 2551–2565.

[16] G. Zissis, The Infrared and Electro-optical Systems Handbook, vol. 1, Sources of Radiation, Infrared Information and Analysis Center, Ann Arbor, MI, 1993383.

[17] D. McGee Thomas, Principles and Methods of Temperature Measurement, Wiley, New York, 1988608.

[18] G. Hetsroni, A. Mosyak, E. Pogrebnyak, R. Rozenblit, Infrared measurements in micro-scale systems, Int. J. Therm. Sci. 50 (2011) 853–868.

[19] T.G. Theofanous, J.P. Tu, A.T. Dinh, T.N. Dinh, The boiling crisis phenomenon. Part I: Nucleation and nucleate boiling heat transfer, Exp. Therm. Fluid Sci. 26 (2002) 775–792.
[20] S.S. Kutateladze, Elements of the hydrodynamics of gas–liquid systems, Fluid Mech.–Soviet Res. 1 (4) (1972) 29–68.
[21] T.G. Theofanous, A.T. Dinh, J.P. Tu, T.N. Dinh, The boiling crisis phenomenon. Part II: Dryout dynamics and burnout, Exp. Therm. Fluid Sci. 26 (2002) 793–810.
[22] G. Hetsroni, M. Gurevich, A. Mosyak, R. Rozenblit, Surface temperature measurement of a heated capillary tube by means of an infrared technique, Meas. Sci. Technol. 14 (2003) 807–814.
[23] P. Carey van, Liquid-Vapor Phase-Change Phenomena, An Introduction to the Thermophysics of Vaporization and Condensation Processes in Heat Transfer Equipment, Hemisphere, 1992, New York.
[24] V.K. Dhir, Boiling heat transfer, Annu. Rev. Fluid Mech. 30 (1998) 365–401.
[25] Y.Y. Hsu, On size range of active nucleation cavities on a heating surface, J. Heat Transfer 84 (1962) 207–216.
[26] T. Sato, H. Matsumura, On the conditions of incipient subcooled-boiling with forced convection, Bull. Jpn. Soc. Mech. Eng. 47 (1964) 392–398.
[27] E.J. Davis, G.H. Anderson, The incipience of nucleate boiling in forced convection flow, AIChE 12 (1966) 774–780.
[28] G.P. Celata, M. Cumo, A. Mariani, Experimental evaluation of the onset of subcooled flow boiling at high liquid velocity and subcooling, Int. J. Heat Mass Transfer 40 (1997) 2879–2885.
[29] W. Qu, I. Mudawar, Prediction and measurement of incipient boiling heat flux in micro-channel heat sinks, Int. J. Heat Mass Transfer 45 (2002) 3933–3945.
[30] S.M. Ghiaasiaan, R.C. Chedester, Boiling incipience in micro-channels, Int. J. Heat Mass Transfer 45 (2002) 4599–4606.
[31] J. Li, P. Cheng, Bubble cavitation in a micro-channel, Int. J. Heat Mass Transfer 47 (2004) 2689–2698.
[32] D. Liu, P.-S. Lee, S.V. Garimella, Prediction of the onset of nucleate boiling in micro-channel flow, Int. J. Heat Mass Transfer 48 (2005) 5134–5149.
[33] H. Ye, N. Naguib, Y. Gogotsi, EM study of water in carbon nanotubes, JEOL News 39 (2004) 2–7.
[34] L.P. Yarin, A. Mosyak, G. Hetsroni, Fluid Flow, Heat Transfer and Boiling in Micro-channels, Springer, Berlin, 2009.
[35] A.E. Bergles, W.W. Rohsenow, The determination of forced-convection surface-boiling heat transfer, ASME J. Heat Transfer 86 (1964) 365–372.
[36] S.G. Kandlikar, V. Mizo, M. Cartwright, E. Ikenze, Bubble nucleation and growth characteristics in subcooled flow boiling water, in: National Heat Transfer Conference HTD-342, ASME, New York, 1997, pp. 11–18.
[37] H.C. Unal, Determination of the initial point of net vapor generation in flow boiling system, Int. J. Heat Mass Transfer 18 (1975) 1095–1099.
[38] I. Hapke, H. Boye, J. Schmidt, Onset of nucleate boiling in micro-channels, Int. J. Therm. Sci. 39 (2000) 505–513.
[39] R. Hino, T. Ueda, Studies on heat transfer and flow characteristics in subcooled flow boiling. Part 1: Boiling characteristics, Int. J. Multiphase Flow 11 (1985) 269–281.
[40] P.C. Lee, F.C. Tseng, C. Pan, Bubble dynamics in micro-channels. Part 1: Single microchannel, Int. J. Heat Mass Transfer 47 (2004) 5575–5589.
[41] A. Mosyak, L. Rodes, G. Hetsroni, Boiling incipience in parallel micro-channels with low mass flux subcooled water flow, Int. J. Multiphase Flow 47 (2012) 150–159.
[42] T.R. Thomas, Rough Surfaces, second ed., Imperial College Press, London, 1999.

[43] J.E. Kennedy, G.M. Roach, M.E. Dowling, S.I. Abdel-Khalik, S.M. Chiaasiaan, S.M. Jeter, Z.H. Quershi, The onset of flow instability in uniformly heated horizontal micro-channels, ASME J. Heat Transfer 122 (2000) 118–125.
[44] G. Hetsroni, A. Mosyak, Z. Segal, Nonuniform temperature distribution in electronic devices cooled by flow in parallel micro-channels, IEEE Trans. Compon. Packag. Technol. 24 (1) (2001) 16–23.
[45] J.G. Collier, J.R. Thome, Convective Boiling and Condensation, third ed., Oxford University Press, Oxford, 1994.
[46] H.J. Lee, S.Y. Lee, Heat transfer correlation for boiling flows in small rectangular horizontal channels with low aspect ratios, Int. J. Multiphase Flow 27 (2001) 2043–2062.
[47] W. Qu, I. Mudawar, Flow boiling heat transfer in two-phase micro-channel heat sinks-2. Annular two-phase flow model, Int. J. Heat Mass Transfer 46 (2003) 2773–2784.
[48] T.-H. Yen, N. Kasagi, Y. Suzuki, Forced convective boiling heat transfer in microtubes at low mass and heat fluxes, Int. J. Multiphase Flow 29 (2003) 1771–1792.
[49] S. Grohmann, Measurement and modeling of single-phase and flow-boiling heat transfer in micro-tubes, Int. J. Heat Mass Transfer 48 (2005) 4072–4089.
[50] J. Lee, I. Mudawar, Two-phase flow in high-heat-flux micro-channel heat sink for refrigeration cooling applications. Part II: Heat transfer characteristics, Int. J. Heat Mass Transfer 48 (2005) 941–955.
[51] R.K. Shah, A.L. London, Laminar flow forced convection in ducts: A source book for compact heat exchanger analytical data, Advances in Heat Transfer, Suppl. 1, Academic, 1978, New York.
[52] M. Steinke, S.G. Kandlikar, An experimental investigation of flow boiling characteristics of water in parallel micro-channels, ASME J. Heat Transfer 126 (2004) 518–526.
[53] G. Hetsroni, A. Mosyak, E. Pogrebnyak, Z. Segal, Explosive boiling of water in parallel micro-channels, Int. J. Multiphase Flow 31 (2005) 371–392.
[54] S.G. Kandlikar, Heat transfer mechanisms during flow boiling in micro-channels, ASME J. Heat Transfer 126 (2004) 8–16.
[55] S.G. Kandlikar, P. Balasubramanian, An extension of the flow boiling correlation to transition, laminar, and deep laminar flows in mini-channels and micro-channels, Heat Transfer Eng. 25 (3) (2004) 86–93.
[56] G. Hetsroni, A. Mosyak, Z. Segal, G. Ziskind, A uniform temperature heat sink for cooling of electronic devices, Int. J. Heat Mass Transfer 45 (2002) 3275–3286.
[57] G. Hetsroni, A. Mosyak, E. Pogrebnyak, Z. Segal, Periodic boiling in parallel micro-channels at low vapor quality, Int. J. Multiphase Flow 32 (2006) 1141–1159.
[58] G. Wang, P. Cheng, A.E. Bergles, Effects of inlet/outlet configurations on flow boiling instability in parallel microchannels, Int. J. Heat Mass Transfer 51 (2008) 2267–2281.
[59] W. Qu, I. Mudawar, Flow boiling heat transfer in two-phase micro-channel heat sinks-1: Experimental investigation and assessment of correlation methods, Int. J. Heat Mass Transfer 46 (2003) 2755–2771.
[60] L. Zhang, E.N. Wang, K.E. Goodson, T.W. Kenny, Phase change phenomena in silicon microchannels, Int. J. Heat Mass Transfer 48 (2005) 1572–1582.
[61] H.Y. Wu, P. Cheng, Liquid/two-phase/vapor alternating flow during boiling in microchannels at high heat flux, Int. Commun. Heat Mass Transfer 30 (2003) 295–302.
[62] H.Y. Wu, P. Cheng, Visualization and measurements of periodic boiling in silicon microchannels, Int. J. Heat Mass Transfer 46 (2003) 2603–2614.
[63] H.Y. Wu, P. Cheng, Boiling instability in parallel silicon microchannels at different heat flux, Int. J. Heat Mass Transfer 47 (2004) 3631–3641.

[64] G. Wang, P. Cheng, H. Wu, Unstable and stable flow boiling in parallel microchannels and in a single microchannel, Int. J. Heat Mass Transfer 50 (2007) 4297–4310.
[65] S.G. Kandlikar, W.K. Kuan, D.A. Willistein, J. Borrelli, Stabilization of flow boiling in microchannels using pressure drop elements and fabricated nucleation sites, J. Heat Transfer 128 (2006) 389–396.
[66] A. Kosar, C.-J. Kuo, Y. Peles, Suppression of boiling flow oscillations in parallel microchannels by inlet restrictors, J. Heat Transfer 128 (2006) 251.
[67] C.J. Kuo, Y. Peles, Pressure effects on flow boiling instabilities in parallel microchannels, Int. J. Heat Mass Transfer 52 (2009) 271–280.
[68] C.M. Diaz, J. Schmidt, Experimental investigation of transient boiling heat transfer in micro-channels, Int. J. Heat Fluid Flow 28 (2007) 95–102.
[69] J. Barber, K. Sefiane, D. Brutin, L. Tadrist, Hydrodynamics and heat transfer during flow boiling instabilities in a single microchannel, Appl. Therm. Eng. 29 (2009) 1299–1308.
[70] D. Krebs, V. Narayanan, J. Liburdy, D. Pence, Spatially resolved wall temperature measurements during flow boiling in microchannels, Exp. Therm. Fluid Sci. 34 (2010) 434–445.
[71] T. Raad, J.E. Myers, Nucleation studies in pool boiling on thin plates using liquid crystals, AIChE J. 17 (5) (1971) 1260–1261.
[72] J.E. Sgheiza, J.E. Myers, Behavior of nucleation sites in pool boiling, AIChE J. 31 (10) (1985) 1605–1613.
[73] D.B.R. Kenning, Wall temperature patterns in nucleate boiling, Int. J. Heat Mass Transfer 35 (1) (1992) 73–86.
[74] D.B.R. Kenning, D.B.R. Youyou, Pool boiling heat transfer on a thin plate: Features revealed by liquid crystal thermography, Int. J. Heat Mass Transfer 39 (15) (1996) 3117–3137.
[75] G. Hetsroni, J.L. Zakin, Z. Lin, A. Mosyak, E.A. Pancallo, R. Rozenblit, The effect of surfactants on bubble growth, wall thermal patterns and heat transfer in pool boiling, Int. J. Heat Mass Transfer 44 (2001) 485–497.
[76] R.F. Gaertner, J.W. Westwater, Population of active sites in nucleate boiling heat transfer, Chem. Eng. Prof. Symp. Ser. 56 (30) (1960) 39048.
[77] C.H. Wang, V.K. Dhir, On the gas entrapment and nucleation site density during pool boiling of saturated water, ASME J. Heat Transfer 115 (3) (1993) 670–679.
[78] R.J. Benjamin, A.R. Balakrishnan, Nucleation site density in pool boiling of saturated pure liquids: Effect of surface microroughness and surface and liquid physical properties, Exp. Therm. Fluid Sci. 15 (1) (1997) 32–42.
[79] G. Kocamustafaogullari, M. Ishii, Interfacial area and nucleation site density in boiling systems, Int. J. Heat Mass Transfer 26 (1983) 1377–1389, 670–679.
[80] I. Hapke, H. Boye, J. Schmidt, Y. Staate, Investigation of flow boiling in mini channels using thermography, in: EUROTHERM Seminar, Microscale Heat Transfer, Proc, Poitiers, France 57 (1998) 159–166.
[81] A. Serizawa, Z. Feng, Z. Kawara, Two-phase flow in micro-channels, Exp. Therm. Fluid Sci. 26 (2002) 703–714.
[82] M.B. Bowers, I. Mudawar, High flux boiling in low flow rate, low pressure drop mini-channel and micro-channel heat sinks, Int. J. Heat Mass Transfer 37 (1994) 321–332.
[83] L. Jiang, M. Wong, Y. Zohar, Phase change in microchannel heat sinks with integrated temperature sensors, J. Microelectromech. Syst. 8 (1999) 358–365.
[84] W. Qu, I. Mudawar, Measurement and correlation of critical heat flux in two-phase micro-channel heat sinks, Int. J. Heat Mass Transfer 47 (2004) 2045–2059.
[85] L. Wojtan, R. Revellin, J.R. Thome, Investigation of saturated critical heat flux in a single, uniformly heated microchannel, Exp. Therm. Fluid Sci. 30 (2006) 765–774.

[86] W. Zhang, T. Hibiki, K. Mishima, Y. Mi, Correlation of critical heat flux for flow boiling of water in mini-channels, Int. J. Heat Mass Transfer 49 (2006) 1058–1072.
[87] D.D. Hall, I. Mudawar, Critical heat flux (CHF) for water flow in tubes—II. Subcooled CHF correlations, Int. J. Heat Mass Transfer 43 (2000) 2605–2640.
[88] R. Revellin, J. Thome, A theoretical model for the prediction of the critical hat flux in heated micro-channel, Int. J. Heat Mass Transfer 51 (2008) 1216–1225.
[89] K. Moriyama, A. Inoue, Thickness of the liquid film formed by a growing bubble in a narrow gap between two horizontal plates, ASME J. Heat Transfer 118 (1996) 132–139.
[90] J.R. Thome, V. Dupont, A.M. Jacobi, Heat transfer model for evaporation in microchannels, Part I: Presentation of the model, Int. J. Heat Mass Transfer 47 (2004) 3375–3385.
[91] H. Sephton, US Patent (1974) 3846254.
[92] W. Frost, C.J. Kippenhan, Bubble growth and heat transfer mechanisms in the forced convection boiling of water containing a surface active agent, Int. J. Heat Mass Transfer 10 (1967) 931–949.
[93] P. Kotchaphakdee, M.C. Williams, Enhancement of nucleate pool boiling with polymeric additives, Int. J. Heat Mass Transfer 13 (1970) 835–848.
[94] H.J. Gannet Jr., M.C. Williams, Pool boiling in dilute non-aqueous polymer solutions, Int. J. Heat Mass Transfer 14 (1971) 1001–1005.
[95] B.H. Shah, R. Darby, The effect of surfactant on evaporative heat transfer in vertical film flow, Int. J. Heat Mass Transfer 16 (1973) 1889–1903.
[96] D.D. Paul, S.I. Abdel-Khalik, Nucleate boiling in drag-reducing polymer solutions, J. Rheol. 27 (1983) 59–76.
[97] Y.M. Yang, J.R. Maa, Pool boiling of dilute surfactant solutions, J. Heat Transfer 105 (1983) 190–192.
[98] Y.L. Tzanand, Y.M. Yang, Experimental study of surfactant effects on pool boiling heat transfer, J. Heat Transfer 112 (1990) 207–212.
[99] W.T. Wu, Y.M. Yang, Enhanced boiling heat transfer by surfactant additives, in: Pool and External Flow Boiling, ASME, New York, 1992361–366.
[100] W.T. Wu, Y.M. Yang, J.R. Maa, Enhancement of nucleate boiling heat transfer and depression of surface tension by surfactant additives, J. Heat Transfer 117 (1995) 526–529.
[101] W.T. Wu, Y.M. Yang, J.R. Maa, Nucleate pool boiling enhancement by means of surfactant additives, Exp. Therm. Fluid Sci. 18 (1998) 195–209.
[102] C.N. Ammerman, S.M. You, Determination of the boiling enhancement mechanism caused by surfactant addition to water, J. Heat Transfer 118 (1996) 429–435.
[103] S.G. Kandlikar, L. Alves, Effects of surface tension and binary diffusion on pool boiling of dilute solutions: An experimental assessment, J. Heat Transfer 121 (1999) 488–493.
[104] V.M. Wasekar, R.M. Manglik, A review of enhanced heat transfer in nucleate pool boiling of aqueous surfactant and polymeric solutions, Enhanc. Heat Transfer 6 (1999) 135–150.
[105] J.L. Zakin, J. Myska, Z. Chara, New limiting drag reduction and velocity profile asymptotes for nonpolymeric additives systems, AIChE J. 42 (1996) 3544–3546.
[106] Y. Katto, S. Yokoya, K. Teraoka, Experimental study of nucleate boiling in case of making interference-plate approach to the heating surface, in: Proceedings of the 3rd International Heat Transfer Conference, 3, 1966, pp. 219–227.
[107] E. Ishibashi, K. Nishikawa, Saturated boiling heat transfer in narrow space, Int. J. Heat Mass Transfer 12 (1969) 863–894.
[108] S.C. Yao, Y. Chang, Pool boiling heat transfer in a confined space, Int. J. Heat Mass Transfer 26 (1983) 841–848.
[109] Y. Fujita, H. Ohta, S. Uchida, K. Nishikawa, Nucleate boiling heat transfer and critical heat flux in narrow space between rectangular surfaces, Int. J. Heat Mass Transfer 31 (1988) 229–239.

[110] K. Nishikawa, Y. Fujita, S. Uchida, H. Ohta, Effect of surface configuration on nucleate boiling heat transfer, Int. J. Heat Mass Transfer 27 (1984) 1559–1571.
[111] J. Bonjour, M. Lallemand, Flow patterns during boiling in a narrow space between two vertical surfaces, Int. J. Multiphase Flow 24 (1998) 947–960.
[112] Y.H. Zhao, T. Masuoka, T. Tsuruta, Unified theoretical prediction of fully developed nucleate boiling and critical heat flux based on a dynamic microlayer model, Int. J. Heat Mass Transfer 45 (2002) 3189–3197.
[113] G. Hetsroni, A. Mosyak, E. Pogrebnyak, Z. Segal, Natural convection boiling of water and surfactants in narrow horizontal annular channels, Int. J. Multiphase Flow 33 (2007) 469–483.
[114] G. Hetsroni, A. Mosyak, R. Rozenblit, E. Pogrebnyak, Z. Segal, Natural convection boiling of water and surfactant solutions having negligible environmental impact in vertical confined space, Int. J. Multiphase Flow 35 (2009) 20–33.
[115] Y.M. Yang, J.R. Maa, Boiling heat transfer enhancement by surfactant additives, in: Proceedings of the 5th Conference Boiling Heat Transfer ICBHT, Montego Bay, Jamaica, May 2003, pp. 4–8.
[116] M.E. Steinke, S.G. Kandlikar, Control and effect of dissolved air in water during flow boiling in micro-channels, Int. J. Heat Mass Transfer 47 (2004) 1925–1935.
[117] D. Klein, G. Hetsroni, A. Mosyak, Heat transfer characteristics of water and APG surfactant solution in a micro-channel heat sink, Int. J. Multiphase Flow 31 (2005) 393–415.
[118] I. Sher, G. Hetsroni, An analytical model for nucleate pool boiling with surfactant additives, Int. J. Multiphase Flow 28 (2002) 699–706.
[119] G. Hetsroni, M. Gurevich, A. Mosyak, R. Rozenblit, L.P. Yarin, Subcooling boiling of surfactant solutions, Int. J. Multiphase Flow 28 (2002) 347–361.
[120] J. Zhang, R.M. Manglik, Additive absorption and interfacial characteristics of nucleate pool boiling in aqueous surfactant solutions, J. Heat Transfer 127 (2005) 684–691.
[121] J. Koo, S. Im, L. Jiang, K.E. Goodson, Integrated micro-channel cooling for three-dimensional electronic circuit architectures, J. Heat Transfer 127 (2005) 56–58.
[122] Guide to the Expression of Uncertainty of Measurements, International Organization for Standardization, 1995, Geneva.
[123] V.A. Patil, V. Narayanan, Spatially resolved temperature measurement in micro-channels, Microfluidics Nanofluidics J. 2 (2006) 291–300.
[124] R.M. Stoddard, A.M. Blasick, S.M. Ghiaasiaan, S.I. Abdel-Khalik, S.M. Jeter, M.F. Dowling, Onset of flow instability and critical heat flux in thin horizontal annuli, Exp. Therm. Fluid Sci. 26 (2002) 1–14.
[125] S. Su, S. Huang, X. Wang, Study of boiling incipience and heat transfer enhancement in forced flow through narrow channels, Int. J. Heat Mass Transfer 31 (2001) 253–260.
[126] B. Thompson, R.V. Macbeth, Boiling Water Heat Transfer—Burnout in Uniformly Heated Round Tubes: A Compilation of World Data with Accurate Correlations, AEEW-R 356, United Kingdom Atomic Energy Authority, Winfrith, UK, 1964.
[127] W.H. Lowdermilk, C.D. Lanzo, B.L. Siegel, Investigation of boiling burnout and flow stability for water flowing in tubes, NACA TN 4382, National Advisory Committee for Aeronautics, Washington, DC, 1958.
[128] K.M. Becker, G. Hernborg, M. Bode, O. Eriksson, Burnout data for flow of boiling water in vertical round ducts, annuli and rod clusters, AE-177, Aktiebolaget Atomenergi, Stockholm, Sweden, 1965.
[129] J. Griffel, Forced convection boiling burnout for water in uniformly heated tubular test sections, USAEC Report NYO 187-7, TID-4500, Columbia University, 1965.
[130] H. Nariai, F. Inasaka, T. Shimura, Critical heat flux of subcooled flow boiling in narrow tube, ASME JSME Therm. Eng. Joint Conf. Proc. 5 (1987) 455–462.

[131] F. Inasaka, H. Nariai, Critical heat flux of subcooled flow boiling with water, Proc. NURETH-4 (1) (1989) 115–120.
[132] F. Inasaka, Critical heat flux of subcooled flow boiling in water under uniform heating conditions, Papers Ship Res. Inst. 30 (4) (1993) 1–69.
[133] G.P. Celata, M. Cumo, A. Mariani, Burnout in highly subcooled water flow boiling in small diameter tubes, Int. J. Heat Mass Transfer 36 (1993) 1269–1285.
[134] C.L. Vandervort, A.E. Bergles, M.K. Jensen, An experimental study of critical heat flux in very high heat flux subcooled boiling, Int. J. Heat Mass Transfer 37 (1994) 161–173.
[135] A.M. Lezzi, A. Niro, G.P. Beretta, Experimental data on CHF for forced convection water boiling in long horizontal capillary tubes, in: G.F. Hewitt (Ed.), Heat Transfer 1994: Proceedings of the 10th International Heat Transfer Conference, Institution of Chemical Engineers, Rugby, 7, 1994, pp. 491–496.
[136] M. Kureta, Critical heat flux for flow boiling of water in small diameter tubes under atmospheric pressure, Kyoto University (in Japanese), Dissertation, 1997.
[137] G.M. Roach, S.I. Abdel-Khalik, S.M. Ghiaasiaan, M.F. Dowling, S.M. Jeter, Low-flow critical heat flux in heated microchannels, Nucl. Sci. Eng. 131 (1999) 411–425.
[138] I. Mudawar, M.B. Bowers, Ultra-high critical heat flux (CHF) for subcooled water flow boiling. I: CHF data and parametric effects for small diameter tubes, Int. J. Heat Mass Transfer 42 (1999) 1405–1428.

CHAPTER THREE

Technology Evolution, from the Constructal Law

Adrian Bejan
Department of Mechanical Engineering and Materials Science, Duke University, Durham, North Carolina, USA

Contents

Abstract

The constructal law of design evolution is the law of physics that expresses the natural tendency of all flow systems, bio and nonbio, to morph into configurations that provide greater flow access over time. River basins, animal locomotion and migratory routes, snowflakes, and turbulence structure illustrate this tendency of design occurrence and evolution over time. The movement and persistence of human life on the landscape (people, goods, construction, and mining) also evolves in accord with the constructal law. Each of us belongs to the "human and machine" species, the evolving design that is better known as technology evolution. In this chapter, I illustrate the technology evolution phenomenon by showing that larger flow components (organs) belong on larger vehicles and animals and that the time arrow of the constructal law points toward smaller sizes over time (miniaturization). This evolutionary direction is further illustrated by the evolution of cooling technology for high-density heat transfer, from natural convection to forced convection and pure conduction, Fourier and non-Fourier. Tree-shaped flow architectures emerged from the same constructal-law tendency when the flow connects one point with an infinity of points (area and volume). At bottom, all technology evolution is about facilitating the movement of the human and machine species on the world's map. This is why the more advanced groups consume proportionally more fuel and why both wealth and fuel consumption continue to rise naturally.

Advances in Heat Transfer, Volume 45
ISSN 0065-2717
http://dx.doi.org/10.1016/B978-0-12-407819-2.00003-7

For all the compasses in the world, there is only one direction, and time is its only measure

***Tom Stoppard,* Rosencrantz and Guildenstern are Dead**

1. TECHNOLOGY EVOLUTION, PREDICTED

A law of physics is a concise statement that summarizes a distinct phenomenon that occurs in nature. A phenomenon is a fact, circumstance, or experience that is apparent to the human senses and can be described. The phenomenon that is the subject of this chapter is the occurrence and evolution of design, or organization in nature: the time direction of the "scenario" of design evolution. The direction is universal, and it is one way—toward configurations and rhythms (designs) that flow and move more easily, for greater access, over time. This phenomenon is summarized by the constructal law [1–4], which governs the time direction in which the scenario evolves [1,2]:

For a finite-size flow system to persist in time (to live), its configuration must evolve in such a way that it provides easier access to the imposed (global) currents that flow through it.

Based on its record, the constructal law is the law of physics that accounts for design in nature in animate, inanimate, and social flow systems [5–13]. The constructal law unites the bio with the nonbio and the human with the nonhuman. The natural phenomenon of design and evolution is universal. For example, on the Earth's surface, the design phenomenon facilitates access for everything that flows, spreads, and is collected [14–16]: river basins, atmospheric and ocean currents, animal life and migration, and our civilization (the evolution of the "human and machine" species).

Technology evolution is just one class of design-in-nature phenomena, and it is no different than animal evolution, river basin evolution, science evolution, or any other kind of evolution. To see this, consider the physics illustrated in Fig. 3.1. A vehicle consumes fuel and moves on the world map. We ask how large one of the components of this vehicle should be, for example, a duct with fluid flowing through it or a heat exchanger surface. Because the size of the component is finite, the vehicle is penalized (in fuel terms) by the component in two ways (Fig. 3.1).

First, the component is a flow system that operates irreversibly, with entropy generation, which means the destruction of useful energy (the consumed fuel), because of currents that flow by overcoming resistances, obstacles, and all kinds of "friction." This fuel penalty is smaller when the component is larger, with wider ducts and larger heat transfer surfaces.

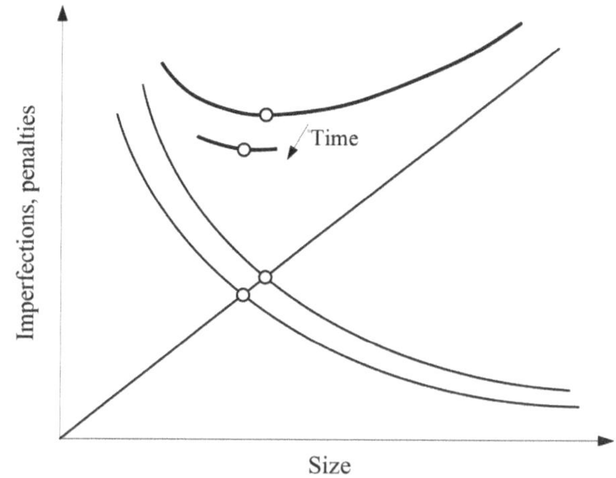

Figure 3.1 Every flow component has a characteristic size, which emerges from two conflicting trends. The useful energy dissipated because of the imperfection of the component decreases as the component size increases. The useful energy spent by the greater system (vehicle and animal) increases with the component size. The sum of the two penalties is minimum when the component size is finite, at the intersection between the two penalties. In time, the component evolves toward smaller sizes, because it improves and its penalty (the descending curve) slides downward.

In this limit, larger is better, because the component poses less resistance to the flow of fluid, heat, mass, and stresses [9,17].

Second, the vehicle must burn fuel in order to transport the component. The fuel penalty for the component is proportional to the weight of the component and teaches that smaller is better. This second penalty is in conflict with the first, and from this conflict emerges the notion—the prediction, the purely theoretical discovery—that the component should have a characteristic size that is finite, not too large, not too small, but just right, for that particular vehicle. At bottom, the total fuel required by the vehicle is proportional to the total weight of the components.

The two penalties are the two asymptotes drawn in Fig. 3.1. The sum of the two penalties is minimal when the component size is essentially the same as the size obtained by intersecting the asymptotes. Locating the design (the component size) in this fashion is one of the more recent examples of the application of the method of intersecting the asymptotes, which is being developed from edition to edition since 1984 in my graduate convection book [18]. "Characteristic size" means that large components (pipes, heat exchangers, etc.) belong on large vehicles, and small components belong on small vehicles. It also means that all components are imperfect, because each has a finite size, not an infinite size.

The whole (the vehicle) is a construct of components that are "imperfect" only when examined in isolation. The vehicle design also evolves in accord with the constructal law, as it becomes a better and better construct for moving the vehicle mass on the world map. Here, "better" means moving more mass farther per unit of fuel consumed.

Everything that I wrote about vehicles in relation to Fig. 3.1 applies unchanged to animal organs and the whole animal. Every organ must have a certain, characteristic size, which is larger when the animal is larger. Every organ is "imperfect" thermodynamically because of its finite size, and this explains the mistake that many scientists make when they marvel that "nature makes mistakes."

No, nature does not make mistakes. Nature has a few universal tendencies, called phenomena, and for these, a few individuals wrote a few laws of physics, which are universal. Nature is faithful to the laws. In the case of the animal, the tendency is to move more animal mass more easily on the world map and move it farther. The whole animal is a truck, a vehicle for animal mass. Finding food, like finding fuel, is a pain.

The crucial difference between the animal and the vehicle, or between the heart and the water pump, is that we cannot witness animal evolution because it occurred over an enormously long time interval. We can, however, witness sports evolution and technology evolution. In fact, the time scale of technology evolution is so short that most of what enables our movement on the map evolved during the past one hundred years: central power plants, electrification, the automobile, and the airplane.

In time, the organ evolves toward designs that flow more easily. This means that in Fig. 3.1, the descending curve migrates downward over time and so does its intersection with the rising line. The minimum of the aggregate penalty curve follows suit and migrates downward and to the left. The discovery is not only that the future component must be better in an evolutionary manner but also that it must be *smaller*. The constructal law calls for the future, and the name for this future is *miniaturization*.

Now we know why miniaturization *should happen*. It is the natural tendency in each of us to move our body, vehicle, and clan more easily and for longer time and space on the world map. In thermal engineering, miniaturization happens, and it did not start with nanotechnology. Before nanotechnology, we had microelectronics, and before microelectronics, we had compact heat exchangers.

There is no "revolution" toward the small and smaller. There is relentless evolution, and we see it in every domain not just in thermal engineering. Just think of writing, from antiquity to our day: clay tablets, slate, and obelisks were displaced by denser written content, by (in order) papyruses, parchments, books, mass printing, and software. There is no end to this movie of evolutionary design. There is just better and better flowing for us the human and machine bodies on the landscape, which means (among many design features) lighter and more efficient artifacts attached to us.

2. EVOLUTION OF COMPACTNESS

The evolution toward greater *density* of volumetric flow, or functionality, is another way to describe the evolution of technology for easier human and machine movement on Earth. The reason is that no matter how small the smallest flow features become (e.g., from micro to nano), the new devices that empower the human and machine species must continue to match the length scale of the human body or the human hand. The smaller the smallest features, the more numerous are the tiniest components of the new device. These tiny flow systems are not poured into the human-scale box like beans in a sack. They must be assembled, connected, and constructed to flow together so that they bathe the available "whole" completely.

The march toward miniaturization is necessarily an evolution toward easier volumetric flow architectures that are more complex because their smallest features become smaller and more numerous. The fascination with the nanophenomenon, the nanoelement, and the nanoperformance misses the big picture, which is the *construction* of the macro device (e.g., lung) that relies on clever elements at the smallest scale and in the largest numbers (e.g., alveoli).

The discovery of the construction is delivered by the constructal law, after invoking it for the flow of the whole. This is illustrated in Fig. 3.2, which is a review of three decades of research on the cooling of electronics packaged on parallel plates in a volume of finite size. The length scale of this volume, L, can vary and has been varying over time. Think of the evolution of electronics, from phone booths to servers and laptops and handheld devices today.

Three cooling technologies are summarized in Fig. 3.2, natural convection (NC), forced convection (FC), and solid-body conduction (C). The

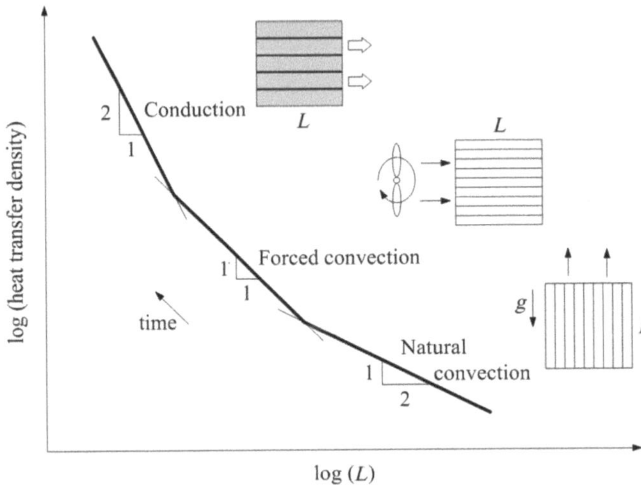

Figure 3.2 Predicted evolution of heat transfer density toward higher values, showing two phenomena: evolution toward smaller sizes (miniaturization) and stepwise changes in cooling technology.

chronological sequence in which these technologies emerged and spread is NC–FC–C. Here is why:

Each cooling technology offers the greatest packing density of functionality when the spacing between the working plates (D) has a certain value. The oldest design of this kind was proposed for NC cooling, based on the intersection of asymptotes method (Ref. [18], problem 11, p. 157; now Section 4.12 in Ref. [19]):

$$\frac{D}{L} \sim Ra^{-1/4} \tag{3.1}$$

where $Ra = g\beta\Delta T L^3/\alpha\nu$, $\Delta T = T_{max} - T_0$, and T_0 is the coolant inlet temperature. Because the packed components generate heat, this design offers the highest density of heat generation $\left(q'''_{NC}\right)$ subject to a temperature ceiling (T_{max}) that cannot be exceeded:

$$q'''_{NC} \sim \frac{k}{L^2}\Delta T Ra^{1/2} \tag{3.2}$$

Because $Ra^{1/2}$ varies as $L^{3/2}$, the density q'''_{NC} varies as $L^{-1/2}$, which is shown in Fig. 3.2. Greater densities of functionality are possible if this elemental package can be made smaller. Notice again the predictive constructal-law message of Section 1: the time arrow of evolution of NC cooling (q'''_{NC} vs. L) points to the left, toward miniaturization.

The second oldest cooling technology is based on FC. The maximum density designs began with Ref. [20], which showed that the spacing should be

$$\frac{D}{L} \sim Be^{-1/4} \tag{3.3}$$

where $Be = \Delta P \cdot L^2/\alpha\mu$ is the pressure difference number, or Bejan number [21], which is analogous to viewing Ra as a temperature difference number [22]. Here, ΔP is the excess pressure maintained by a fan across the package. The highest heat generation density offered by this D spacing subject to the same excess temperature ceiling ($\Delta T = T_{max} - T_0$) is

$$q'''_{FC} \sim \frac{k}{L^2} \Delta T Be^{1/2} \tag{3.4}$$

Because $Be^{1/2}$ varies as L, we discover that q'''_{FC} varies as L^{-1}, as shown in Fig. 3.2.

Greater cooling densities by FC are possible in the direction of smaller elements L, toward miniaturization. This is not new relative to the evolution of NC cooling technology. New is the prediction (correct, in hindsight) that in the pursuit of greater densities in smaller elements, there must be a *transition* from natural to FC cooling and not from forced to NC.

The evolution of volumetric cooling does not end with FC through properly sized spacings, parallel plates, or other packed elements (cylinders, spheres, staggered or aligned, etc.): for a review, see Ref. [9]. It is possible to cool the L-scale body by pure conduction, as shown in the upper detail of Fig. 3.2. This time, the body generates heat volumetrically at the rate q'''_C, its peak temperature is T_{max}, and its thermal conductivity is k_0. To facilitate the flow of heat from the volume to one or more points on the side, channels (blades, pins, and trees) of a second material with much higher conductivity (k_p) are inserted in the k_0 material. The volume fraction occupied by the k_p blades in the L-scale body is $\phi \ll 1$. The ratio of conductivities is $\tilde{k} = k_p/k_0 \gg 1$.

In sum, conduction cooling is facilitated by designing the body as a composite material with two components (k_0 and k_p) and with *design*, or *organization*, which means the configuration of the k_p paths on the k_0 background. The composition is described by ϕ and $\tilde{k}$, while the design can evolve such that the heat generated by the volume flows more and more easily to the heat sink on the boundary. All such designs tend toward maximal densities of heat generation of order (Ref. [9], problem 5.5)

$$q'''_{\mathrm{C}} \sim k_0 \frac{\Delta T}{L^2} \left(\widetilde{k}\phi\right)^{1/2} \tag{3.5}$$

For the design with parallel high-conductivity blades shown in the upper part of Fig. 3.2, this q'''_{C} level is achieved when the spacing between blades has the scale $L\left(\widetilde{k}\phi\right)^{-1/2}$.

Equation (3.5) shows that higher densities of functionality continue on the path toward smaller scales over time. However, because q'''_{C} increases as L^{-2} when L decreases, and because the path toward greater densities is more direct than in designs with FC cooling, and certainly greater than in NC cooling, we discover that there must be a transition from FC to volumetric conduction. This step of technology evolution is one way, from q'''_{FC} to q'''_{C}, not the other way around.

The conclusion that follows from invoking the constructal law in designs that facilitate the flow of volumetric heat from a body to the outside is that the technology should evolve in time in two ways:

i. Toward smaller and smaller scales (miniaturization)

ii. Through stepwise changes (transitions) in heat-flow mechanisms

These changes should occur in a particular direction over time, NC → FC → *C*, in the same time direction as the morphing toward miniaturization (*i*). These changes should not end: Fourier conduction (*C*) is not an "end design," that is, not an ultimate technology. The stepwise evolution continues, for example, from Fourier conduction cooling (*C*) to conduction at nanoscales [23].

Interesting as this discovery with the constructal law may appear, it is deeply ironic because in my work, it did not happen this way. Conduction cooling with high-conductivity inserts was not discovered with the constructal law. It happened the other way around [1]: the constructal law was discovered with the conduction cooling designs (Fig. 3.3). Here is how:

I imagined the necessity of the evolution toward smaller scales for higher q''', and into that future, I saw the appearance of solid-body conduction with evolving heat-flow architectures from volume to point. To summarize this vision, at the end of my 1996 paper, I discovered the constructal law of design in nature (Ref. [1], p. 815). My 1996 paper began with these passages (from Ref. [1], p. 799), before I knew the constructal law:

International Journal of Heat and Mass Transfer

Volume 40, *No.* 4 (Published November 1, 1996) *March* 1997

CONTENTS

Figure 3.3 The back cover of the journal issue in which the volume-to-point flow problem and the constructal law were published in 1996 [1]. That year, the *International Journal of Heat and Mass Transfer* was experiencing an overflow of accepted papers. The overflow was accommodated by assigning 1997 numbers to published issues that were made available on the library shelves in 1996. This figure is needed because many authors mistake 1997 as the year of publication of the constructal law. (For color version of this figure, the reader is referred to the online version of this chapter.)

This paper is about one of those fundamental problems that suddenly appear 'obvious', but only after considerable technological progress has been made on pushing the frontier. The technology in this case is the cooling of electronics (components and packages), where the objective is to install a maximum amount of electronics (heat generation) in a fixed volume in such a way that the maximum temperature does not exceed a certain level. The work that has been done to devise cooling techniques to meet this objective is enormous and is continuing at an accelerated pace [. . .]. In brief, most of the cooling techniques that are in use today rely on convection or conjugate convection and conduction, where the coolant is either a single phase fluid or one that boils.

The frontier is being pushed in the direction of smaller and smaller package dimensions. There comes a point where miniaturization makes convection cooling *impractical, because the ducts through which the coolant must flow take up too much space. The only way to channel the generated heat out of the electronics package is by conduction. This conduction path will have to be very effective (of high thermal conductivity, k_p), so that the temperature difference between the hot spot (the heart of the package) and the heat sink (on the side of the package) will not exceed a certain value.*

Conduction paths also take up space. Designs with fewer and smaller paths are better suited for the miniaturization evolution. The fundamental problem addressed in this paper is this:

Consider a finite-size volume in which heat is being generated at every point and which is cooled through a small patch (heat sink) located on its boundary. A finite amount of high-conductivity (k_p) material is available. Determine the optimal distribution of k_p material through the given volume such that the highest temperature is minimized.

I will show that the solution to this problem is astonishingly simple and beautiful, with far reaching implications in physics, mathematics and the natural evolution of living systems.

The design evolution that began with formulating the volume-point heat-flow problem is discussed in the next section and in review articles and books [9,13,24].

3. TREE-SHAPED DESIGNS: CONDUCTION, FLUID FLOW, AND CONVECTION

The constructal law is a general statement of physics. It does not use words such as *tree*, *complex* versus *simple*, or *natural* versus *engineered*. There are several classes of flow configurations in nature, and each class can be derived from the constructal law in several ways: analytically or numerically, approximately or more accurately, blindly (random search) or using strategy (shortcuts), and so on. Classes that our group treated in detail, and by several methods, are the cross-sectional shapes of ducts, the cross-sectional shapes of rivers, internal spacings, and tree-shaped architectures in animate and inanimate flow systems.

Regarding the tree architectures, we treated them not as "models" but as fundamental problems of access to flow: volume to point (e.g., exhaling), area to point (e.g., river basin), line to point (e.g., electronics cooling), and the respective reverse flow directions. Important is the geometric notion that the "volume," the "area," and the "line" represent *infinities* of points.

The theoretical discovery of trees (in Ref. [1], and later) came from the decision to connect one point (source or sink) with an *infinity* of points (volume, area, and line). It is the reality of the continuum (the infinity of points) that is routinely discarded by modelers who approximate the space with a finite number of discrete points and then cover the space with drawings made of sticks, which cover the space incompletely (and from this, fractal geometry). The reality of nature requires a study of the interstitial spaces between the tree links. The interstices can only be bathed by high-resistivity diffusion (an invisible, disorganized flow), whereas the tree links serve as conduits for low-resistivity organized flow (visible streams and ducts).

The two modes of flowing with imperfection (irreversibility), the interstices and the links, must be *balanced* so that together they ease the global flow. The drawing—the flow architecture—is the graphical expression of the balance between channels and their interstices. The deduced architecture (tree, duct shape, spacing, etc.) is the *distribution* of imperfection over the available flow space. It is the architecture for access into and out of the flow space, which is finite. Those who model natural trees and then draw them as black lines on white paper (while not struggling to discover the layout of every black line on its allocated white patch) miss half of the drawing. The white is as important as the black.

The constructal-law discovery of tree-shaped flow architectures was based on three approaches. It started in 1996 with an analytic shortcut [1,2] based on several simplifying assumptions: 90° angles between stem and tributaries, a construction sequence in which smaller optimized constructs are retained, constant-thickness branches, and so on.

Months later, we solved the same problem numerically [25] by abandoning most of the simplifying assumptions (e.g., the compounding construction sequence) used in the first papers. In 1998, we reconsidered the problem in an area-to-point flow domain with randomly moving low-resistivity blocks embedded in a high-resistivity background [26,27] with Darcy flow (permeability instead of conductivity and resistivity) (Fig. 3.4). Grains of high resistivity were identified and replaced with grains of low resistivity in such a way that the global resistance of the area-to-point flow decreased in every frame of the evolutionary design. Along the way, we found better performance and trees that look more "natural" as we progress in time, that is, as we endowed the flow structure with more freedom to morph.

Figure 3.5 shows the most recent tree design for conduction in a heat-generating medium with high-conductivity channels that are the most free to morph [28]. Darcy fluid flow is one form of "diffusion," that is, the same physics phenomenon as thermal diffusion (Fourier conduction) and

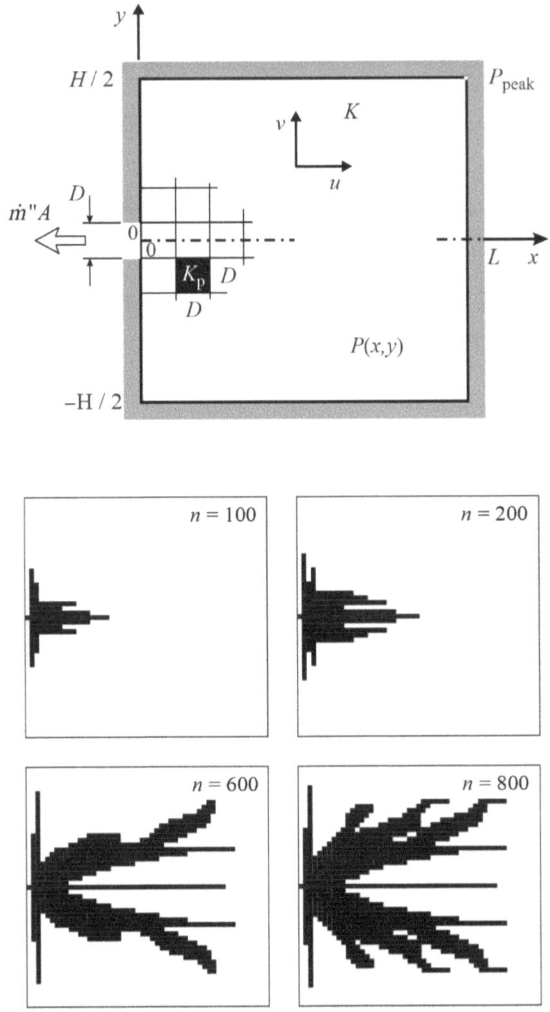

Figure 3.4 Darcy flow on a square domain with low permeability (K) and high permeability (K_p). In time, K grains are searched and replaced by K_p grains such that the overall area-to-point flow access is increased faster [26,27].

electrical diffusion (Ohm conduction). Yet, a disturbing trend in publishing is to copy the original work [26,27], change a few key words (i.e., "translate"), and publish the same idea brazenly as "new" after replacing the Darcy diffusion notation with thermal diffusion notation (as Guo *et al.* [29] have done; a related critique of Guo *et al.* [29] was just published by Grazzini *et al.* [30]).

The constructal literature is expanding rapidly in the domain of tree-shaped designs for conduction [31–34], fluid flow [35–43], and convective

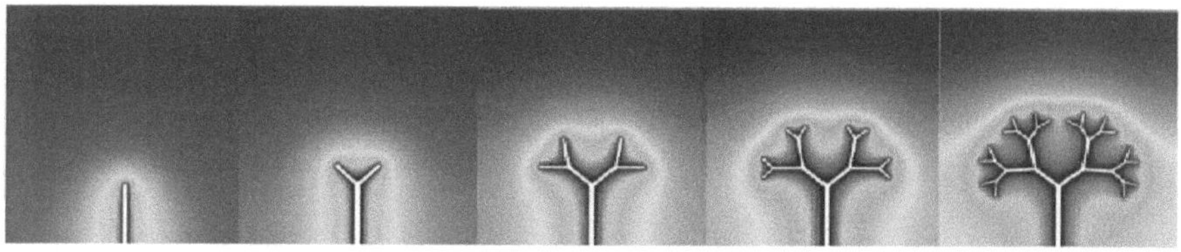

Figure 3.5 Constructal invasion of a conducting tree into a conducting body [28]. (For color version of this figure, the reader is referred to the online version of this chapter.)

heat transfer [44–48]. A central feature of these designs is the notion that when channels bifurcate or merge, their diameters should change by precise factors, so that the overall flow through the architecture is facilitated. The best known design rule of this kind is the Hess–Murray rule ($D_1/D_2 = 2^{1/3}$) for selecting the ratio of channel diameters at a bifurcation [9]. This rule was extended by constructal design in several directions: to junctions with n branches ($D_1/D_2 = n^{1/3}$), to bifurcations with two identical branches (length L_2, diameter D_2) and one stem (L_1, D_1) on an area of fixed size ($L_1 \times 2L_2$) in fully developed laminar flow [9],

$$\frac{D_1}{D_2} = 2^{1/3}, \quad \frac{L_1}{L_2} = 2^{1/3} \tag{3.6}$$

and fully developed turbulent flow [9],

$$\frac{D_1}{D_2} = 2^{3/7}, \quad \frac{L_1}{L_2} = 2^{1/7} \tag{3.7}$$

and to bifurcations with unequal branches (L_2, D_2 and L_3, D_3) (problem 4.4 in Ref. [9]). All these developments come from evolving the flow configuration in accord with the constructal law, toward providing greater access, which led analytically to minimal flow resistance in T- and Y-shaped (and more complicated) constructs of tubes and other channels, as in the trees matched canopy to canopy of Ref. [49].

The constructal design of the T- or Y-shaped constructs of tubes (Eqs. 3.6 and 3.7) is the same as the uniform distribution of fluid residence time in the channels. This means that the time (t_1) spent by the fluid in the D_1 tube is the same as the time (t_2) spent in the D_2 tube. The residence time in any tube is $t \sim L/U \sim LD^2/\dot{m}$, where U and $\dot{m}$ are the mean fluid speed and mass flow rate. Next, at a bifurcation, we note $\dot{m}_2/\dot{m}_1 = 1/2$, and with the laminar flow architecture of Eq. (3.1), we obtain

$$\frac{t_1}{t_2} \sim \frac{L_1}{L_2}\frac{\dot{m}_2}{\dot{m}_1}\frac{D_1^2}{D_2^2} = 2^{1/3}\frac{1}{2}\left(2^{1/3}\right)^2 = 1 \tag{3.8}$$

Similarly, for the turbulent flow design of Eq. (3.7), we obtain

$$\frac{t_1}{t_2} \sim \frac{L_1}{L_2}\frac{\dot{m}_2}{\dot{m}_1}\frac{D_1^2}{D_2^2} = 2^{1/7}\frac{1}{2}\left(2^{3/7}\right)^2 = 1 \tag{3.9}$$

Svelteness [9] is a new property of flow architectures that was brought to light by the constructal design of flow architectures. The svelteness Sv is the ratio of the external length scale of the flow design (e.g., $A^{1/2}$, if the area of the flow layout is A) divided by the internal length scale of the flow design (e.g., $V^{1/3}$, if the total volume occupied by the flow is V). A flow architecture has three main characteristics: sizes, aspect ratios (shapes), and svelteness, that is, the relative thinness of the lines of its drawing. Svelteness is intimately tied to the flow performance of the architecture [9,50].

It is important to keep in mind the actual constructal-law statement and its phenomenon (design in nature), because there is a tendency in the literature to confuse the constructal law with "tree networks." While it is true that the first flow designs derived from the constructal law were tree-shaped [1], the design that emerged was a "tree" only because the flow was between one point and an area or volume (an infinity of points). At about the same time, we showed (with high-conductivity inserts, as in Ref. [1]) that if the conducting body is cooled from one side surface, not from one point, the inserts must be parallel plates, or needles, as in Figs. 3.2 (top) and 3.6 [51].

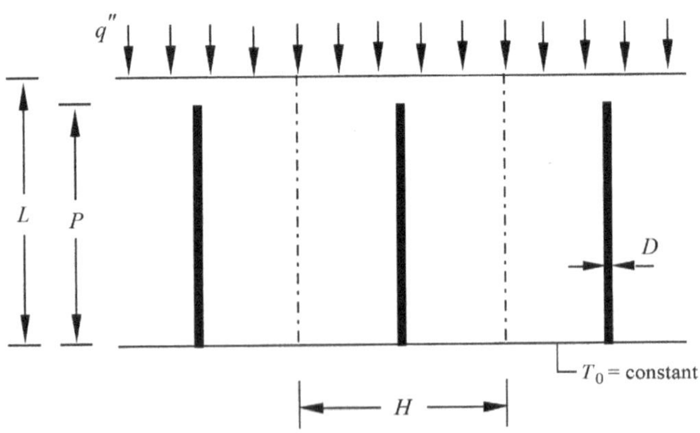

Figure 3.6 When the heating or cooling of a conducting body is from a side surface, the high-conductivity inserts are parallel plates or needles, not trees [51].

Table 3.1 How, not what: the balancing of high-resistivity flow with low-resistivity flow in a wide diversity of flow systems

		How	
Application	**What**	**Interstices High resistance at the Smallest Scale**	**Channels Low resistance at Larger Scales**
Electronics packages	Heat	Low-conductivity substrate	High-conductivity inserts (blades and needles)
River basins	Water	Darcy flow, porous medium	Rivulets and rivers
Lungs	Air	Diffusion in alveoli, tissues	Bronchial passages
Circulatory systems	Blood	Diffusion in capillaries, tissues	Blood vessels, capillaries, arteries, veins
Turbulent flow	Momentum	Laminar, viscous diffusion	Streams, rolls, eddies
Urban traffic	People	Walking in urban structure	Street traffic
Economics	Goods	Hand delivery and collection	Freight, rail, truck, air, ship

Courtesy of Mr. Stephen Perin.

In drawings such as Figs. 3.4–3.6, we discover patterns, trees, and needles, because they are the architectures that facilitate flow access. Every detail of the tree geometry is the result of invoking the constructal law. The discovery of the tree as the flow architecture for greater access between one point and an infinity of points is general—it is not restricted to trees of streets and trees of high-conductivity inserts. The generality of the tree discovery is stressed by Table 3.1, which shows that "how" unites and "what" divides. How the tree is generated (through a balance between high-resistivity flow and low-resistivity flow) is the same in many classes of flow systems, regardless of the diversity of the currents that flow through them. To see the principle that unites is harder than to see the "diversity" of flow systems in nature.

4. FREE CONVECTION: AN ENGINE + BRAKE SYSTEM

Technology evolution is about us, about the evolutionary design of all the flows and movements that facilitate the persistence of human flow (life) on the Earth's surface (people, goods, material, etc.). Nothing moves by itself.

Everything that moves does so because it is pushed or pulled. The force times the distance of movement is the work dissipated (destroyed) by the movement.

No design, no movement is "for free." This may come as a surprise to us in the thermal sciences, where we often speak of free fall and free convection. Here, I show that although invisible, the driving mechanism (the engine) for NC is present, and it is a work-producing engine like all the engines that drive the fans and pumps of FC.

A simple configuration with NC is in Fig. 3.7. Think of a body of temperature T_0 and height H immersed in a fluid of temperature T_∞. For a more meaningful discussion, think of a heat-generating body immersed in a cold fluid reservoir, such as an old-fashioned stove in the middle of a room. Since air at constant pressure expands upon being heated, the air layer adjacent to the wall expands (becomes lighter and less dense) and rises. At the same time, the cold reservoir fluid is displaced downward. Thus, the wall–reservoir temperature difference drives the all-familiar "natural circulation" or free-convection cell sketched in Fig. 3.7. In view of the constructal law, it is appropriate to ask the question: Who or what is responsible for the movement known as NC?

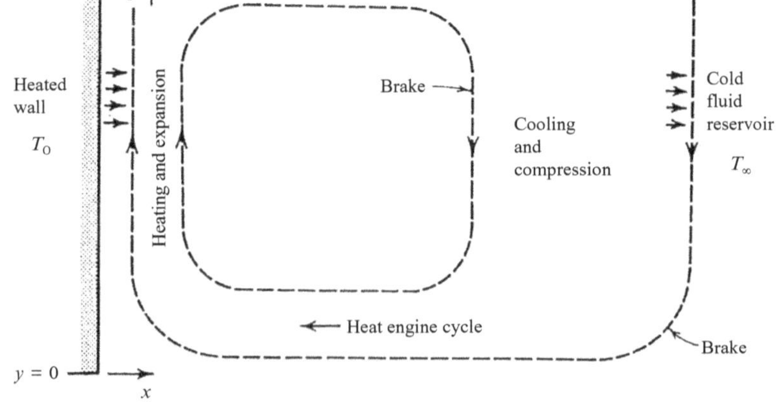

Figure 3.7 Heat engine responsible for driving natural convection.

To answer this question, while writing my 1984 convection book [18], I proposed to follow the evolution of a fluid packet through the imaginary duct that holds the flow. Starting from the bottom of the heated wall, the packet is heated by the wall and expands as it rises to lower pressures in the hydrostatic pressure field maintained by the reservoir. Later on, along the downflowing branch of the cycle, the fluid packet is cooled by the reservoir and compressed as it reaches the depths of the reservoir. From the circuit executed by each fluid packet, we learn that the flow is the succession of four processes:

$$\text{heating} \rightarrow \text{expansion} \rightarrow \text{cooling} \rightarrow \text{compression}$$

In sum, the fluid packet traveling along the flow loop of Fig. 3.7 is equivalent to the cycle executed by the working fluid in a heat engine modeled as a closed system. This heat engine cycle should be capable of delivering work, for example, if we insert a suitably designed propeller in the stream: this is the origin of the "wind power" discussed in connection with the harnessing of solar work indirectly from the atmospheric heat engine loop. In the absence of work-collecting devices (e.g., windmill wheels), the heat engine cycle drives its working fluid fast enough so that its entire work output is destroyed because of irreversibilities due to friction between adjacent fluid layers and heat transfer along finite temperature gradients. The entire circulation pattern of Fig. 3.7 is an infinity of nested heat engine wheels with friction between them.

The legacy of inanimate flow systems on Earth is the same as that of animate flow systems. All move mass by destroying exergy that originates from the sun. Rivers and animals use and destroy exergy in proportion with the moved weight (Mg) times the horizontal displacement (L). The same holds for our vehicles, on land, in air, and in water. The spent fuel is proportional to the weight of the vehicle times the distance traveled.

River and animal designs morphed and perfected themselves over millions of years. Engineered designs are evolving right now, in our minds, on design tables, and in factories. The vehicle consumes fuel in proportion with the high-temperature heating rate Q_H. This stream of heat is converted partially into mechanical power delivered to the wheels (W), while the remainder (Q_L or $Q_H - W$) is dissipated into the environment. Not mentioned in thermodynamics until recently [13] is the fact that W is itself dissipated into the ambient, because of the movement of the vehicle weight (Mg) over the horizontal distance L (Fig. 3.8).

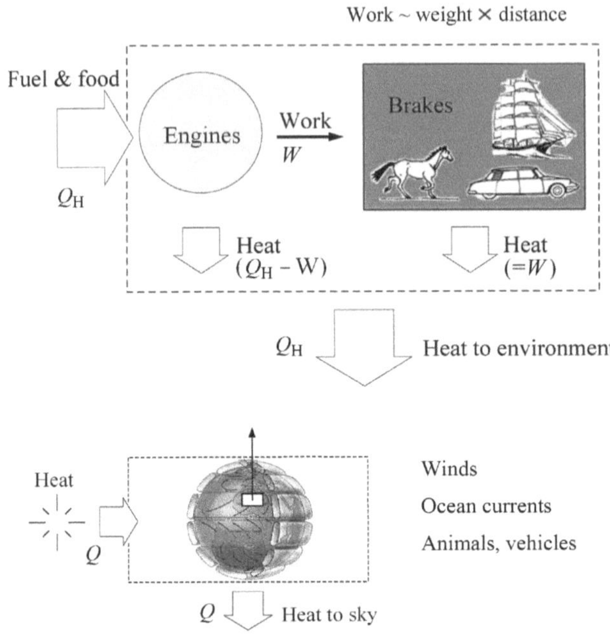

Figure 3.8 The whole Earth is an engine+brake system, containing innumerably smaller "engine+brake" systems (winds, ocean currents, animals, and human and machine species). (For color version of this figure, the reader is referred to the online version of this chapter.)

To summarize, all the high-temperature heating that comes from burning fuel (Q_H) is ultimately transmitted to lower temperature into the environment, all of it, 100%. This is evident in Fig. 3.8 and any other "heat transfer" configuration. The need for higher efficiencies in power generation (greater W/Q_H) is the same as the need to have more W, that is, the need to move more weight over greater distances on the surface of the Earth. This is the natural phenomenon (the tendency) summarized in the constructal law.

At the end of the day, when all the fuel has been burned, and all the food has been eaten, this is what animate flow systems have achieved. They have moved mass on the surface of the Earth (they have "mixed" the Earth's crust) more than in the absence of animate flow systems.

The moving animal or vehicle is the equivalent to an engine connected to a brake (Fig. 3.8), first proposed in my book *Solved Problems in Thermodynamics* (1976) and later [2]. The power generated by muscles and motors is ultimately and necessarily dissipated by rubbing against the environment. There is no taker for the W produced by the animal and vehicle. This is why in Fig. 3.9, the GDP of a country is roughly proportional to the amount of fuel burned in that country [13].

Figure 3.8 holds also for the whole Earth, as a closed thermodynamic system (note: closed does not mean isolated). Earth, with its solar heat input, heat rejection, and wheels of atmospheric and oceanic circulation, is a heat engine without shaft. Its mechanical power output cannot be delivered to an extraterrestrial system. Instead, the Earth engine must dissipate through air, water, and solid friction and other irreversibilities (e.g., heat leaks and mass diffusion) all the mechanical power that it produces. It does so by "spinning in its brake" as fast as is necessary (and from this follow the winds and the ocean currents, which proceed along easier and easier routes, and flow at finite, characteristic speeds, never getting out of hand). This is the constructal-law basis of all NC phenomena.

In the human and nonhuman biosphere (power plants, animals, vegetation, and water flow), the engines have shafts, rods, legs, and wings that

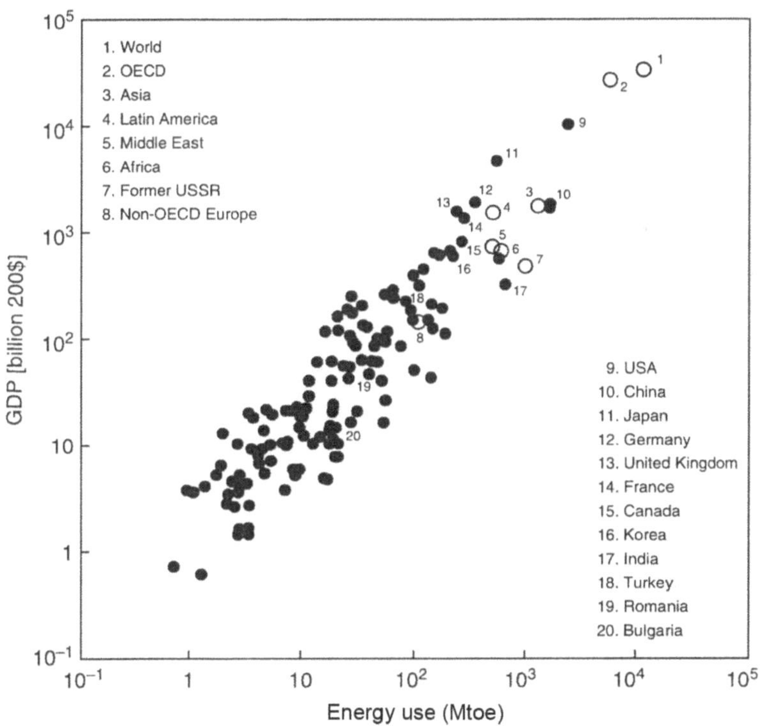

Figure 3.9 Economic activity means fuel that is being burned. The annual gross domestic product of regions and countries all over the globe versus their annual consumption of fuel.

deliver the mechanical power to external entities that use the power (e.g., vehicles and animal bodies needing propulsion). Because the engines of engineering and biology are constructal, they morph in time toward easier flowing configurations. They evolve toward producing more mechanical power (under finiteness constraints), which, for them, means a time evolution toward *less* dissipation or greater efficiency.

Outside the engineering or biology engine, all the mechanical power is destroyed through friction and other irreversibility mechanisms (e.g., transportation and manufacturing for humans, animal locomotion, and body heat loss to ambient). The engine and its immediate environment (the "brake"), as one thermodynamic system, are analogous to the entire Earth (cf. Fig. 3.8). The flowing Earth (with all its engine + brake components, rivers, fish, turbulent eddies, etc.) accomplishes as much as any other flow architecture, animate, or inanimate: It mixes the Earth's crust most effectively—more effectively than in the absence of constructal phenomena of generation of flow configuration.

The movement of animals—the flow of animal mass on Earth—is analogous to other moving and mixing designs such as the turbulent eddies in rivers, oceans, and the atmosphere. It is not an exaggeration to regard animals as self-driven packs of water, that is, motorized vehicles of water mass, which spread and mix the Earth like the eddies in the ocean and the atmosphere.

Irrefutable evidence in support of this unifying view is that all these moving packs of biological matter have morphed and spread over larger areas, depths, and altitudes, in this remarkable sequence over time, that is, fish in water, walking fish and other animals on land, flying animals in the atmosphere, human and machine species in the air, and human and machine species in the outer space, and not in the opposite time direction. The balanced and intertwined flows that generate our engineering, economics, and social organization are not different than the natural flow architectures of biology (animal design) and geophysics (river basins and global circulation) [1–5].

Returning to Fig. 3.7, we see the fundamental difference between FC and NC. In FC, the engine that drives the flow is external, whereas in NC, the engine is built into the flow itself.

5. CONSTRUCTAL LAW, DESIGN IN NATURE, AND COMPLEXITY

This chapter focused on the phenomenon of technology evolution in order to illustrate how the constructal law governs design and evolution in nature. Technology evolution is about the evolving design of the human movement on the Earth's surface: people, goods, material, construction, mining, etc. As the whole vehicle or animal evolves its architecture to become better at moving mass on the landscape, the rising line of Fig. 3.1 rotates downward as shown in Fig. 3.10. The intersection between the two competing trends shifts downward and to the right, and the total penalty decreases. In the direction of a more efficient whole, the whole vehicle or animal is larger [17], lives longer, and covers a greater distance [52]; the component is relatively larger (Fig. 3.10); and the scaling rule that larger components belong on larger vehicles (animals) is preserved.

More broadly, the chapter points to the unifying power of the constructal law and its applications in all the domains of design generation and evolution, from biology and geophysics to globalization, engineering, sustainability, and security. This growing activity covers the board, from physics and biology to social organization and technology evolution [3–17].

The constructal law has generated a worldwide movement toward design as science, that is, design as a physics phenomenon as captured in 1996 by the constructal law (Fig. 3.3): "For a finite-size flow system to persist in time (to live), its configuration must evolve in such a way that it provides easier access to the imposed (global) currents that flow through it."

Life is movement and the persistent morphing of the configuration of this movement [3]. The constructal law (a) identifies the design and evolution (changes in configuration) as a physical phenomenon and (b) captures the time direction of design generation and evolution.

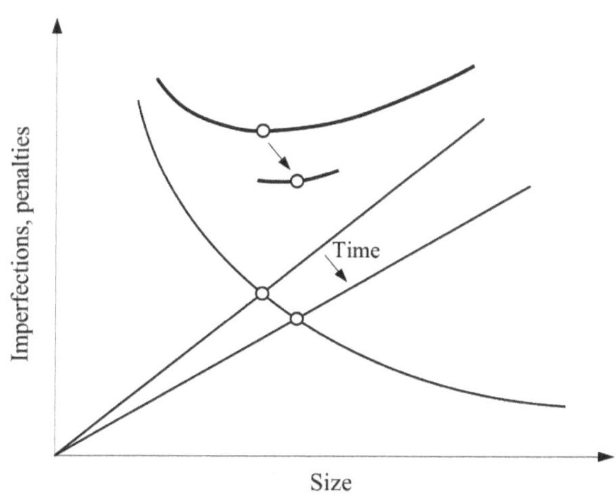

Figure 3.10 Technology evolution also occurs at the vehicle (animal) level. As the vehicle improves over time, the fuel penalty associated with the component decreases.

To place the constructal law and its field in the greater framework of scientific inquiry, it is timely to review what we mean by design in nature and by other words that refer to design in nature: complexity, networks, diversity, chance, turbulence, etc. These words are old and numerous because the fascination with the surroundings has inspired human curiosity and creativity throughout history. Science is only the latest and most powerful mental construct that came out of this natural tendency to understand and use the surroundings in order to move more easily, farther, and more persistently in time through the surroundings. Science is an evolutionary design [15,53,54].

Design has two meanings in English. The first is the noun, which means organization, shape, structure, configuration, pattern, drawing, figure, rhythm, motif, architecture, and many more words that represent the mental viewing of an image—sharp lines on a background of a different color. Design in nature is about this. Science began with images: from cave paintings to geometry (the science of figures) and mechanics (the science of contrivances made out of moving figures). We think, we create, and we speak in terms of images. Design in nature is about this, the images. The very fact that these images have names—river basin, lung, snowflake—means that we all know what they are individually even though they all look like trees.

The second meaning is the verb "to design," which is about the human activity of creating images and contrivances that are useful. This verb refers strictly to what people do on a design project, for example, in engineering, where along with the verb "to design" comes "the designer" as one or many. This second meaning is not the object of the constructal law. Design in nature is not about "to design" and "the designer."

The constructal law is about predicting the design (the flow configuration) and its evolution in time. The constructal law is about why organization and arrangement happen. Constructal "theory" is not the same as constructal "law." Constructal *theory* is the view that the constructal *law* is correct and reliable in a predictive sense in a particular flow system. For example, reliance on the constructal law to predict the evolving architecture of the snowflake is the constructal theory of rapid solidification. Using the constructal law to predict the architecture of the lung and the rhythm of inhaling and exhaling is the constructal theory of respiration.

The law is one, and the theories are many—as many as the phenomena that the thinker wishes to predict by invoking the law.

The constructal law is predictive, not descriptive. This is the big difference between the constructal law and other views of design in nature. Previous attempts to explain design in nature are based on empiricism:

observing first and explaining after. They are backward-looking, static, descriptive, and at best explanatory. They are not predictive theories even though some are called "theory," for example, complexity theory, network theory, chaos theory, power laws (allometric scaling rules), general models, and optimality statements (minimum, maximum, and optimum).

With the constructal law, complexity and scaling rules are discovered, not observed. Complexity is finite (modest) and is part of the description of the constructal design that emerges. If the flows are between points and areas or volumes, the constructal designs that are discovered are tree-shaped networks. The "networks" are discovered, not observed, and not postulated. Networks, scaling rules, and complexity are part of the description of the world of constructal design that emerges predictively from the constructal law.

The constructal law is not about optimality, destiny, or end design. It is about the fact that the generation and evolution of design never ends. The phenomenon of design in nature is dynamic, forever changing. With the constructal law, we anticipate the evolving design and its direction in time.

> *We cross our bridges when we come to them and burn them behind us, with nothing to show for our progress except a memory of the smell of smoke, and a presumption that once our eyes watered.*
>
> ***Tom Stoppard,* Rosencrantz and Guildenstern are Dead**

REFERENCES

[1] A. Bejan, Constructal-theory network of conducting paths for cooling a heat generating volume, Int. J. Heat Mass Transfer 40 (1996) 799–816 (published 1 Nov. 1996, cf. Fig. 3).

[2] A. Bejan, Advanced Engineering Thermodynamics, second ed., Wiley, New York, 1997.

[3] T. Basak, The law of life: the bridge between physics and biology, Phys. Life Rev. 8 (2011) 249–252.

[4] A. Bejan, J.P. Zane, Design in Nature. How the Constructal Law Governs Evolution in Biology, Physics, Technology, and Social Organization, Doubleday, New York, 2012.

[5] A.H. Reis, Constructal theory: from engineering to physics, and how flow systems develop shape and structure, Appl. Mech. Rev. 59 (2006) 269–282.

[6] D. Queiros-Conde, M. Feidt (Eds.), Constructal Theory and Multi-Scale Geometries: Theory and Applications in Energetics, Chemical Engineering and Materials, Les Presses de L'ENSTA, Paris, 2009.

[7] P. Kalason, Le Grimoire des Rois: Théorie Constructale du Changement, L'Harmattan, Paris, 2007.

[8] P. Kalason, Épistémologie Constructale du Lien Cultuel, L'Harmattan, Paris, 2007.

[9] A. Bejan, S. Lorente, Design with Constructal Theory, Wiley, Hoboken, 2008.

[10] A. Bachta, J. Dhombres, A. Kremer-Marietti, Trois Ètudes sur la Loi Constructale d'Adrian Bejan, L'Harmattan, Paris, 2008.

[11] L. Rocha, Convection in Channels and Porous Media. Analysis, Optimization, and Constructal Design, VDM Verlag, Saarbrücken, 2009.
[12] G. Lorenzini, S. Moretti, A. Conti, Fin Shape Optimization Using Bejan's Constructal Theory, Morgan & Claypool Publishers, San Francisco, 2011.
[13] A. Bejan, S. Lorente, The constructal law and the evolution of design in nature, Phys. Life Rev. 8 (2011) 209–240.
[14] A. Bejan, S. Lorente, The constructal law origin of the logistics S curve, J. Appl. Phys. 110 (2011) 024901.
[15] A. Bejan, S. Lorente, The physics of spreading ideas, Int. J. Heat Mass Transfer 55 (2012) 802–807.
[16] E. Cetkin, S. Lorente, A. Bejan, The steepest S curve of spreading and collecting: discovering the invading tree, not assuming it, J. Appl. Phys. 111 (2012) 114903.
[17] A. Bejan, S. Lorente, B.S. Yilbas, A.S. Sahin, The effect of size on efficiency: power plants and vascular designs, Int. J. Heat Mass Transfer 54 (2011) 1475–1481.
[18] A. Bejan, Convection Heat Transfer, Wiley, New York, 1984.
[19] A. Bejan, Convection Heat Transfer, fourth ed., Wiley, Hoboken, 2013.
[20] A. Bejan, E. Sciubba, The optimal spacing of parallel plates cooled by forced convection, Int. J. Heat Mass Transfer 35 (1992) 3259–3264.
[21] S. Bhattacharjee, W.L. Grosshandler, The formation of a wall jet near a high temperature wall under microgravity environment, ASME HTD 96 (1988) 711–716.
[22] S. Petrescu, Comments on the optimal spacing of parallel plates cooled by forced convection, Int. J. Heat Mass Transfer 37 (1984) 1283.
[23] L. Gosselin, A. Bejan, Constructal trees at micro and nanoscales, J. Appl. Phys. 96 (2004) 5852–5859.
[24] A. Bejan, S. Lorente, Constructal law of design and evolution: Physics, biology, technology, and society, J. Appl. Phys. 113 (2013) 151301.
[25] G.A. Ledezma, A. Bejan, M. Errera, Constructal tree networks for heat transfer, J. Appl. Phys. 82 (1997) 89–100.
[26] M.R. Errera, A. Bejan, Deterministic tree networks for river drainage basins, Fractals 6 (1998) 245–261.
[27] A. Bejan, Shape and Structure from Engineering to Nature, Cambridge University Press, Cambridge, UK, 2000.
[28] H. Kobayashi, S. Lorente, R. Anderson, A. Bejan, Trees and serpentines in a conducting body, Int. J. Heat Mass Transfer 56 (2013) 488–494.
[29] Z.-Y. Guo, H.-Y. Zhu, S.-G. Liang, Entransy—a physical quantity describing heat transfer ability, Int. J. Heat Mass Transfer 50 (2007) 2545–2556.
[30] G. Grazzini, R. Borchiellini, U. Lucia, Entropy versus entransy, J. Non-Equilib. Thermodyn. 38 (2013) 259–271.
[31] X. Xianghua, L. Xingang, R. Jianxun, Optimization of heat conduction using combinatorial optimization algorithms, Int. J. Heat Mass Transfer 50 (2007) 1675–1682.
[32] M.M. Fyrillas, Heat conduction in a solid slab embedded with a pipe of general cross-section: shape factor and shape optimization, Int. J. Eng. Sci. 46 (2008) 907–916.
[33] M. Eslami, K. Jafarpur, Optimal distribution of imperfection in conductive constructal designs of arbitrary configurations, J. Appl. Phys. 112 (2012) 104905.
[34] L. Combelles, S. Lorente, A. Bejan, Leaflike architecture for cooling a flat body, J. Appl. Phys. 106 (2009) 044906.
[35] B.V.K. Reddy, P.V. Ramana, A. Narasimhan, Steady and transient thermo-hydraulic performance of disc with tree-shaped micro-channel networks with and without radial inclination, Int. J. Therm. Sci. 47 (2008) 1482–1489.
[36] A. Sciacovelli, V. Verda, Entropy generation minimization for the optimal design of the fluid distribution system in a circular MCFC, Int. J. Thermodyn. 14 (2011) 167–177.

[37] H.R. Williams, R.S. Trask, P.M. Weaver, I.P. Bond, Minimum mass vascular networks in multifunctional materials, J. R. Soc. Interface 5 (2008) 55–65.
[38] P. Bieupoude, Y. Azoumah, P. Neveu, Optimization of drinking water distribution networks: computer-based methods and constructal design, computers, Environ. Urban Syst. 36 (2012) 434–444.
[39] A.F. Miguel, Dendritic structures for fluid flow: laminar, turbulent and constructal design, J. Fluids Struct. 26 (2010) 330–335.
[40] C. Bai, L. Wang, Constructal structure of nanofluids, J. Appl. Phys. 108 (2010) 074317.
[41] J. Fan, L. Wang, Constructal design of nanofluids, Int. J. Heat Mass Transfer 53 (2010) 4238–4247.
[42] C. Bai, L. Wang, Constructal design of particle volume fraction in nanofluids, J. Heat Transfer 131 (2009) 112402.
[43] F. Wu, L. Chen, A. Shu, X. Kan, K. Wu, Z. Yang, Constructal design of stack filled with parallel plates in standing-wave thermo-acoustic cooler, Cryogenics 49 (2009) 107–111.
[44] P. Xu, X.Q. Wang, A.S. Mujumdar, C. Yap, B.M. Yu, Thermal characteristics of tree-shaped microchannel nets with/without loops, Int. J. Therm. Sci. 48 (2009) 2139–2147.
[45] H. Ghaedamini, M.R. Salimpour, A. Campo, Constructal design of reverging microchannels for convective cooling of circular disc, Int. J. Therm. Sci. 50 (2011) 1051–1061.
[46] S. Tescari, N. Mazet, P. Neveu, Constructal theory through thermodynamics of irreversible processes framework, Energy Convers. Manag. 52 (2011) 3176–3188.
[47] A. Nakayama, F. Kuwahara, W. Liu, A macroscopic model for countercurrent bioheat transfer in a circulatory system, J. Porous Media 12 (2009) 289–300.
[48] H. Kobayashi, S. Lorente, R. Anderson, A. Bejan, Freely morphing tree structures in a conducting body, Int. J. Heat Mass Transfer 55 (2012) 4744–47523.
[49] S. Lorente, W. Wechsatol, A. Bejan, Tree-shaped flow structures designed by minimizing path lengths, Int. J. Heat Mass Transfer 45 (2002) 3299–3312.
[50] H. Ghaedamini, M.R. Salimpour, A.S. Mujumdar, The effect of svelteness on the bifurcation angles role in pressure drop and flow uniformity of tree-shaped microchannels, Appl. Therm. Eng. 31 (2011) 708–716.
[51] M. Neagu, A. Bejan, Constructal placement of high-conductivity insert in a slab: optimal design of "roughness" J. Heat Transfer 123 (2001) 1184–1189.
[52] A. Bejan, Why the bigger live longer and travel farther: animals, vehicles, rivers and the winds. Nat. Sci. Rep. 2 (2012) 594, http://dx.doi.org/10.1038/srep00594.
[53] A. Bejan, Science and technology as evolving flow architectures, Int. J. Energy Res. 33 (2009) 112–125.
[54] A. Bejan, Two hierarchies in science: the free flow of ideas and the academy, Int. J. Des. Nat. Ecodyn. 4 (2009) 386–394.

CHAPTER FOUR

Recent Advances in Vapor Chamber Transport Characterization for High-Heat-Flux Applications

Justin A. Weibel, Suresh V. Garimella
Cooling Technologies Research Center, An NSF IUCRC, School of Mechanical Engineering and Birck Nanotechnology Center, Purdue University, West Lafayette, Indiana, USA

Contents

Abstract

Owing to their high reliability, simplicity of manufacture, passive operation, and effective heat transport, flat heat pipes and vapor chambers are used extensively in the thermal management of electronic devices. The need for concurrent size, weight, and

Advances in Heat Transfer, Volume 45
ISSN 0065-2717
http://dx.doi.org/10.1016/B978-0-12-407819-2.00004-9

performance improvements in high-performance electronics systems, without resort to active liquid-cooling strategies, demands passive heat-spreading technologies that can dissipate extremely high heat fluxes from small hot spots. In response to these daunting application-driven trends, a number of recent investigations have focused on the design, characterization, and fabrication of ultrathin vapor chambers for proximate heat spreading away from these hot spots. The predominant transport mechanisms and operational limits have been found to be different under these conditions relative to conventional low-power heat pipes. Noteworthy advances in the fundamental understanding of evaporation and boiling from porous microstructures fed by capillary action and improvements in vapor chamber characterization, modeling, design, and fabrication techniques are reviewed. Characterization of evaporation and boiling from idealized and realistic wick structures, observation of vapor formation regimes as a function of operating conditions, assessment of fluid dryout limitations, design of novel multiscale and nanostructured wicks for enhanced transport, and incorporation of these high-heat-flux transport phenomena into device-level models are discussed. These recent developments have successfully extended the maximum operating heat flux limits of vapor chambers.

NOMENCLATURE

$\boldsymbol{A}$ area
$\boldsymbol{A_{c,p}}$ characteristic pore area ($\pi D_p^2/4$)
$\boldsymbol{C}$ constant
$\boldsymbol{D}$ particle diameter
$\boldsymbol{D_p}$ pore diameter ($0.42D$)
$\boldsymbol{h}$ height
$\boldsymbol{h_{lv}}$ latent heat of vaporization
$\boldsymbol{k}$ thermal conductivity
$\boldsymbol{k_{eff}}$ effective thermal conductivity
$\boldsymbol{K}$ permeability
$\dot{m}_i''$ mass flux
$\bar{\boldsymbol{M}}$ molecular weight
$\boldsymbol{P}$ pressure
$\boldsymbol{\Delta P}$ capillary pressure
$\boldsymbol{P_p}$ characteristic pore perimeter (πD_p)
q'' heat flux
$\boldsymbol{R}$ thermal resistance, radius of curvature
$\bar{\boldsymbol{R}}$ universal gas constant
$\boldsymbol{t}$ time
$\boldsymbol{T}$ temperature
$\boldsymbol{T_{ref}}$ vapor reference temperature
$\boldsymbol{T_{sat}}$ saturation temperature
$\boldsymbol{T_{substrate}}$ substrate temperature
$\boldsymbol{\Delta T_{sl}}$ surface to liquid–vapor saturation temperature drop
$\boldsymbol{\Delta T^*}$ nucleation temperature drop ($4\sigma T_{sat}/\rho_g h_{lg} D_p$)
$\boldsymbol{L_e}$ characteristic length $[(D_p(A_{c,p}/P_p)^4)^{0.2}]$

GREEK SYMBOLS

$\boldsymbol{\delta}$ thickness
$\boldsymbol{\varepsilon}$ porosity
$\boldsymbol{\Phi}$ wedge angle
$\boldsymbol{\mu}$ dynamic viscosity
$\boldsymbol{\theta}$ contact angle
$\boldsymbol{\rho}$ density
$\boldsymbol{\sigma}$ surface tension
$\hat{\boldsymbol{\sigma}}$ accommodation coefficient
$\boldsymbol{\upsilon}$ kinematic viscosity

SUBSCRIPTS

Cu copper
i interface
l liquid
v vapor

1. INTRODUCTION

The high effective thermal conductance offered by heat pipes and other passively driven multiphase fluid loop heat transport devices has spurred their utilization across a range of thermal management applications for purposes of heat conveying and spreading, temperature control and equalization, and energy harvesting and storage. The advantageous device heat transport characteristics in all such applications stem from the thermodynamic cycle of vaporization and condensation of an enclosed working fluid that serves as the primary energy carrier.

Such devices are evacuated and charged with an internal working fluid as part of the assembly process. During operation, the working fluid is vaporized via evaporation at the location of heat input, which results in a local increase in the vapor pressure. The elevated internal pressure forces vapor to travel toward the condenser region where heat is rejected via condensation. By taking advantage of the highly effective phase-change heat transfer across the liquid–vapor interfaces, this heat transport process can occur under near-isothermal conditions at a small saturation pressure gradient. Continuous cyclical operation is passively achieved since the condensed liquid is returned to the heat input section by capillary action in a porous wick material that connects the vaporization and condensation interfaces, by gravity, or by other means. The capillary pressure generated by the pores of a wick material must overcome the cumulative viscous and inertial pressure drops along the vapor and liquid flow paths to sustain operation at any given heat transport rate. When this condition and several other operating limits are satisfied, such as avoiding entrainment of liquid into the vapor flow and choking at sonic vapor velocities, order-of-magnitude gains in overall effective thermal conductance can be realized compared to solid heat-spreading materials of the same shape.

The term thermosyphon typically refers to devices where the liquid return occurs by gravity, while the term heat pipe is used to refer to other liquid return schemes. The most common heat pipe configuration is a sealed cylindrical tube with axial capillary grooves or porous wick material lining the inner wall and is intended for bulk transport of heat along the device length. A wide variety of nonconventional heat pipe types have been developed due to the need for different operational characteristics and form factors. Examples including variable conductance gas-loaded heat pipes for constant temperature control, capillary pumped loops for wickless transport over long working distances, rotating heat pipes in which liquid return occurs by centrifugal force, micro heat pipes, thermal diodes, oscillating heat pipes, and others are described in detail in Refs. [1–4]. Recent reviews summarize research on emerging heat pipe applications [5,6], novel working fluid developments [7], advances in modeling approaches [8,9], and performance of archetypal device geometries, such as loop heat pipes [10,11], micro heat pipes [12], and oscillating heat pipes [13]. A comprehensive summary of all such device types, applications, and analysis methodologies is outside the scope of the current review; instead, the focus is on the burgeoning interest and developments in heat pipes targeted at spreading of very high heat fluxes in ultrathin form factors, as discussed in the following section.

1.1. Ultrathin vapor chambers for thermal management of electronics

The most widespread application of heat pipes is in the thermal management of electronics. Heat pipes for this application are typically mass-produced for integration into consumer devices such that heat may be dissipated efficiently from integrated circuits that must be kept below specified temperature limits. In most conventional low-power configurations, the primary purpose is efficient transport of heat to a remote location where a larger, more effective heat sink can be accommodated, resulting in a net reduction of junction-to-ambient thermal resistance; however, the need for concurrent size, weight, and performance improvement in high-performance commercial and military electronics over the last decade has given rise to an increasing number of thermally limited systems that demand low-thermal-resistance ultrathin heat spreaders (<1 mm thick) that can dissipate extremely high heat fluxes (>500 W/cm^2) from small areas (<1 cm^2). The thermal packaging infrastructure of electronics systems, such as radar power amplifiers, engine digital control systems, and high-performance graphics processing units, requires passive heat removal directly from microelectronic devices, typically achieved by die attach to moderate-conductivity copper alloy substrates. Their proven reliability, hermetic packaging, passive operation, and high effective conductance offer heat pipe devices as a potential solution for these applications.

A vapor chamber is a flat heat pipe with rectangular cross section employed for heat spreading (Fig. 4.1). While flat heat pipes have long been utilized in electronics cooling applications and are often produced simply by pressing a cylindrical heat pipe into a flat shape, the development of ultrathin vapor chambers for high-heat-flux applications requires a paradigm shift in terms of the transport mechanisms coming into play at this different form factor. In relative terms, conventional heat pipes are thicker devices intended to transport heat over longer working distances from larger, lower-heat-flux sources. This mode of operation results in low surface superheat temperatures, and evaporation occurs from a continually receding meniscus up to

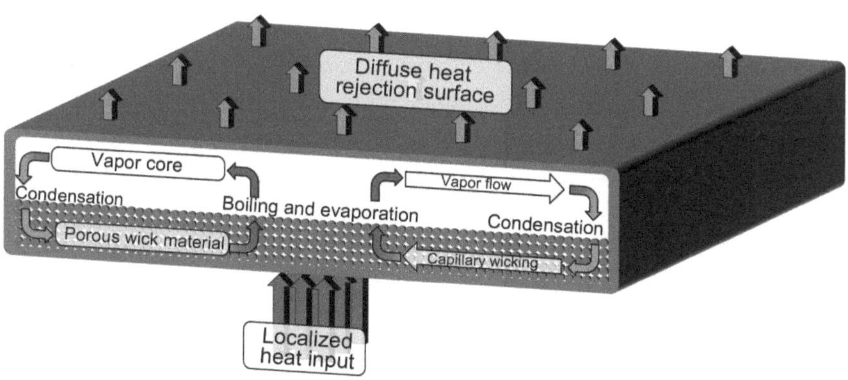

Figure 4.1 Schematic diagram of the operation and form factor of a flat heat pipe or vapor chamber heat spreader (scaled in the thickness direction to show details of operation). (For color version of this figure, the reader is referred to the online version of this chapter.)

the capillary-limited heat input level. The dominant system thermal resistance is conduction between the external surfaces and the liquid–vapor interfaces [14]; figures of merit for working fluid selection (see Merit Number in Ref. [4]) are established based on this capillary operating limit. Conversely, ultrathin vapor chamber heat spreaders have a comparatively small working length and total input power, albeit at a high local heat flux. Therefore, a capillary limit is not expected to be encountered until the total heat input gives rise to an extremely high local heat flux, large surface superheat, and potential incipience of boiling in the wick structure. Accurate quantification of the pressure drop in the vapor space at ultrathin form factors and the temperature differential associated with internal phase-change processes at high heat fluxes is necessary for prediction of overall device thermal resistance.

1.2. Nucleate boiling in porous wick structures

Conventional wisdom calls for nucleate boiling to be avoided in heat pipes having longitudinal groove wick structures. In these wicks, nucleation of vapor bubbles completely obstructs the noncommunicating individual paths of capillary liquid return to the evaporator section; a boiling limit is imposed in this case based on the conventional nucleation incipience superheat criterion. Alternatively, sintered screen mesh, sintered powder, and fibrous wick structures affixed to the wall of a heat pipe can continue to feed liquid to the heat source during boiling via the inherently stochastic network of interconnected pores. While it is commonly accepted that these structures may support nucleate boiling, conventional heat pipe applications discussed in Section 1.1 have not necessitated comprehensive assessment of the boiling behavior from wick structures during heat pipe operation. The recent development of ultrathin vapor chambers for high-heat-flux dissipation has altered this perspective.

Submerged pool boiling from porous surfaces has been extensively characterized for a number of surface geometries, working fluids, and flooded porous wick structures often found in heat pipes [15–28]. The fundamental mechanism of heat transfer during boiling from a porous surface submerged in a liquid pool differs from that from a wick that is passively fed by capillary action. The schematic diagram in Fig. 4.2 illustrates the primary differences in heat transfer regimes between capillary-fed and pool boiling. In pool boiling, heat transport initially occurs by natural convection at low heat fluxes and is followed by nucleate boiling until the critical heat flux (CHF) is

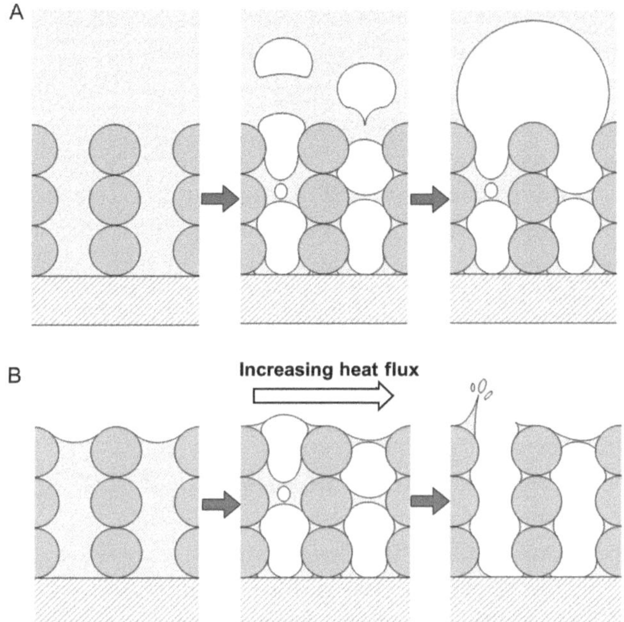

Figure 4.2 Schematic representation of the vapor formation characteristics for (A) submerged pool boiling and (B) capillary-fed boiling conditions. (For color version of this figure, the reader is referred to the online version of this chapter.)

reached. In contrast, a liquid–vapor free surface exists at the top layer of the liquid-saturated porous structure in a heat pipe, and heat transport first occurs in an evaporation regime prior to boiling incipience. Also, during capillary-fed boiling, the liquid–vapor free surface boundary in the pore layer causes drastic differences in the bubble departure characteristics. The observed heat transfer regimes (e.g., natural convection vs. evaporation at low heat fluxes), vapor formation characteristics, and transition criteria for incipience/dryout are all different under capillary-feeding conditions and require targeted investigation under representative conditions that is different from pool boiling.

1.3. Recent advances

In response to the application-driven demands on vapor chamber design and functionality, a number of recent studies have developed testing and analysis capabilities intended specifically for investigation of capillary-fed evaporation and boiling phenomena for a variety of porous wick structures. Recent experimental investigations that isolate the capillary-fed evaporation/boiling process for dedicated characterization are presented in Table 4.1.

Recent advances in the experimental investigation and analysis of ultrathin vapor chambers for high-heat-flux applications are reviewed in this chapter. The discussion focuses on

1. the observation of capillary-fed boiling regimes, visualization of vapor formation characteristics, and parametric performance quantification of vapor chamber wick structures subjected to high heat fluxes (Section 2.1);
2. the development and fabrication of novel heterogeneously structured and patterned vapor chamber wicks for reduced capillary-fed boiling thermal resistance (Section 2.2);
3. the prediction of heat transfer coefficients, exploration of incipience transition criteria, and assessment of heat input area dependencies of dryout mechanisms for capillary-fed boiling (Sections 2.3–2.5);
4. the vapor chamber modeling developments, viz., wick microstructure characterization and methodologies for incorporation into device-level models, implementation of boiling in models at high heat fluxes, and optimization for ultrathin form factors (Sections 3.1–3.3);
5. the comparison of capillary-fed boiling experiments to, and validation of transport models against, ultrathin vapor chamber device-level tests (Section 3.4);
6. the analytical and experimental assessment of nanostructures for improved capillary wicking/wetting behavior and enhancement of capillary-fed evaporation/boiling (Section 4).

2. EXPERIMENTAL EVALUATION OF CAPILLARY-FED EVAPORATION AND BOILING

It is well known that the overall effective thermal resistance of a heat pipe is impacted by the response of the wick structure to the evaporator heat load and the internal pressure distribution, which in turn dictate the liquid–vapor interface shape, level of recession into the porous structure, and mechanism by which phase change occurs at the location of heat input. One approach to deducing the impact of these phenomena is by testing actual heat pipe devices and measuring external heat loads and temperatures, according to the procedures reviewed in Refs. [2] and [3]; however, the ability to infer detailed information about the phase-change mechanisms

Table 4.1 Summary of recent experimental investigations on capillary-fed evaporation/boiling

Authors	Wick structure	Wick thickness	Test fluid	Heat source details	Maximum heat flux
Homogeneous wick structures					
Brautsch and Kew [29]	Clamped 304-L stainless steel screen mesh Wire diameter 40–190 μm Porosity 0.313–0.469	1–5 layers	Water ($T_{sat}=100\ °C$)	Variable area (not reported) Vertical rise height 55–79 mm	8.24–21.39 W/cm^2
Hanlon and Ma [30]	Monoporous sintered copper powder Particle diameter 635 μm Porosity 0.43	1.9–5.7 mm	Water ($T_{sat}=100\ °C$)	10 $mm^3 \times 20\ mm^2$ Horizontal	~35–55 W/cm^2
Davis and Garimella [31]	Monoporous sintered copper powder Particle dia. 45–355 μm Porosity 0.46–0.60	1 mm	Water ($T_{sat}=47–52\ °C$)	384 mm^2 Horizontal	~30–33 W/cm^2
Li *et al.* [32]	Sintered copper screen mesh Wire diameter 56 μm Porosity 0.693–0.737	0.21–0.74 mm	Water ($T_{sat}=100\ °C$)	8 $mm^3 \times 8\ mm^2$ Horizontal	150–367.9 W/cm^2
Li and Peterson [33]	Sintered copper screen mesh Wire dia. 56–191 μm Pore size 119.3–232.8 μm Porosity 0.409–0.692	0.37 mm	Water ($T_{sat}=100\ °C$)	8 $mm^3 \times 8\ mm^2$ Horizontal	150–367.9 W/cm^2
North *et al.* [34]	Biporous sintered copper powder Particle diameter 66–89 μm Cluster diameter 250–710 μm Porosity ~0.70	0.64 mm	Water ($T_{sat}=20\ °C$)	30 $mm^3 \times 8\ mm^2$ Horizontal	27–70 W/cm^2

Continued

Table 4.1 Summary of recent experimental investigations on capillary-fed evaporation/boiling—cont'd

Authors	Wick structure	Wick thickness	Test fluid	Heat source details	Maximum heat flux
Semenic *et al.* [35]	Monoporous sintered copper powder Particle diameter 53–710 μm Porosity 0.318–0.371	1 mm	Water (T_{sat} = 40–48 °C)	70.9 mm^2 Horizontal	209.5–223 W/cm^2
Semenic *et al.* [35]	Biporous sintered copper powder Particle diameter 29–63 μm Cluster dia. 250–710 μm Porosity 0.597–0.631	1–4 mm	Water (T_{sat} = 40–48 °C for 1 mm thick)	70.9 mm^2 Horizontal	150–494 W/cm^2
Semenic and Catton [36]	Biporous sintered copper powder Particle diameter 41–83 μm Cluster diameter 302–892 μm Porosity 0.51–0.68	0.8–3 mm	Water (T_{sat} = 36–46 °C)	32.2 mm^2 Horizontal	244–990 W/cm^2
Weibel *et al.* [37]	Monoporous sintered copper powder Particle diameter 60–302.5 μm Porosity 0.635–0.657	0.6–1.2 mm	Water (T_{sat} = 100 °C)	5 mm^3 × 5 mm^2 Vertical rise height 8.25 mm	500–596.5 W/cm^2
Weibel [38]	Monoporous sintered copper powder Particle diameter 100 μm Porosity 0.5	0.2 mm	Water (T_{sat} = 100 °C)	25–100 mm^2 Vertical rise height 8.25 mm	160–430 W/cm^2

Heterogeneous and patterned wick structures					
Zhao and Chen [39]	Sintered copper powder with vertical microgrooves Particle dia. 50 μm Groove width 150–500 μm Bank width 250–500 μm	2 mm	Water (T_{sat} = 100 °C)	5 mm^3 × 5 mm^2 Vertical rise height 7–20 mm	~200–350 W/cm^2
Hwang *et al.* [40]	Oxidized sintered copper powder with capillary artery posts Particle diameter 60 μm Artery pitch 3.52 mm	0.06 mm	Water (T_{sat} = ~40 °C)	10 mm^3 × 10 mm^2 Distributed liquid return	387 W/cm^2
Hwang *et al.* [41]	Sintered copper powder with converging lateral arteries Particle diameter 60 μm	0.06 mm	Water (T_{sat} = ~40 °C)	10 mm^3 × 10 mm^2 Vertical rise height ~45 mm	580 W/cm^2
Ju *et al.* [42]	Sintered copper powder with converging lateral arteries Particle diameter 60 μm Sixteen to thirty-two 1 mm wide arteries	0.06 mm Artery 1.5 mm	Water (T_{sat} = ~40 °C)	10 mm^3 × 10 mm^2 Vertical rise height ~45 mm	~350–400 W/cm^2
Weibel and Garimella [43]	Sintered copper powder with radial and square grid converging arteries Particle diameter 100 μm Porosity 0.5 0.5 mm wide arteries	1 mm	Water (T_{sat} = 100 °C)	5 mm^3 × 5 mm^2 Vertical rise height 8.25 mm	437–558 W/cm^2

Continued

Table 4.1 Summary of recent experimental investigations on capillary-fed evaporation/boiling—cont'd

Authors	Wick structure	Wick thickness	Test fluid	Heat source details	Maximum heat flux
Coso *et al.* [44]	Biporous silicon pin fins with vertical microgrooves Pin diameter 3.1–29 μm Pin pitch 4.9–28 μm Groove width 30–61 μm Pin cluster width 156–288 μm	0.135–0.243 mm	Water ($T_{sat}=100\ °C$)	6.25–100 mm^2 Vertical rise height 5 mm	73.6–733.1 W/cm^2
Nanostructured wicks					
Cai and Chen [45]	Carbon nanotube (CNT) biwick with vertical microgrooves CNT diameter <20 nm Groove width 50 μm CNT cluster width 100 μm	0.250 mm	Water ($T_{sat}=50–100\ °C$)	2 $mm^3 \times 2\ mm^2$ Vertical rise height ~5 mm	~600 W/cm^2
Cai and Bhunia [46]	CNT biwick with vertical microgrooves or pillar clusters CNT diameter <20 nm Groove width 50 μm CNT cluster width 100–250 μm	0.250 mm	Water ($T_{sat}=100\ °C$)	4–100 mm^2 Vertical rise height 9–10 mm	130–770 W/cm^2
Cai and Chen [47]	CNT biwick with microgrooves or pillar clusters Groove width 50 μm CNT cluster width 100–150 μm	0.210–0.300 mm	Water	4–100 mm^2 Vertical height ~2–3 mm	195–938 W/cm^2

Nam *et al.* [48]	Copper oxide (CuO) nanostructured copper microposts Post diameter 50 μm Post pitch 20–50 μm Details in Ref. [49]	0.100 mm	Water ($T_{sat} = 44\ °C$)	4–25 mm^2 Vertical height 15 mm	~125–800 W/cm^2
Weibel *et al.* [43,50]	Cu-functionalized CNT-coated sintered copper powder with square grid converging arteries CNT diameter ~100 nm Particle diameter 100 μm Porosity 0.5	0.2–1 mm	Water ($T_{sat} = 100\ °C$)	5 $mm^3 \times 5\ mm^2$ Vertical rise height 8.25 mm	~450–530 W/cm^2
Kousalya *et al.* [51]	Cu-functionalized CNT-coated sintered copper powder Cu-coated CNT diameter ~85–275 nm Particle diameter 100 μm Porosity 0.5	0.2 mm	Water ($T_{sat} = 100\ °C$)	5 $mm^3 \times 5\ mm^2$ Vertical rise height 8.25 mm	~350–450 W/cm^2

by this means, and to quantify the local thermal resistance across only the evaporator section, is limited.

Several studies have devised novel testing approaches that attempt to characterize the thermal resistance across a wick structure under simulated heat pipe operating conditions [29–31,52–54]. The form factors (working length and wick thickness) and operating conditions (heat input area and heat flux) investigated are not strictly applicable to thin vapor chamber heat spreaders; however, these studies represent the first sets of such investigations that offered measurement approaches that informed subsequent studies.

Mughal and Plumb [52] and Williams and Harris [53] both placed an open heat pipe (i.e., a porous wick layer attached to a surface that provided heated, adiabatic, and condenser sections) into a large vapor containment vessel for the purpose of measuring the capillary limit as a function of wick properties, independent of vapor-domain confinement effects. The saturated vapor temperature was measured just above the wick surface and provided a temperature differential for direct assessment of the evaporation/boiling thermal resistance as a function of heat input. Even though differing working lengths, wick structures, and test fluids were considered, similar trends were observed in both studies. At low heat inputs, the slope of the boiling curve (heat flux vs. excess temperature) was linear, consistent with the dominance of the conduction thermal resistance through the saturated wick layer. It was presumed that the evaporation resistance at the liquid–vapor interface was comparatively small and the measured resistance represented the saturated wick effective thermal conductivity [53]. The heat flux was slowly increased until a large surface temperature excursion associated with wick dryout and the capillary limit was observed; however, in both studies a sharp reduction in the evaporator thermal resistance was measured prior to dryout. This is attributed to the onset of nucleate boiling, which provides significant improvement in performance, but simultaneously causes premature starvation of the porous structure for the effective working lengths ($\sim$100 mm) and heat input areas ($\gg$1 cm^2) tested. Channels cut into the porous wick structure (1.6 mm wide) reduced the thermal resistance during capillary-fed boiling and extended the maximum dryout heat flux by providing a low-resistance path for vapor escape from the surface [52].

An alternative approach for measuring the capillary limit and effective evaporator thermal resistance was developed by Brautsch and Kew [29], who placed a screen mesh wick sample vertically into a large sealed vapor containment vessel. The lower edge of the sample was submerged into a liquid pool, and the working fluid was wicked up the sample surface a distance of 55–79 mm by capillary action to the location of the heat input. The evaporator thermal resistance was directly measured based on a temperature differential between the heated wick base and vapor space. The trend of variation in thermal resistance with respect to increasing wick thickness was reported, and conduction through the saturated wick layer was found to dominate the thermal resistance in an evaporation mode, consistent with the findings in Refs. [52] and [53]. Boiling was visually observed in the wick structure prior to complete dryout but caused an increase in the thermal resistance in all cases. While the single-phase capillary limit and boiling limit have been previously described and treated as distinct phenomena in the literature, the critical conclusion was drawn that they are fundamentally interrelated [29]. The presence of vapor in the wick structure affected both the capillary pressure generated in the pores and the overall pressure drop, resulting in an altered effective dryout heat flux that cannot necessarily be predicted by either of the discrete limiting mechanisms.

More recently, the level of liquid charge in a heat pipe (a parameter that could not be varied in the studies discussed earlier) has been found to play a critical role in determining the evaporator resistance and the modes of phase change that occur. Studies by different groups of researchers—Wong and Kao [55] and Wong *et al.* [56,57]—fabricated heat pipe devices specifically intended for measuring the thermal resistance and visualizing the internal saturated wick structure at the evaporator section. Wong and Kao [55] tested a 150 mm long, 6 mm diameter heat pipe with various screen mesh wick sizes and levels of fluid charge. A range of regimes was observed within the evaporator as a function of heat load. These regimes included evaporation from a saturated wick layer, nucleate boiling in the wick structure, suppression of boiling by recession of the meniscus, and evaporation from a thin-liquid film at the base of the wick structure. While distinct operating conditions that resulted in each regime were not identified, finer mesh sizes and larger fluid charges tended to result in boiling in the wick structure. Wong and coworkers fabricated 120 mm long, 7 mm thick heat pipes with both screen mesh [56] and sintered powder wick structures [57]. For both wick structures, boiling was never observed, even at heat fluxes >100 W/cm^2. The decreasing evaporator resistance measured with increasing heat flux was attributed to the meniscus recession observed within the porous structure and the reduced conduction resistance offered by the portion of the wick that remains saturated with liquid.

Despite the observation of boiling incipience in simulated heat pipes in multiple studies, the occurrence of capillary-fed boiling in actual heat pipe devices remains a point of contention in the literature, as exemplified by the comments of Wong *et al.* [57]. They concluded that the absence of boiling in their studies, using <1 mm thick sintered powder and sintered screen mesh wick structures, proved that the conditions required for boiling are not relevant to electronics cooling applications and that only overly thick wick structures, fluid overcharge, or fluids other that water fluids other than water might lead to boiling [57]. The suppression of boiling in highly porous structures is indeed an important observation (as discussed further in Section 2.4), but these authors recognized [56] that conditions giving rise to a large evaporator superheat prior to dryout would undoubtedly result in boiling. While capillary-fed boiling may indeed not have occurred for the device geometry and wick thicknesses tested in Ref. [57], which reached a capillary limit at moderate heat fluxes, this conclusion does not consider vapor chamber heat spreaders that feature very short working lengths (and do not suffer from capillary performance limitations) for which boiling has been shown to be prevalent (see Section 3.4).

In the studies described earlier [52,53,55–57], although the thermal resistance is measured across the evaporator wick structure, the performance is inextricably related to the test facility geometry and operating conditions. Specifically, the range of heat fluxes investigated was limited by the long capillary wicking length that gave rise to dryout at moderate heat fluxes (~100 W/cm^2). The experimental investigations reviewed in the rest of this section aim to decouple the evaporator performance from overall heat pipe performance. The capillary-fed evaporation/boiling regimes, conditions for transition, and thermal performance of a variety of porous wick structures are discussed.

2.1. Homogeneous wick structures: morphological, pore size, porosity, and thickness effects

Observing the capillary-fed evaporation/boiling regimes and determining the effect of parametric wick structure variations in response to a range of heat fluxes from small hot-spot heat sources (<1 cm^2) is critical to improved design and operation of ultrathin vapor chamber heat spreader devices. To evaluate the behavior of specific components and regions of the vapor chambers, test facilities have been devised to feed a wick structure by capillary action under conditions where the effective wicking length is on the same order as the size of the heat input area. Capillary liquid supply is virtually unlimited under these conditions; the viscous pressure drop due to single-phase flow to the heat input area is negligible at all operating conditions investigated.

Several early efforts developed novel facilities of this type but were targeted at studying the evaporation regime at lower heat fluxes and from larger-heat-input areas. Hanlon and Ma [30] oriented a sintered powder wick horizontally and maintained constant lateral capillary feeding to the base of the wick by controlling the height of a separate, hydraulically coupled liquid reservoir. A model was developed to predict the thin-film evaporation performance as a function of wick thickness, but comparison to model predictions was limited due to the observation of boiling during a majority of the experimental data points. The incipience heat flux was described as a critical boiling limit (independent of the capillary limit) that starves the wick of liquid supply; however, the measured performance was only marginally reduced in the boiling regime, and the magnitude of heat transfer coefficients was not consistent with complete dryout. Davis and Garimella [31] investigated a copper powder layer wick sintered onto a pedestal with its top surface sitting just above a standing liquid pool at the base of a thermosyphon. The thin wick layer on top of the pedestal drew liquid on demand from the pool with negligible capillary resistance, and the thermal resistance was measured across the wick layer. Testing was limited to the evaporation regime.

The following subsections focus on similar studies in which the performance and vapor formation regimes are measured up to the point of wick dryout for small-heat-input areas. Most importantly, these comprehensive investigations considered multiple, independently varied wick parameters for three basic classes of homogeneous wick structures (Fig. 4.3): sintered copper screen mesh, monoporous sintered copper powder, and biporous sintered copper powder (where two characteristic pore sizes are formed by sintering large clusters of small particles).

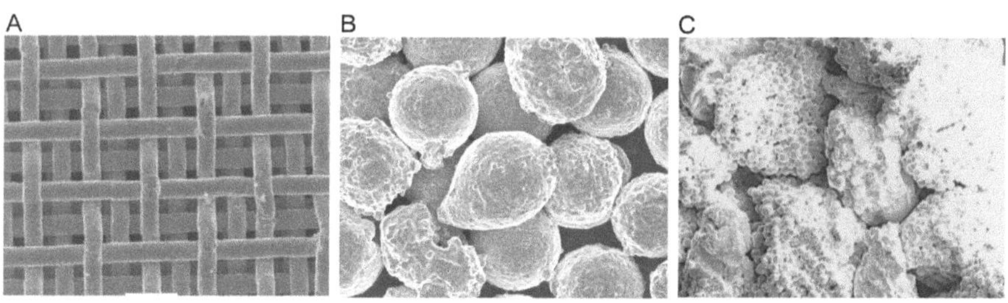

Figure 4.3 Scanning electron microscopy (SEM) images of (A) a 56 μm wire diameter sintered copper screen mesh [32, *reprinted with permission from ASME*], (B) 100 μm diameter sintered copper particles, and (C) a biporous wick with 600 μm diameter clusters of 60 μm diameter particles [35, *reprinted with permission from Elsevier*].

2.1.1 Sintered screen mesh

A series of studies by Peterson and coworkers [32,33,58] investigated the parameters affecting capillary-fed evaporation/boiling heat transfer and dryout heat flux for sintered copper screen meshes. The dependence on overall wick thickness, mesh size, and porosity was independently evaluated using consistent sample fabrication procedures. Based on the collective measurements and observations made, a new boiling curve was proposed for capillary wicking structures.

The range of wick thicknesses and mesh sizes investigated (see Table 4.1) was chosen based on critical length scales according to conventional bubble nucleation theory. The bubble departure diameter for a heated wall, based on the required buoyant forces required to overcome surface tension-driven adhesion to the surface, is on the order of millimeters for water [59]. Wick thicknesses from 0.21 to 0.82 mm were chosen under the hypothesis that this would prevent formation of discrete rising vapor bubbles within the porous structure. Therefore, bubbles formed in the wick structure would condense at the liquid–vapor interface at an elevated frequency while being constantly replenished by capillary action, effectively increasing the heat transfer rate at a given superheat [32]. Direct sintering to the heated surface can interrupt and prevent formation of a large lateral vapor film that would induce CHF. An 8 mm × 8 mm sample of the desired wick parameters was directly sintered to a copper heater block of the same dimensions for testing.

The experimental test facility used by Peterson and coworkers [32,33,58] is shown in Fig. 4.4. Degassed water flows into a sealed chamber that contains the sample and heater block. The sample was fed only by capillary action (and did not become submerged) by controlling the liquid level with an outlet overflow system. A heated glass window enabled observation of the sample under test. The facility and data reduction procedures were validated by testing a polished copper surface in a pool boiling mode and comparing results against the literature.

The dependence on wick thickness was first evaluated in Ref. [32]. During testing, after a critical incipience superheat was reached, it was clearly observed that both evaporation and boiling occurred in the wick structure (in the form of small bubbles that easily escaped to the vapor space). In contrast to prior studies using thicker wicks [29,30], capillary-fed boiling in the thin wick structures yielded improved performance compared to an evaporation regime; weak dependence of the thermal resistance on the wick thickness suggests that the boiling process, which occurs at the base of the wick structure independent of thickness, governs overall performance (compared to the conduction-dominated evaporation regimes described in Refs. [30] and

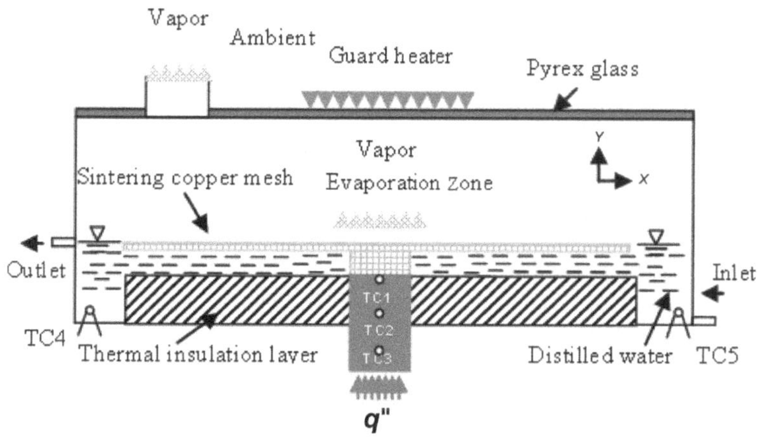

Figure 4.4 Schematic diagram of experimental capillary-fed evaporation/boiling facility used for testing sintered copper screen mesh [32, *reprinted with permission from ASME*]. (For color version of this figure, the reader is referred to the online version of this chapter.)

[53]). In regard to the limiting heat input, it was noted that a CHF was not induced by vapor blanketing of the surface, but dryout occurred at the center of the heat source due to capillary starvation in the presence of departing vapor [32]. Hence, this could be considered a capillary limit that is strongly interdependent on the pore-scale vapor departure behavior, as previously concluded by Brautsch and Kew [29]. An increase in the wick thickness delayed dryout via an increase in the available flow area (Fig. 4.5A).

Subsequent investigation [33] revealed the dependence of performance on mesh size, which controls the effective capillary pore size. For all pore sizes investigated from 119 to 232 μm, the superheat dependence on heat flux was linear in the boiling regime, as observed previously for pool boiling from porous surfaces [19]. A definitive decrease in the capillary-fed boiling thermal resistance was observed with decreasing pore size and was attributed to an increase in the effective heat transfer area [33]. Conversely, a smaller pore size reduced the measured dryout heat flux (Fig. 4.5B). The menisci were observed to recede to the lowest mesh layer prior to dryout; under this condition, a qualitative analysis of a single meniscus formed between a mesh wire and plane wall showed that larger pore sizes are favored by the trade-off between capillary pressure and liquid permeability.

From visualizations acquired for the samples investigated [58], and physical insights obtained from the parametric trends and conventional boiling theory, a new set of capillary-fed boiling curve regimes was proposed for

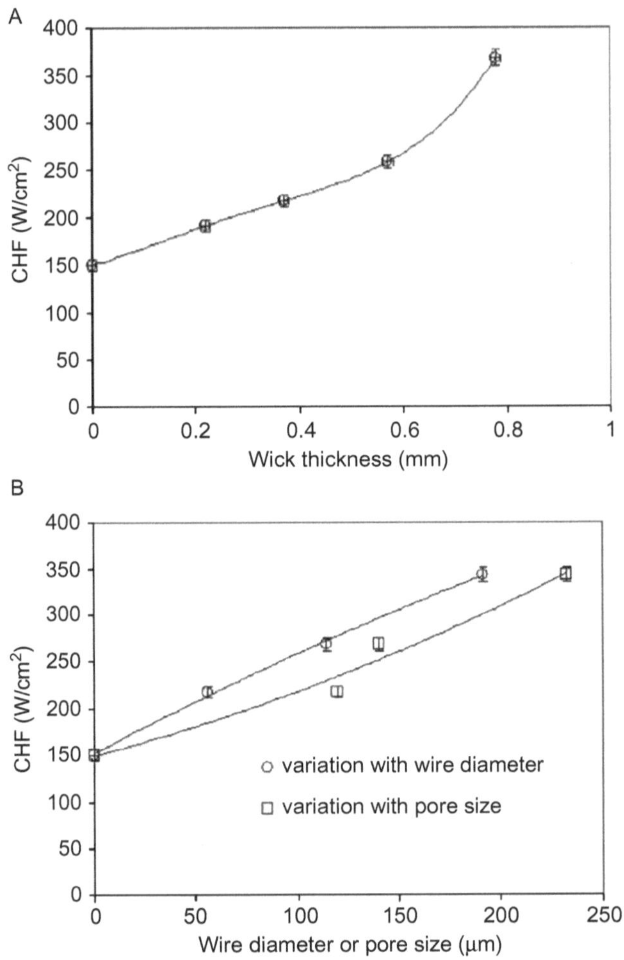

Figure 4.5 Dryout heat flux (CHF) as a function of (A) wick thickness [32, *reprinted with permission from ASME*] and (B) wire diameter or pore size [33, *reprinted with permission from ASME*] for sintered copper screen mesh.

sintered screen mesh wicks (Fig. 4.6). At low heat fluxes, heat transfer primarily occurs by conduction/convection to the evaporating liquid meniscus at the top of the wick structure. Under capillary-fed conditions, a sharp reduction in the surface superheat is observed concurrent with boiling incipience. Performance is improved, and vapor formation does not block capillary supply, as pore-scale diameter bubbles break up and condense at a high frequency (~250 Hz) due to close proximity of the liquid–vapor interface [33,58]. As the heat flux is increased further, the liquid meniscus recedes and the bubble departure frequency increases; at a critical transition point, the

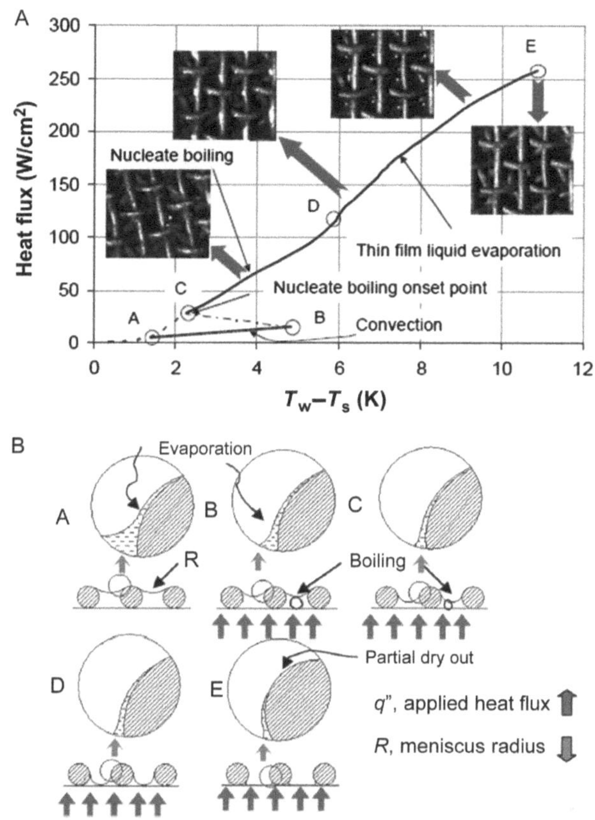

Figure 4.6 Typical (A) capillary-fed evaporation/boiling curve of heat flux versus wall superheat (T_W–T_s) for sintered copper screen mesh [58, *reprinted with permission from ASME*] and (B) local phase-change mechanisms as a function of heat transfer regime [33, *reprinted with permission from ASME*]. (For color version of this figure, the reader is referred to the online version of this chapter.)

evaporating meniscus becomes increasingly thin such that the available superheat cannot support nucleation and boiling is suppressed [33]. The heat transfer coefficient is maximized in this thin-film evaporation regime just prior to dryout.

2.1.2 Monoporous sintered powder

The phase-change regimes, associated thermal resistances, and dependence of performance on wick parameters were investigated by Weibel *et al.* [37] for monoporous sintered copper wicks fed by capillary action. While a number of other studies had considered homogeneous sintered copper powder samples as a baseline for comparison against various enhancement structures, a robust evaluation of the evaporation and boiling processes from this foundational porous structure was lacking in the literature. Hence, a set of sintered powder wick samples were fabricated by commercial sintering techniques for investigation. Three particle size distributions and three wick thicknesses (0.6, 0.9, and 1.2 mm) were selected for being commonly used in commercial heat pipes and vapor chambers.

A schematic diagram of the test facility used in Ref. [37], and in subsequent studies by Weibel and coworkers [43,50], is shown in Fig. 4.7. Copper powder was sintered directly to the surface of a copper substrate to produce a uniform layer of the desired thickness. The backside of the copper

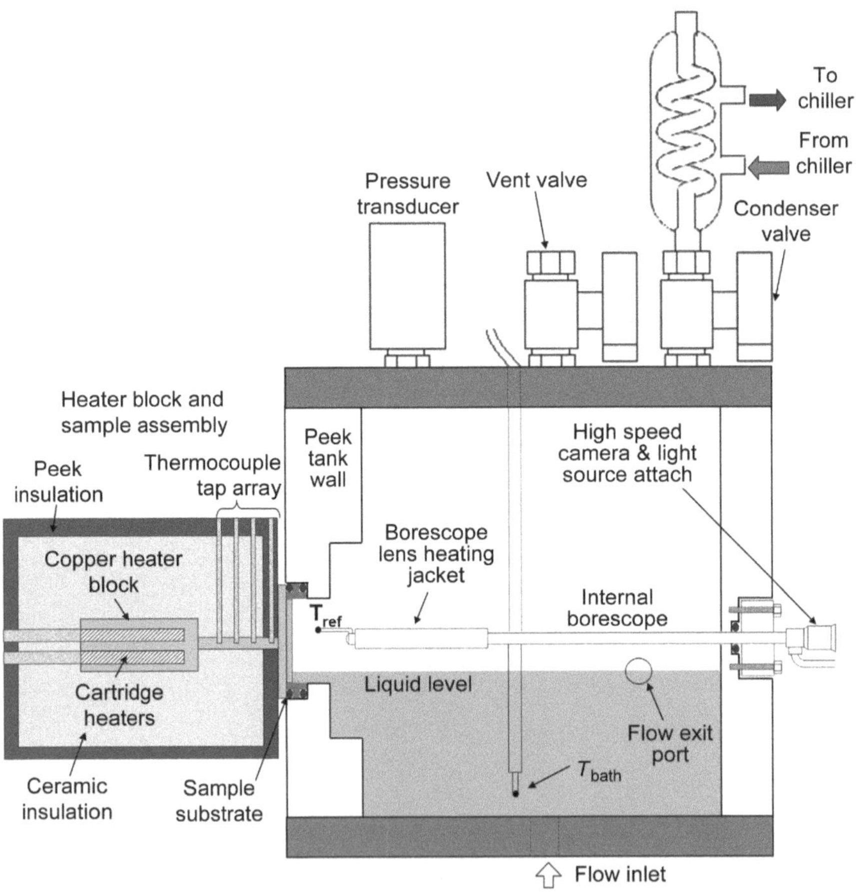

Figure 4.7 Schematic diagram of experimental capillary-fed evaporation/boiling facility used by Weibel *et al.* [37,43,50]. (For color version of this figure, the reader is referred to the online version of this chapter.)

substrate was then soldered directly to an insulated copper heater block with a heat input area of 5 mm × 5 mm. The porous sintered copper surface is sealed into a saturated vapor test chamber. Degassed water is continuously fed to the test chamber at the saturation temperature, and the liquid level in the chamber is maintained at a fixed height by means of an overflow drain on the side wall. To ensure liquid feeding exclusively by capillary action, the liquid level is chosen so that only the lower edge of the vertically oriented wick surface is submerged into the pool below. Unlike similar previous facilities [29], the results are independent of orientation and single-phase liquid pressure drop: The distance from the liquid free surface in the chamber to the center of the heat input area is only 8.25 mm, and the liquid pressure drop was calculated to be negligible at this effective working length [37]. The thermal resistance and surface superheat were measured as a function of heat flux for each sample tested.

A test case was presented with fine resolution of heat flux data points and simultaneous *in situ* visualization to identify the boiling curve regimes and associated phase-change mechanisms for sintered powder wicks (Fig. 4.8A). At low heat inputs, approximately less than 75 W/cm^2, visualization confirmed that heat transfer occurred by evaporation from the liquid meniscus at the top of the saturated wick structure [37]. As was found in previous studies [29,52,53], the superheat was linearly dependent on heat flux in this regime, dominated by conduction through the saturated wick layer. At a critical incipient heat flux and wall superheat, boiling in the wick structure caused a sudden transient drop in the surface superheat (in a manner analogous to pool boiling incipience but at an elevated heat flux due to the low thermal resistance of the evaporation regime compared to natural convection). For the small-heat-input area investigated, boiling was sustained in the wick structure. Once boiling was initiated, a significant reduction in thermal resistance was measured, with the magnitude of reduction corresponding to the thermal conduction resistance through the wick layer. This reduction was attributed to a transition in the mode of heat transfer from evaporation at the free surface of a liquid-saturated wick to bubble nucleation at the substrate–wick interface, which effectively eliminated the wick conduction resistance (Fig. 4.8B) [37]. This behavior is consistent with previous observations by Mughal and Plumb [52] and Williams and Harris [53] and illustrated the phenomenon more clearly without the confounding effects of a capillary limiting condition. As the heat flux was further increased in the capillary-fed boiling regime, the bubble departure rate increased dramatically (>500 Hz); however, unlike the observations for

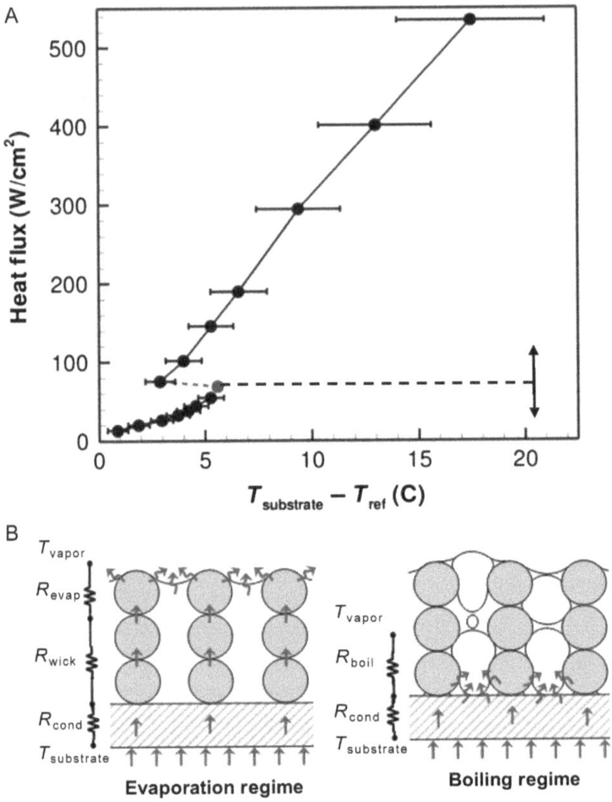

Figure 4.8 (A) Capillary-fed evaporation/boiling curve of heat flux versus substrate superheat ($T_{substrate}-T_{ref}$) for a 0.6 mm thick sintered copper powder sample (250–355 μm) and (b) schematic illustrations of the visualized two-phase regimes. (For color version of this figure, the reader is referred to the online version of this chapter.)

sintered screen meshes made by Li and Peterson [33], the liquid meniscus did not significantly recede into the wick structure; boiling was sustained at a relatively constant thermal resistance up to the maximum heat flux tested [37]. Monoporous sintered powder wicks tested in Ref. [37] were able to support local heat fluxes of greater than 500 W/cm^2 without the occurrence of dryout.

The effect of particle size on the average thermal resistance within the capillary-fed boiling regime is shown in Fig. 4.9. For the thicker wick samples, a minimum thermal resistance was achieved at an intermediate particle size, suggesting an inherent performance trade-off with decreasing particle size. Li and Peterson [33] previously observed and argued that a decreasing

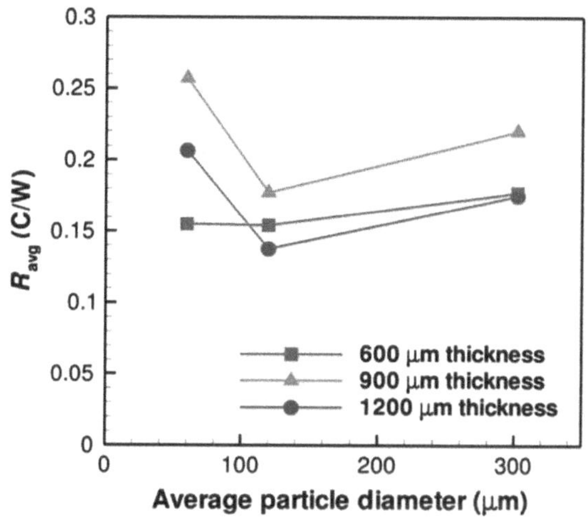

Figure 4.9 Average thermal resistance (R_{avg}) in the capillary-fed boiling regime (~200–500 W/cm^2) as a function of the average sintered copper powder particle size [37, *reprinted with permission from Elsevier*]. (For color version of this figure, the reader is referred to the online version of this chapter.)

particle size increased the effective surface area for heat transfer; however, Weibel *et al.* [37] observed a reversal in this performance trend for the smaller particle sizes. They attributed the thermal resistance increase at the lowest particle sizes to a reduction in wick permeability that caused an increased drag resistance impeding vapor escape from the wick. This implies that the optimum size is dependent on wick thickness and is consistent with reduced optimum particle size for the thinnest wick tested, which would pose the shortest path for vapor escape. Similar observations of optimum thickness-to-particle size ratio have been made in studies of submerged pool boiling from wick structures [60,61]. The observed influence of wick thickness on overall thermal resistance in the boiling regime was negligible. This agrees with the conclusions drawn in Ref. [33] and mechanistic arguments presented in Section 2.1.1.

2.1.3 Biporous sintered powder

Unlike monoporous structures, a biporous (or bidisperse) wick is characterized by two principal pore sizes and is typically homogeneously composed of large clusters of smaller size particles. Monoporous structures typically remain saturated with liquid and are limited by conduction to the liquid–vapor interface during evaporation. By contrast, the liquid meniscus preferentially recedes in the larger pores of the biporous structure at moderate heat fluxes, allowing increased area for thin-film evaporation from cluster surfaces. Further, during operation in the boiling regime, the large pores allow a path for vapor escape from the surface while liquid is pumped by the small capillary pores.

The capillary-fed evaporation/boiling performance of biporous wick structures was first measured by North *et al.* [34]. For the comparatively large heat input area (30 mm × 8 mm) and long working distance (17 mm) investigated, the menisci in the large pores were observed to continuously recede until dryout was observed at less than 100 W/cm^2. Later, Cao *et al.* [62] also measured the evaporation performance of biporous wick structures but used a test facility that was intended to replicate the liquid feeding and vapor removal mechanisms that occur in a capillary-pumped loop. In both cases, performance was augmented relative to monoporous structures.

Catton and coworkers [35,36,63] were the first to study the capillary-fed boiling process from a small hot-spot source (<100 mm^2), independent of single-phase capillary wicking limits, in the test facility shown in Fig. 4.10. A biporous wick sample was sintered to a horizontal copper pedestal that was fed by capillary action from a liquid pool below with minimal

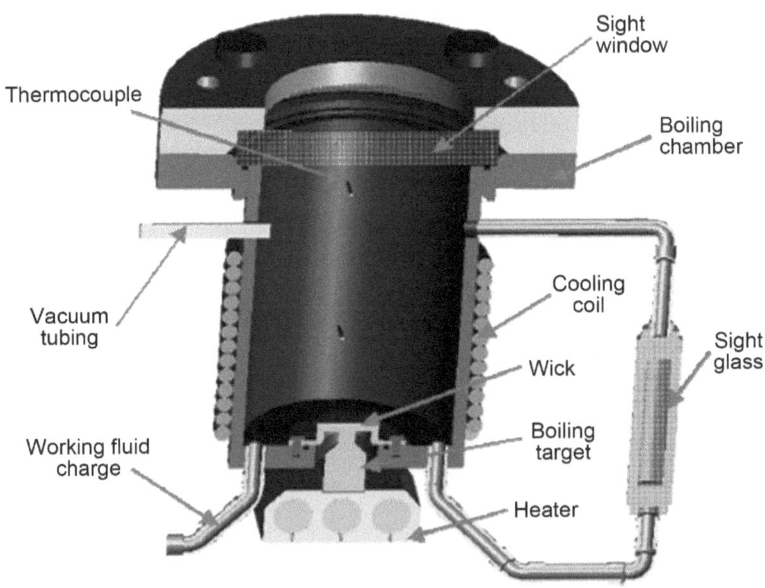

Figure 4.10 Experimental test rig used by Catton *et al.* for evaluation of capillary-fed evaporation/boiling from biporous sintered copper powder [35, *reprinted with permission from Elsevier*]. (For color version of this figure, the reader is referred to the online version of this chapter.)

flow resistance. The sample sat in a sealed thermosyphon chamber and was heated via an insulated copper block (the size of the heat input area was reduced in successive studies as indicated in Table 4.1). This approach is similar to the experimental facility devised by Davis and Garimella [31] but incorporated an auxiliary charging port and sight glass to monitor and precisely control the liquid level [35]. The steady-state temperature drop across the wick structure was acquired as a function of heat input for a condenser temperature of 40 °C.

In the initial study by Semenic *et al.* [35], monoporous and biporous wicks were compared at a thickness of 1 mm. At a constant particle size of 60 μm, the maximum measured heat transfer coefficients were larger for the biporous wicks, demonstrating the effectiveness of the liquid–vapor flow separation mechanism. The relatively low dryout heat fluxes (<200 W/cm^2) that were measured for monoporous wicks compared to the results of Weibel *et al.* [37] can be attributed to (1) the significantly lower porosity considered, which has been previously shown to reduce the maximum achievable heat flux [33], and (2) the larger-heat-input area (see Section 2.5).

The objective of a subsequent study by Semenic and Catton [36] was to manufacture copper biporous wicks with a range of different particle diameters, cluster diameters, and wick thicknesses, to inform the design of optimal biporous wick structures. A detailed description of the geometric and thermophysical properties of biporous wicks tested was presented in Ref. [64]. In Ref. [36], CHF was defined as the heat flux at which the minimum thermal resistance was measured. Large measured critical superheats (>50 °C) indicated that partial film boiling likely occurred in the wick structure at the minimum resistance value. CHF measurements from the data set did not reveal a clear mechanistic relationship between reduced thermal resistance and wick parameters. Correlation of the surface heat flux to wick structure yielded an optimum set of wick parameters at a constant surface superheat of 90 °C (Fig. 4.11). This agrees with previous observations of optimal particle size for capillary-fed boiling in the literature [37]. An additional set of different thickness wick structures (0.8–3 mm) was tested at these optimal cluster and particle diameters; decreasing wick thickness led to a lowered thermal resistance at a constant dryout heat flux [36]. While this disagreed with earlier observations for thin monoporous wick layers, the overall biporous wick thicknesses investigated were all significantly larger than those tested in Refs. [32] and [37].

The observed vapor formation regimes and mechanisms [36] associated with boiling from thin biporous layers (<1 mm) are consistent with the

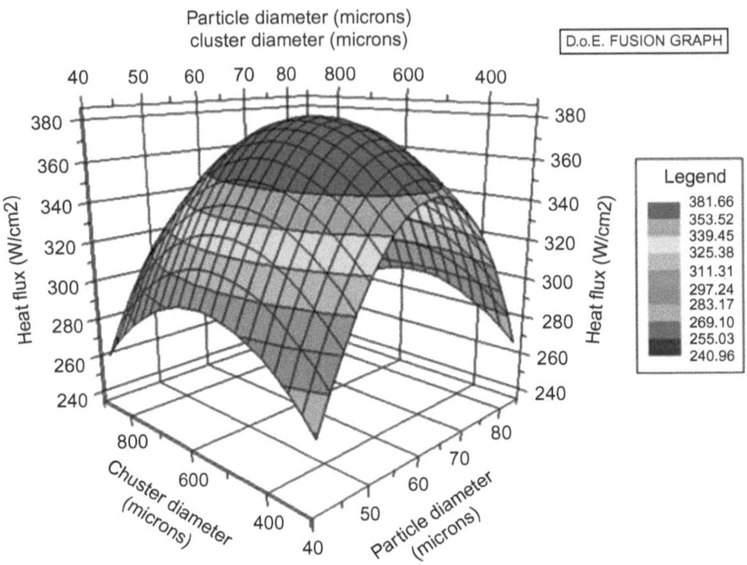

Figure 4.11 Empirical fit of the surface heat flux as a function of biporous wick parameters at 90 °C superheat [36, *reprinted with permission from Elsevier*]. (For color version of this figure, the reader is referred to the online version of this chapter.)

proposed analytical models [65]. The wick is initially saturated, and the menisci in the larger capillary pores recede with increasing heat flux, exposing additional liquid–vapor interfacial area. Performance in this regime is governed by the total area for thin-film evaporation from the menisci formed in the interstices of the smaller particles [36], as depicted in Fig. 4.12. Dryout is induced when the smaller capillary pores cannot sustain liquid feeding in the presence of the departing vapor and occurred at heat fluxes in excess of 600 W/cm^2. In contrast, for thicker wick structures, bubble nucleation occurs at discrete sites to form vapor columns within the wick structure. A unique regime was observed to exist in which film boiling occurs in the base of the wick structure, but sufficient liquid is provided to the evaporating menisci above the vapor layer to achieve a minimum thermal resistance at heat fluxes up to 990 W/cm^2, albeit at a large critical superheat of 147 °C [36].

2.1.4 Summary

Comprehensive investigations have been performed for a variety of homogeneous wick structures that isolate and identify the mechanisms of capillary-fed evaporation/boiling from a small hot-spot heat source. Heat

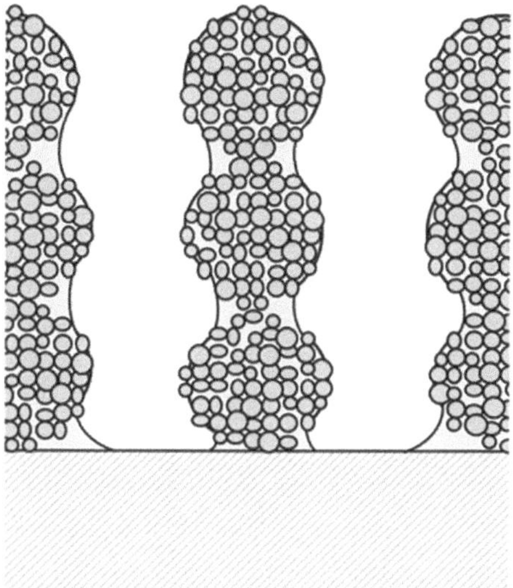

Figure 4.12 Schematic diagram of thin-film evaporation from the interstitial menisci within a biporous wick. (For color version of this figure, the reader is referred to the online version of this chapter.)

transfer regimes that have an increased liquid–vapor interfacial area for phase change, due to the presence of a vapor phase within the wick structure, provide a reduced thermal resistance compared to conduction-dominated evaporation from the surface of a saturated wick layer. Such regimes may be manifested as pseudovapor columns above active nucleation sites with a high departure frequency or as continuous vapor columns in comparatively larger pore spaces that cannot sustain capillary liquid supply at the given heat flux.

Due to similarities between the dominant heat transfer mechanisms of pseudovapor and continuous vapor columns, a number of common enhancement mechanisms are identified. Optimized structures result from pore sizes, porosities, and wick thicknesses that promote efficient vapor removal from the wick structure (which in turn prevents vapor layer formation on the surface, allows for liquid replenishment, and reduces the effective saturation vapor temperature at the surface) as well as an increased area for interstitial phase change via evaporation or boiling.

The remainder of this section focuses on enhancement and prediction of these common capillary-fed evaporation/boiling mechanisms and reviews the following: wick structure modification techniques employed to reduce thermal resistance and extend dryout (Section 2.2), prediction of boiling regime thermal resistance as a function of wick parameters (Section 2.3), prediction of transition to boiling regimes (Section 2.4), and prediction of dryout (Section 2.5).

2.2. Efficient liquid feeding and vapor extraction features

Improved heat transfer coefficients in capillary-fed evaporation/boiling were demonstrated to be closely linked to increased pore-scale interstitial evaporation area as discussed in Section 2.1; however, excess pressure drop in both the vapor leaving the structure and the liquid feeding the site of evaporation/boiling may increase the effective surface superheat and reduce the dryout heat flux, respectively. While homogeneous structures suffer from the inherent trade-off between these effects, multiscale heterogeneous wicks have been proposed to ease vapor escape from the structure with minimal effect on the liquid supply capacity. This multiscale design approach is analogous to the incorporation of device-scale graded wick structures [4] for reduced liquid-phase pressure drop but instead focuses on separation of the liquid feeding and vapor extraction flow paths to reduce the vapor-phase pressure drop during high-heat-flux operation. Multiple recent studies have designed and tested patterned wick structures to explore potential performance enhancement (Fig. 4.13).

Zhao and Chen [39] investigated capillary-fed boiling from sintered powder structures with and without patterned microgrooves. The performance of the proposed structures was evaluated in a saturated chamber wherein the lower edge of the wick was submerged in a liquid pool set at a fixed working distance below a 5 mm × 5 mm heat source. The microgrooves were aligned in the direction of liquid feeding and provided open

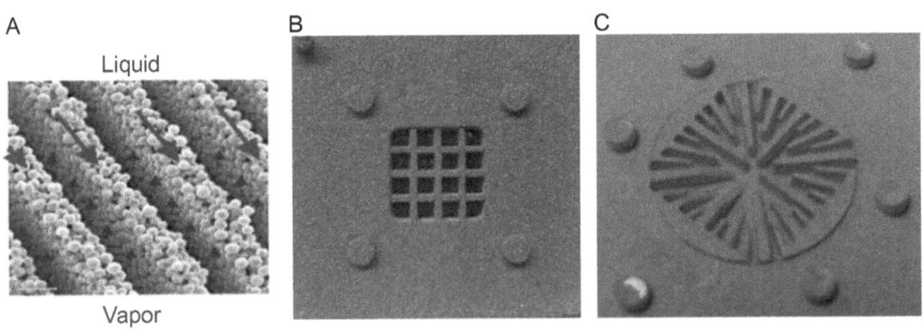

Figure 4.13 Images of (A) a microgrooved sintered powder wick [39, *reprinted with permission from ASME*], (B) a grid patterned sintered powder wick tested in Ref. [43], and (C) a copper particle monolayer fed by lateral converging arteries [42, *reprinted with permission from Elsevier*]. (For color version of this figure, the reader is referred to the online version of this chapter.)

areas to ease vapor escape during boiling. Due to this enhancement mechanism, a 3 mm thick microgrooved wick structure exhibited a dryout heat flux of 350 W/cm^2, compared to 0.5 mm and 3 mm thick monoporous wick structures that dried out soon after boiling incipience at or below 100 W/cm^2 [39]. This dryout heat flux of 100 W/cm^2 is lower than other monoporous sintered wicks in the literature [35,37], which was likely due to the extremely small constituent particle size (13 μm effective pore radius). Further testing showed that 500 μm grooves outperform narrower 150 μm grooves that tend to remain partially filled with liquid. This liquid poses an additional resistance to vapor removal, as evidenced by liquid expulsion from the wick structure observed at higher heat fluxes [39].

In a later study by Weibel and Garimella [43], patterns were fabricated into the sintered powder wick structures over the location of heat input to create a network of capillary-feeding arteries; millimeter-scale gaps between the arteries provide increased permeability to vapor exiting the wick. The wick structures were experimentally evaluated in the capillary-fed evaporation/boiling facility described in Ref. [37]. High-speed *in situ* visualizations supported the explanation of the relative performance differences between the monolithic and patterned structures as a function of the prevalent heat transfer regime (e.g., evaporation vs. boiling) for 1 mm thick wicks, as summarized in Fig. 4.14.

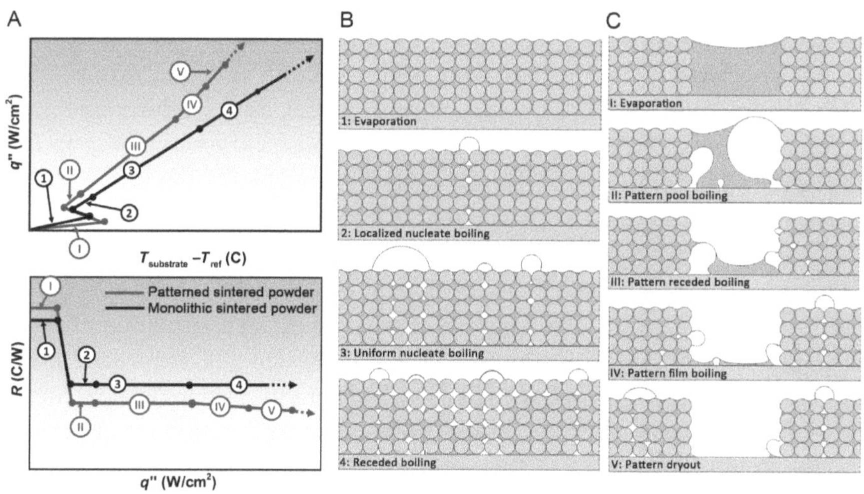

Figure 4.14 Schematic diagram of typical vapor formation regimes along the (A) boiling curve for (B) monolithic and (C) patterned sintered powder wicks. (For color version of this figure, the reader is referred to the online version of this chapter.)

At low heat fluxes, even the largest millimeter-scale pores in the patterned wick structures generated the necessary capillary pressure to maintain liquid saturation. In this evaporation-dominant regime, the patterning replaced high-conductivity copper regions with water and increased the conduction resistance between the substrate and the evaporating menisci, relative to the monolithic structure. At intermediate heat fluxes, the onset of boiling was observed. In the patterned structures, vapor preferentially departed the surface through the large pores, whereas it was forced to depart through the low-permeability smaller pores of the sintered powder in the monolithic structures. This reduced vapor-phase pressure drop due to the patterning yielded a performance enhancement throughout the boiling regime and became more effective as the liquid menisci recede into the larger pores of the patterned structures [43]. As liquid in the patterned pores further receded, a thin film was formed over the exposed substrate and nucleation occurred primarily from the corners and sides of the patterned area. Film boiling in this regime was observed to provide the largest comparative enhancement (28.5% reduction in the boiling thermal resistance) versus monolithic structures [43]. It was later shown that patterning of thinner wicks of 200 μm thickness [50] did not provide enhancement due to their inherently lower total frictional resistance to vapor exiting through the structure. It may be inferred that the optimum pattern or groove size is dependent on the wick thickness.

A series of studies by Hwang *et al.* [40,41] investigated the use of novel liquid return structures to sustain evaporation from a thin monolayer of 60 μm particles for cooling a 10 mm × 10 mm hot spot. The proposed structures were tested in a vapor chamber device with a condenser surface temperature fixed at 40 °C. The first type of novel liquid return structures investigated was columnar cylindrical arteries distributed over the heated area that fed fluid directly from the opposing condenser surface [40]. The artery geometry tested was optimized using a fluid thermal resistance network model [66]. While the columnar arteries act to provide local fluid delivery, the performance regimes are determined by evaporation from the thin monolayer of particles. The thermal resistance remained constant for a majority of the heat fluxes up to ~150 W/cm^2, in accordance with models assuming a constant saturated wick thickness. At higher heat fluxes, the thermal resistance varied as the meniscus receded, prior to complete dryout at 387 W/cm^2. Modeling efforts failed to capture the experimental performance, which was attributed to the use of an idealized liquid meniscus profile in the stochastic monolayer [40]. Nucleate boiling was suppressed in the 60 μm particle monolayer and was not observed.

A second novel liquid-feeding geometry was evaluated by Hwang *et al.* [41]. Unlike the columnar arteries, a set of lateral arteries converged toward the center of the wick structure, again with the purpose of providing distributed liquid flow to increase the evaporation regime dryout limit of a thin sintered monolayer. For this feeding geometry, the thermal resistance was observed to be relatively constant up to a heat flux of ~350 W/cm^2, attributed to a relatively constant liquid level in the monolayer at which the meniscus shape provides maximum capillary pressure [41]. Unlike the columnar arteries, for which sudden dryout was observed, operation after partial dryout was maintained up to 580 W/cm^2 via lateral arteries (albeit at an increased thermal resistance). This was due to the increased lateral inlet flow area to the monolayer provided by the lateral arteries [41]. Further design iterations revealed that an increased number of distributed liquid arteries provided improved performance [42]. Ultimately, a significant increase in the dryout heat flux was obtained, greater than would otherwise be possible for a thin particle monolayer with low evaporative thermal resistance.

2.3. Prediction of capillary-fed boiling thermal resistance

In a capillary-fed evaporation regime, it is well known that the limiting thermal resistance is imposed by conduction through the saturated wick layer. The performance under these conditions is amenable to prediction by assuming an effective thermal conductivity of the wick structure and depends on the ability to accurately characterize the porous properties. Despite the number of recent experimental studies, the development of correlations and methods for predicting the thermal performance during capillary-fed boiling is less explored; however, the inherent similarities between the vapor formation regimes and the governing heat transfer mechanisms observed across a range of homogeneous wick morphologies (see Section 2.1.4) suggest that generalized correlations are possible.

Weibel and Garimella [43] first assessed correlations developed for pool boiling for their ability to predict capillary-fed boiling. Expectedly, because they do not account for enhancement by the porous surface, established smooth surface pool boiling correlations [67–70] overpredicted the surface superheat and failed to capture the boiling curve trends. While there are several existing correlations for pool boiling from porous surfaces [16,61,71], the empirical dependence on the specific porous morphology, surface material, and test fluid prevented generalized evaluation against a range of wick structures. A semiempirical method was identified in Ref. [43] that predicted boiling performance for generalized porous media and was validated against experimental data. Smirnov [72] described capillary-fed boiling as evaporation heat transfer from an elementary unit cell of the porous structure represented by a fin coated with a thin-liquid film. The model was later verified against experimental data [73,74]. This model [72] was shown to compare favorably against capillary-fed boiling data for the sintered powder particle size and porosity investigated by Weibel and Garimella [43].

The simplified conduction model postulated by Smirnov [72] is applicable for both pseudovapor columns above active nucleation sites and continuous vapor columns. The thin-liquid layer formed around a vapor column is depicted in Fig. 4.15. Under an assumption that the liquid film thickness is smaller than the vapor column diameter, the saturated wick volume can be modeled as an extended surface. Solving for heat transfer by evaporation from this film-coated fin yields [72]

$$q'' = C\left(\frac{h_{\mathrm{lv}}\sigma k_{\mathrm{l}}^{3}}{\upsilon_{\mathrm{l}}}\right)^{1/6}(1-\varepsilon_{\mathrm{l}})^{1/6}k_{\mathrm{eff}}^{1/3}\left(\frac{\Delta T_{\mathrm{sl}}-\Delta T*}{L_{\mathrm{e}}}\right)^{5/6} \tag{4.1}$$

A fitted empirical constant of $C=0.094$ for water was recommended based on experimental data obtained using stainless steel meshes and monel balls. Measured experimental data fell within 50% of the predicted value [72].

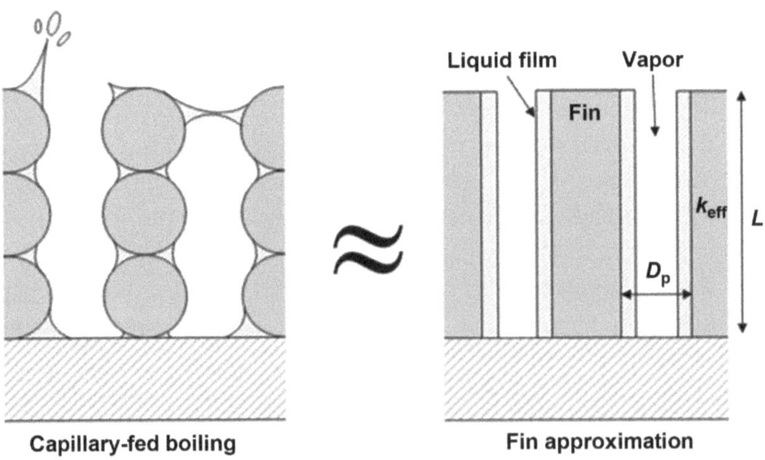

Figure 4.15 Illustration of the wetted-fin model proposed by Smirnov [72] for prediction of the heat transfer rate under capillary-fed boiling conditions from [38]. (For color version of this figure, the reader is referred to the online version of this chapter.)

A survey of the available experimental data reveals that the studies by Weibel *et al.* [37,43], Li *et al.* [32,33], and Semenic and Catton [36] provide sufficient wick property data to apply Smirnov's correlation [72]. As recommended by the correlation, the effective thermal conductivity for all wick structures was assumed to be a function of the porosity according to

$$k_{\mathrm{eff}} = k_{\mathrm{Cu}} \frac{(1-\varepsilon)}{(1+\varepsilon/2)} \tag{4.2}$$

The pore size was assumed to be 40% of the average particle size for the monoporous sintered powder wicks, also as recommended by the correlation. The cluster diameter, which governs the wick permeability to exiting vapor [64], was used to estimate the approximate pore size for the biporous wick structures. The predicted heat flux for a given superheat is evaluated and compared against 428 experimental data points, as shown in Fig. 4.16A. A single empirical constant of $C=0.127$ fitted to the data set yields a mean average error of 39.4% for the collective data set. Even with a simplified modeling approach, this suggests an improved error bound compared to the advertised correlation accuracy of 50% [72] over a broad range of wick parameters.

Interrogation of the predicted values obtained for each specific data set reveals several trends that are not captured by the model. The different particle sizes tested by Weibel *et al.* [37] fall into distinct groupings (Fig. 4.16B). The correlation exaggerates the trend of increasing performance with decreasing particle size, thereby overpredicting performance for smaller particle sizes. This is likely due to the assumption that vapor columns are maintained in each individual pore. The observations presented in Ref. [43] show that the nucleation sites actually engulf multiple pores for small particles, reducing the actual liquid film area for evaporation. This discrepancy with varying pore/particle size is not observed for the sintered screen mesh [33] and biporous wick [36] data, for which pore-scale vapor columns are indeed observed.

For the data of Li *et al.* [32,33], the correlation overpredicts the heat flux at a given superheat for experimental values below $\sim$100 W/cm^2 (Fig. 4.16C). This heat flux coincides with transition to film boiling on the phase-change regime map proposed in Ref. [58]. Above this heat flux, liquid recedes into the wick structure and forms pore-scale vapor columns that are mechanistically similar to the proposed correlation; the data points

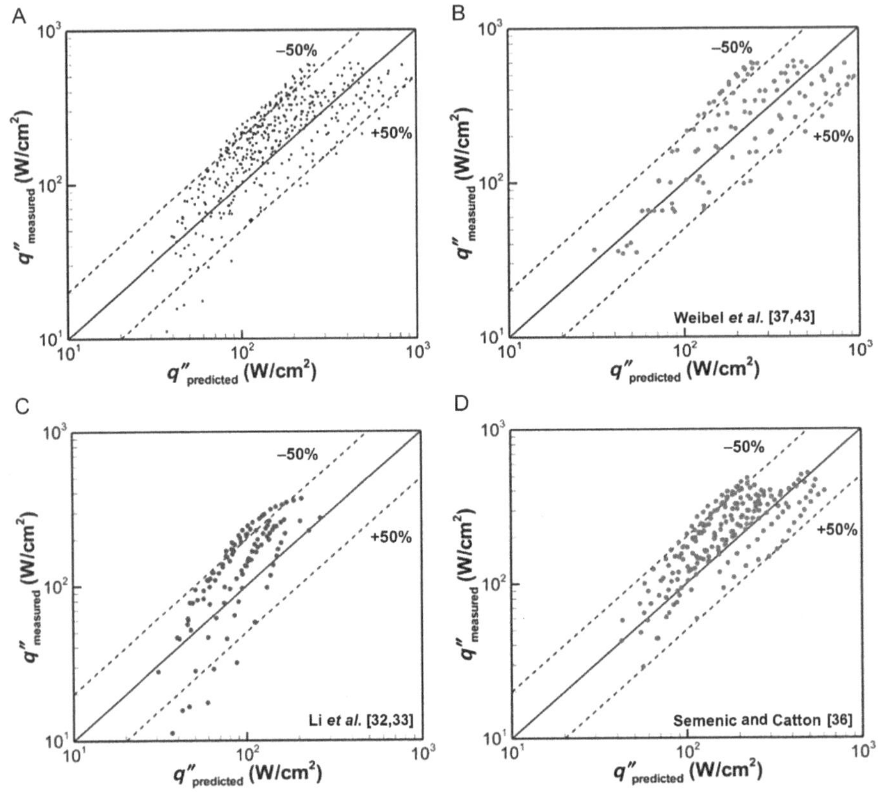

Figure 4.16 (A) Correlation accuracy of the wetted-fin model proposed by Smirnov [72] for data from the literature [32,33,36,37,43]. Detailed comparisons of the model predictions with data from (B) Weibel *et al.* [37,43], (C) Li *et al.* [32,33], and (D) Semenic and Catton [36] from Ref. [38]. (For color version of this figure, the reader is referred to the online version of this chapter.)

begin to converge above 200 W/cm^2. The correlation best agrees with the experimental data of Semenic and Catton [36]. These data collapse into a single grouping (Fig. 4.16D) with no obvious deviation for a particular cluster size or wick thickness. The relatively large cluster diameters in this work (300–800 μm) suit the model assumption that the pore size should be much larger than the film thickness. Additionally, the smaller-diameter particles in the biporous wick are expected to sustain a thin evaporating liquid film surrounding the vapor core.

The model for capillary-fed boiling proposed by Smirnov [72] is seen to match reasonably well with recent experimental data in the literature reviewed here; however, the density of the vapor columns formed in the wick structure and the effective liquid film area surrounding the vapor columns are expected (and observed) to vary uniquely as a function of wick morphology and input heat flux. This leads to the large error observed for some operating conditions in Fig. 4.16. In order to more accurately predict the heat flux versus wall superheat for a given wick structure, a more accurate representation of the vapor flow structures and interstitial liquid film thickness as a function of operating conditions is required.

2.4. Incipience of boiling under capillary-fed conditions

Understanding and predicting incipience of nucleate boiling in porous wick structures fed by capillary action is critical to the design of vapor chambers that are subjected to high heat fluxes. As is further explained in this section, bubble nucleation is suppressed under capillary-fed evaporation compared to pool boiling, and the sudden onset of nucleate boiling may cause a larger incipient overshoot in this case. Design of devices for sustained operation in evaporation or boiling regimes may serve to avoid such repeated fast thermal transients that may damage sensitive electronics.

It is well documented that large incipience superheats are observed for pool boiling from smooth surfaces using highly wetting fluids due to the vapor embryo entrapment process [59]. Surface superheats for pool boiling of FC-72 have been observed in the range of 20–40 °C from platinum surfaces [75] and 10–20 °C from copper surfaces [76,77]. The incipience superheat may be reduced through creation of high-cavity-angle sites for vapor embryo trapping [76]. Water has a comparatively higher surface tension and therefore tends to form stable nucleation sites at lower superheats than highly wetting fluids. This translates into measured incipience superheats of ~10 °C from smooth surfaces [76] and less than 1.5 °C for pool boiling of water from sintered porous surfaces [19,78].

Under capillary-feeding conditions, an increased substrate superheat temperature requirement has routinely been observed for incipience of boiling, even for highly porous structures with a wide distribution of cavity radii. Relatively thick wick structures [29,52,53] yielded boiling incipience at superheats ranging from 1 to 6 °C. Li and Peterson [33] reported superheats of up to 10 °C for the range of sintered screen wick geometries tested and noted that the required superheat increased with decreasing wick thickness. Wong and coworkers [56,57] tested thin heat pipe devices with sintered screen mesh and sintered copper powder structures and found that superheats up to 13.7 °C did not induce boiling up to the capillary-limited heat fluxes investigated. Boiling is also suppressed entirely for 60 μm particle monolayers up to superheats of 15–20 °C [40]. The most extreme example was provided by Altman *et al.* [79], who performed testing of ultrathin vapor chamber devices containing sintered powder wicks and observed incipience at superheats of 20–70 °C. The increased incipience superheat due to reduced vapor chamber operating saturation pressure can be calculated using a nucleation parameter [80] that predicts the relative required incipience superheat based on thermophysical fluid properties. Clearly, reduced saturation pressure alone does not account for the large incipience superheat discrepancy between capillary-fed boiling in vapor chambers and pool boiling at atmospheric pressure. Due to the expected mechanistic differences, conventional pool boiling incipience criteria are not suitable for the analysis of capillary-fed boiling.

All of the studies earlier based their observations on a single trial using different wick structures; however, it has been previously observed that the inherently variable boiling incipience process yields a large standard deviation in measured superheat heat values, even for exacting replication of test conditions [81]. Incipience criteria are therefore more appropriately defined statistically. The required incipience superheat can be characterized by repeated testing of nominally identical surfaces to develop a probabilistic representation. This approach is typically used to predict pool boiling incipience for highly wetting fluids that are prone to similarly large temperature overshoots [75,76].

Weibel *et al.* [50] characterized the probability of incipience versus wall superheat for capillary-fed 200 μm thick sintered powder wicks. Boiling was expected to be suppressed for this very thin wick structure compared to thicker structures, based on previous experimental observations in the literature [33,40]. Following a specified surface aging protocol, after which there was no discernible change in trend in the measured incipience superheat, a total of 18 boiling curve trials were obtained using six nominally identical samples. Multiple samples were used to account for sample-to-sample variability as well as variation between multiple trials. The measured substrate superheat at incipience ranged from 2.1 to 19.8 °C, and the data set is represented statistically in Fig. 4.17. The plotted relative cumulative frequency depicts the percentage of total test trials for which onset of boiling began below a given superheat. The slope provides insight into the wide distribution and limited predictability of the incipience phenomenon. It is concluded that capillary-fed thin sintered powder structures can sustain evaporation at superheats greater than 10 °C [50].

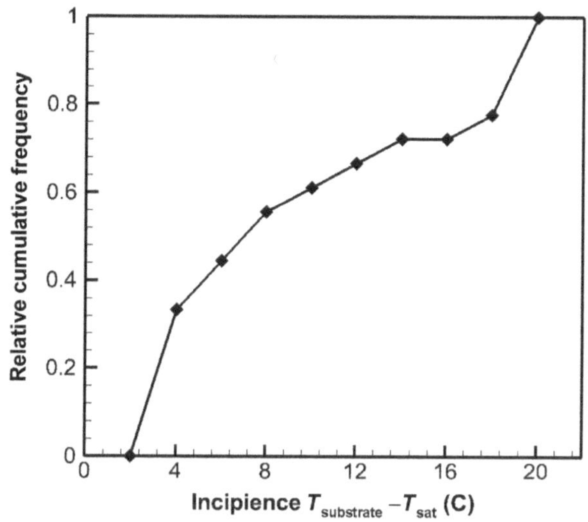

Figure 4.17 Relative cumulative frequency of superheat-dependent boiling incipience for a 200 μm thick sintered copper powder wick. *Adapted from Ref. [50].*

Several mechanisms that result in sustained evaporation from capillary-fed structures have been proposed. The classical model of Hsu [82] recognizes that the required incipience superheat is based on a limiting thermal boundary layer thickness in the superheated liquid. This model is widely used to understand the nucleation process and agrees with experimental trends for pool boiling. In capillary-fed evaporation/boiling, the liquid menisci recede and form thin-liquid films over the particle surfaces with increasing heat input. It was postulated by Weibel *et al.* [50] that these thin-liquid films dictate a thin thermal boundary layer that prevents bubble growth and increases the threshold superheat for nucleation compared to pool boiling. This is analogous to the effect of bulk liquid convection that suppresses nucleation in flow boiling. Li and Peterson [33] similarly attributed boiling suppression to the formation of thin-liquid films formed in the structure that do not have the necessary superheat to sustain nucleation. Further mechanistic modeling and experimental validation is required to ensure reliable prediction of incipience superheat under capillary-feeding conditions.

2.5. Dryout mechanisms and heater size dependency

Dryout of the vapor chamber wick at the location of heat input presents a critical limiting condition that results in a rapid surface temperature rise at which the device ceases to operate. Hence, accurate prediction of the dryout conditions is a primary focus for vapor chamber development. For thin vapor chambers intended to operate at a high heat flux but with the heat spread over only a small area, incipience of boiling or formation of vapor columns in the wick structure is expected to occur before a liquid-phase capillary pressure limitation is reached. Prediction of dryout under these multiphase flow conditions is critical. Experimental studies of capillary-fed evaporation/boiling often describe dryout mechanisms on a case-by-case basis [29,52] and sometimes as a function of the porous structure [33,36], but an investigation of the sensitivity to heater size has been limited to inferences drawn by comparison between disparate experimental facilities and porous structures.

In pool boiling, the CHF is characterized by transition to film boiling and formation of a vapor layer over the surface, most notably due to instability in the vapor jets leaving the surface and starvation of the liquid at the surface [59]. It has been shown that CHF is partially dependent on the size of the heated surface area due to a change in the proportional perimeter length through which ambient fluid can be drawn to the sites of nucleation. Lienhard *et al.* [83] first proposed that CHF would increase for a finite heater size compared to the assumption of an infinite size for the heated plate. Several subsequent experimental studies [84–87] investigated pool boiling from decreasing heat input areas and confirmed an increase in CHF.

Similar dependence is expected for capillary-fed wick structures due to the distance the liquid must travel to the center of the heat input, especially in the presence of departing vapor. To prevent dryout, the capillary pressure generated under these conditions must overcome the maximum two-phase cross flow pressure drop for flow to the center of the heated area. Two-phase flow in porous media is generally classified into countercurrent and crosscurrent flow arrangements [88]. Simplified modeling and empirical correlation of two-phase pressure drop is limited to these arrangements [89–92]. Further, pressure drop is highly dependent on the flow regime, and mechanistic predictions cannot be formulated without extensive visualization of the interstitial vapor phase. While comprehensive two-phase pressure drop and flow regime visualization data are not available, several recent studies have focused on studying the influence of heater size on dryout under capillary-feeding conditions, as described in the succeeding text.

A strong dependence of the maximum capillary-fed boiling dryout heat flux on heater size was first reported by Nam *et al.* [48]. In common with the testing approaches in all studies discussed in the rest of this section, the wick samples were oriented vertically inside a saturated test chamber. The lower edges of the samples were submerged in a liquid pool below to allow capillary self-feeding. Nam *et al.* [48] fabricated serpentine resistor heaters of two different areas, 4 and 25 mm^2, to evaluate the surface superheat response to input heat flux. The wick structures evaluated were composed of 50 μm diameter copper microposts with nanoscale surface roughness features. For all samples, boiling was observed in the wick structure after a superheat of ~25 °C was reached. The maximum supported heat flux increased from ~150 to ~800 W/cm^2 solely due to the heater size area reduction. While a small percentage of this increase is attributable to heat-spreading effects in the substrate, the authors ascribe this dependence to capillary supply to the heated area under vigorous boiling conditions [48].

Several studies reported similar findings and quantified novel dryout mechanisms and heater size dependence metrics. Cai and Chen [47] evaluated the performance of carbon nanotube-based wick structures using heat input areas of 4 mm^2 and 100 mm^2. In the most extreme case, the dryout heat flux increased from 195 to 938 W/cm^2 due to the reduction in heater area. Due to the high vapor departure velocity at such heat fluxes, a new physical mechanism of liquid loss due to spraying of droplets from the surface was identified [47]. Quantitative measurements revealed that this mechanism accounted for as much as 12% of the total liquid supplied to the hot spot. Ćoso *et al.* [44] fabricated and tested silicon-pillared wick structures and observed similar ejection of liquid from the surface at high heat fluxes during nucleate boiling. By testing multiple pillar heights, a critical transition geometry governing the dryout mechanism was observed. For short pillars, the evaporation mode is sustained until dryout; taller pillars produce a higher surface superheat at lower heat fluxes that allows for bubble nucleation [44]. Based on a measured increase in the dryout heat flux from ~160 to 733 W/cm^2 for a heater size decrease from 100 to 6.25 mm^2, the authors proposed a wick-to-heater area ratio metric to capture the qualitative relationship between dryout and heater size [44], but quantitative correlations were not provided.

Weibel [38] proposed a methodology for quantitatively evaluating the effect of heater size. The dryout heat flux was measured for 200 μm thick sintered copper powder wick structures with heat input areas of 25 and 100 mm^2. As shown in Fig. 4.18, a significant reduction in the dryout heat flux was measured for the larger heater area. Vigorous boiling was observed inside the wick structure at dryout. Based on the available capillary pressure head, the approximate two-phase radial pressure drop can be calculated for

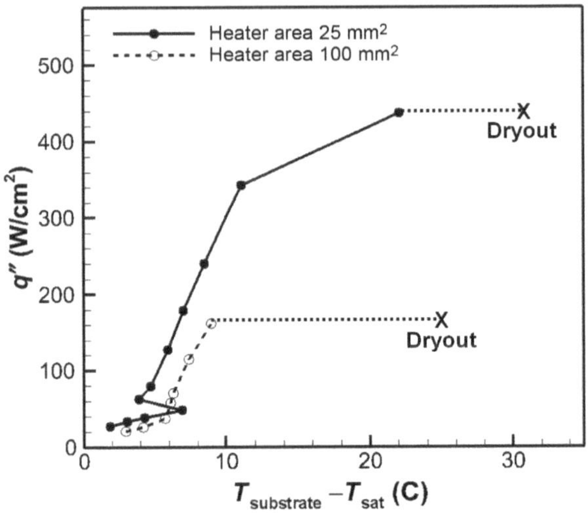

Figure 4.18 Boiling curve comparing the effect of heat input area for a 200 μm thick sintered copper powder wick. *Adapted from Ref. [38].*

the superficial vapor–liquid velocities corresponding to the dryout heat flux and used to extrapolate the dryout limit for intermediate heater sizes. This approach, while limited to a specific wick structure in Ref. [38], can be emulated to correlate the two-phase pressure drop at dryout over a wider parametric range.

The dryout heat flux under capillary-fed boiling has been demonstrated to be extremely sensitive to heater size; however, few quantitative prediction methods have been proposed. Additional investigation, and detailed visualization of the vapor–liquid regimes inside the porous structure, is required to develop methodologies for correlating and modeling the complex dryout mechanisms.

3. DEVICE-LEVEL MODELING, TESTING, AND DESIGN FOR HIGH-HEAT-FLUX APPLICATIONS

In order to develop vapor chamber heat spreaders with significant enhancements in the maximum heat dissipation and effective thermal conductance within a reduced-thickness package, it is becoming increasingly important to optimize the design of the internal wick and vapor core structures. Optimization via validated device-level models may avoid time-consuming parametric testing approaches, which are confounded by variability due to fabrication, fluid charge, and experimental uncertainties.

Modeling of vapor chambers is complicated by the coupled heat, mass, and momentum transport mechanisms prevalent in the device, such as evaporation and condensation at the liquid–vapor interfaces; however, a number of modeling approaches have been developed that approximate transport with various levels of fidelity. Garimella and Sobhan [8] and Faghri [9] provided reviews of the modeling approaches for a variety of heat pipe device types. Modeling of the transport in thin vapor chambers (i.e., flat heat pipes) is briefly reviewed first, to provide a context for the state-of-the-art advances in modeling approaches discussed in detail in Sections 3.2 and 3.3.

3.1. Flat heat pipe and vapor chamber models

3.1.1 Analytical modeling approaches

Simplified one-dimensional thermal resistance network models, such as those presented by Faghri [3] and Prasher [14], account for the dominant conduction and vapor pressure drop resistances in the vapor chamber. While these models only predict steady-state performance and do not capture spatial flow characteristics and operating limits, they enable first-order design for multiple working fluids and different power dissipation levels, and selection of favorable vapor chamber form factors [93], at reasonable accuracy compared to complex, three-dimensional analytical solutions [94]. Additional predictive capabilities can be incorporated into these modeling schemes as necessary. When the vapor core resistance is negligible, Zuo and Faghri [95] predicted one-dimensional transient behavior by assuming that conduction in the wick and wall dominated the thermal response and showed good agreement with transient, multidimensional numerical modeling approaches [96]. Chang *et al.* [97] modified a one-dimensional vapor chamber model to account for the thermal resistance reduction due to liquid meniscus recession at high-heat-flux inputs. These thermal resistance network models typically assume a given transfer coefficient for the evaporator section of the vapor chamber.

Higher-order analytical models provide additional insight into the vapor- and liquid-phase flow profiles and spatial pressure distribution. A series of studies by Vafai and coworkers [98–100] offered detailed analytical models for a specific asymmetric heating profile applied to a flat heat pipe geometry. Vafai and Wang [98] first established the nondimensional governing equation formulation for pseudo-three-dimensional incompressible vapor flow and solved for device steady-state pressure, temperature, and velocity fields using an integral method. This modeling approach was extended to transient modeling of start-up and shutdown by Zhu and

Vafai [99] and Wang and Vafai [100], respectively. Transient conduction was solved in the saturated wick and wall domains (which govern the overall system time constant) and thermally coupled to a quasisteady incompressible vapor domain. Analytical modeling results were in good agreement with experimental investigations of the same device [101,102]. An analytical model capable of simulating steady-state vapor chamber operation under a more generalized asymmetric heating profile (having multiple hot spots on one side) was formulated by Lefèvre and Lallemand [103]. To achieve this, a two-dimensional hydrodynamic model for the liquid and vapor domains was coupled to three-dimensional thermal diffusion in the wall. Most recently, Aghvami and Faghri [104] developed a two-dimensional analytical model that solved mass, momentum, and energy equations in the vapor chamber wall, wick, and vapor domains. A variety of canonical heater/condenser configurations was investigated, and the importance of considering axial variation in the evaporation and condensation rates was noted for high-thermal-conductivity wall materials. By necessity, each of the individual analytical vapor chamber modeling approaches in the preceding text represents a trade-off between simplifying thermal fluid operating assumptions, limiting investigation to a set or range of external heating/cooling configurations, and reducing temporal/spatial dimensionality.

3.1.2 Numerical modeling approaches

Numerical domain-discretization approaches offer the potential for complete description of transient, three-dimensional conjugate heat and mass transfer within the vapor chamber for improved model functionality compared to simplified analytical solutions. Zhu and Vafai [105] expanded their prior analytical model [99] using a numerical approach to solve for three-dimensional vapor flow in the vapor chamber and revealed the vapor chamber aspect ratios for which a three-dimensional model was necessary to resolve the viscous pressure drop caused by the side walls. Sonan *et al.* [106] implemented a discretized form of the analytical steady-state model developed by Lefèvre and Lallemand [103] to enable transient analysis of a vapor chamber subjected to multiple discrete heat sources.

A number of additional transient, three-dimensional vapor chamber models have been developed independently and are described in the literature. Vadakkan *et al.* [107] developed a three-dimensional numerical model that accounted for vapor core pressurization to study the transient and steady-state response of a vapor chamber with multiple discrete sources. The effects of heat source strength and separation were studied, and the importance of accounting for axial thermal diffusion was noted. Carbajal *et al.* [108] used a numerical finite volume analysis to obtain the temperature distribution on the backside of a vapor chamber subjected to a nonuniform heating profile. To reduce computational cost, a simplified transient, three-dimensional model intended for implementation into system-level models was developed by Chen *et al.* [109]. The linear model assumed negligible thermal resistance due to vapor convection and treated the vapor as a common interface between the evaporator and condenser wicks; the utility and accuracy of the model were later demonstrated through analysis of a complete thermal heat sink package with an embedded vapor chamber [110]. Additional detailed, three-dimensional numerical models that solved for flow and heat transport in vapor chambers were presented by Koito *et al.* [111] and Xiao and Faghri [112].

In all of the studies mentioned earlier, the wick–vapor interface was assumed to be flat, and the effect of dynamic meniscus curvature on capillarity was not considered. Tournier and El-Genk [113] developed a two-dimensional axisymmetric model for transient analysis of cylindrical heat pipes. In addition to numerically solving for flow and temperature in the wall, wick, and vapor domains as in the studies mentioned earlier, the analysis approach determined the local radius of curvature of the liquid–vapor interface in the axial direction. Based on experimental observations, a liquid pooling model was incorporated. It was assumed that the equivalent volume of convex menisci formed by transient thermal expansion of working liquid was entrained into the vapor stream and pulled to the condenser section as shown in Fig. 4.19. Rice and Fahgri [114] presented a numerical analysis of the transient operation of a heat pipe with given screen mesh wick structures. This model captured the effects of local evaporative mass flux on the meniscus curvature and therefore yielded a more accurate approximation of the capillary limit as a function of heat load.

3.1.3 Summary

Direct representation of the wick structure at a device scale is not computationally tractable, and therefore, all modeling approaches assume a representative porous medium domain for the wick structure. While this approach captures device-level transport phenomena, its accuracy hinges upon adequate characterization of the transport properties for realistic wick structures, viz., effective thermal conductance, permeability, and capillarity. Furthermore, most modeling approaches do not consider the implications of

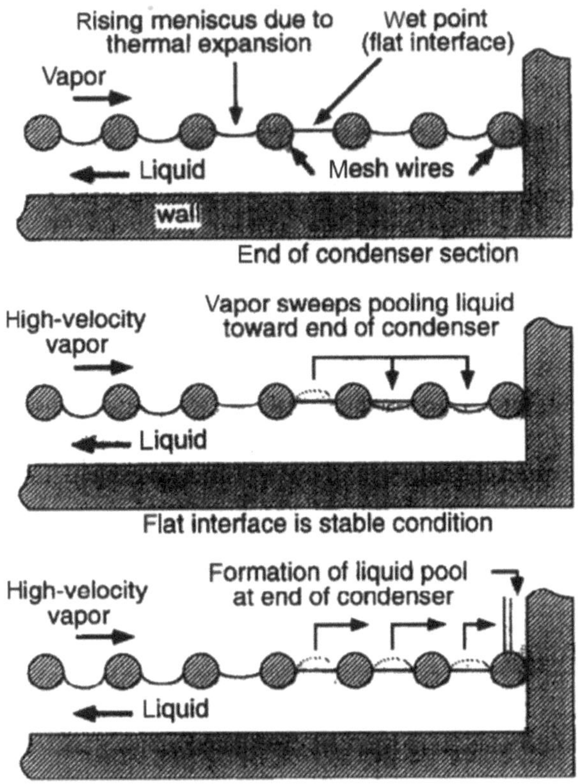

Figure 4.19 Illustration of liquid pooling at the condenser end of a heat pipe [113, *reprinted with permission from Elsevier*].

wick microstructure topology at the liquid–vapor interface on evaporation and heat transfer rates, which become increasingly important in shrinking form factors where surface area-to-volume ratios increase. The numerical models discussed earlier are also not applicable to high-heat-flux conditions for which boiling occurs in the wick structure.

Recent advances in the characterization of transport in wick structures are reviewed in Section 3.2, and numerical models that can account for the effects of the wick microstructure at the liquid–vapor interface and boiling are discussed in Section 3.3. Optimization of vapor chambers for high-heat-flux operation has become possible due to the development of these models. Informed by intensive modeling efforts and subcomponent testing described in Section 2, the recent design, development, and testing of nascent ultrathin vapor chamber technologies are then described.

3.2. Wick thermophysical properties and pore-scale evaporation characteristics

To enable device-scale analysis of transport in wick materials, pore-scale transport mechanisms are typically averaged over a suitably larger-scale representative elementary volume for which macroscale governing equations can be formulated [115]. To employ this approach, geometry-specific relations are required for estimating the representative transport properties of interest, viz., thermal conductivity, permeability, capillary pressure, and evaporation/condensation mass transfer. The effective wick transport properties can then be represented on a simplified basis in device models (see Section 3.1).

3.2.1 Simplified analytical prediction

Several approaches commonly employed for prediction of transport properties are based on simplified representations of the wick geometry. The effective thermal conductivity of saturated porous media is often represented as a function of the porosity and solid-/liquid-phase thermal conductivities. At a given porosity, commonly suggested expressions [3,4] represent meshes as a square array of uniform cylinders [116] and sintered powder beds as having randomly dispersed liquid-phase spheres [117].

The pressure drop imposed to flow through a porous medium is correlated to the superficial velocity using the permeability of the medium in Darcy's law [88], given by

$$\frac{\nabla P = -\mu \vec{V}}{K} \tag{4.3}$$

For a collection of common wick structures [3], expressions for permeability are derived assuming Hagen–Poiseuille flow through passages in the porous structure having a hydraulic diameter representative of the pore size and geometry. The Carmen–Kozeny theory uses a hydraulic radius to predict the friction factor of a porous matrix [88].

The capillary pressure difference generated by surface tension across a spherical curved liquid–vapor interface in a capillary tube is given by the Young–Laplace equation as

$$\Delta P = \frac{2\sigma}{R} \tag{4.4}$$

where R is the radius of curvature. For nonspherical menisci formed in actual wick structures, an effective radius of curvature (or pore radius) is typically defined based on simplifying geometric approximations or empirical observations, as summarized in Ref. [3]. The evaporation/condensation resistance from these menisci can be estimated locally based on the interfacial mass flux as described using the kinetic theory of gases by Schrage [118]:

$$\dot{m}_{\mathrm{i}}'' = \frac{(2\hat{\sigma})}{(2-\hat{\sigma})}\sqrt{\frac{\bar{M}}{2\pi\bar{R}}}\left(\frac{P_{\mathrm{v}}}{T_{\mathrm{v}}^{1/2}} - \frac{P_{\mathrm{i}}}{T_{\mathrm{i}}^{1/2}}\right) \tag{4.5}$$

This formulation can be used to predict the detailed evaporation characteristics of an extended liquid film [119] or a meniscus in a simplified geometry [120], but generalized analytical expressions for evaporation from saturated wick structures are not available.

3.2.2 Empirical characterization

The simplifying assumptions employed in analytical predictive methods prevent accurate portrayal of transport characteristics for realistic wick structures over a broad range of geometries. In order to improve prediction accuracy and enable validation of device-level models, effective transport properties must typically be determined experimentally. Studies that review empirical correlation of heat pipe wick transport properties are briefly summarized in the succeeding text.

Effective thermal conductivity is typically measured by maintaining a one-dimensional temperature gradient across a wick sample at a known heat flux [121]. Atabaki and Baliga [122] reviewed several thermal conductivity models for two-phase mixtures and developed a modified empirical correlation based on experimental data for sintered metal powders [121,122]. Similar testing of biporous and monoporous sintered copper powders performed by Catton and coworkers [64,123] revealed that the effective medium theory model [124] discussed by Carson *et al.* [125], which assumes a random dispersion of both material phases, provides an upper bound on effective thermal conductivity. Li and Peterson [126] reviewed models for predicting the effective conductivity of layers of wire screen and performed experiments using sintered screen to validate a proposed analytical model as a function of the mesh number and wire diameter. Zhao *et al.* [127] formulated a more generalized screen mesh thermal conductivity model in comparison to previous efforts [126,128] that accounted for arbitrary weave patterns.

The permeability of a wick can be found by forcing liquid through a representative sample at a known flow rate and measuring the resultant pressure drop, in either a linear [129] or a radial orientation [130]. Kozai *et al.* [129] investigated the permeability of metal screen wicks as functions of the number of layers, mesh size, and screen compression. An empirical correlation for permeability was developed, and geometries were identified for which accuracy was improved compared to the modified Blake–Kozeny equation [1]. A number of additional comparisons have been drawn against the Carmen–Kozeny theory based on experimental measurement of felt [131], metal screen [132], sintered powder [123], and composite [133] wick permeability.

Several methods are available for evaluating the capillary pressure generated in the pores of a realistic wick structure. In the bubble-point method [134], the capillary pressure is determined based on the minimum pressure required to penetrate gas through the sample; a bubble will first pass through the largest pore, providing a conservative estimate. Conversely, the maximum capillary pressure can be determined by measuring the pressure head sustained by a plug of wick material holding a liquid column against gravity [64,123,135]. An alternative method is to measure the transient rate of rise of liquid in a sample to predict the capillary pressure and permeability based on Washburn's equation [136], as first described with respect to heat pipe wick materials by Adkins and Dykhuizen [130]. Holley and Faghri [137] improved upon the viability of this technique by measuring the liquid mass uptake instead of relying on visualization of a uniform liquid front and by quantifying mass loss due to evaporation. The technique has been recently utilized to characterize a variety of monoporous [38,138] and biporous [139] powder, composite [133], foam [140], micropillared [141], and nanostructured copper post wicks (Fig. 4.20) [49].

Simplified analytical correlations relating evaporation characteristics to wick microstructure do not exist, and most modeling approaches assume evaporation from a flat liquid–vapor interface. Numerous experimental efforts [29–31,53–57], as described in Section 2, have measured the combined conduction and evaporation resistances of saturated wick structures; however, direct comparison of local evaporative mass flux predictions against experiments is usually limited to more simplified geometries [142–148].

3.2.3 Advances in characterization methods

Simplified analytical approaches can readily predict transport properties as a function of wick geometry and are convenient for use in device-scale models, but they lack accuracy. To avoid the need for empirical correlation

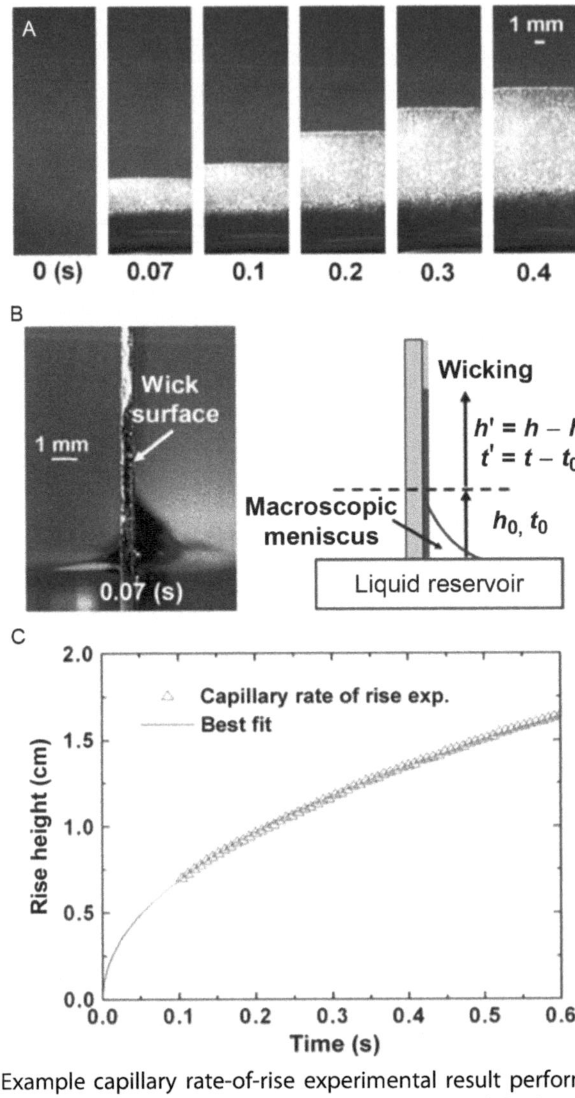

Figure 4.20 Example capillary rate-of-rise experimental result performed for a nanostructured copper post wick showing (A) selected images of the liquid front, (B) side view of the sample under test, and (C) best-fit Washburn's equation to experimental rise height versus time [49, *reprinted with permission from IEEE*]. (For color version of this figure, the reader is referred to the online version of this chapter.)

of transport for each desired wick structure, several numerical prediction approaches have recently been employed and are reviewed here. These recent advances have enabled the design of novel wick microstructures for ultrathin vapor chambers.

With the advent of microtomography techniques that provide submicron resolution, it is possible to obtain intricate three-dimensional details of stochastic porous wick structures to enable direct simulation of transport. A review of this characterization approach with respect to sintered powder wicks is reviewed in detail in Ref. [149]. Direct simulation of transport yields improved accuracy compared to experimental methods that have practical measurement uncertainties. Bodla *et al.* [150] performed computations to resolve the effective thermal conductivity and permeability of commercial sintered wicks composed of multiple particle sizes. The computations showed that the effective thermal conductivity of sintered wicks was a strong function of sintering conditions and the resulting morphology, viz., the interparticle necking, which could not be captured by available predictive methods [125]. Similarly, simplified analytical expressions for permeability [88] significantly overpredicted the computed values due to nonspherical particle shape. Correlated functions of thermal conductivity and permeability for realistic wick structures were proposed [150].

In order to develop a methodology for reverse-engineering capillary wick structures, multiple recent studies have performed numerical analyses to predict the capillary pressure and evaporation characteristics for three-dimensional representative unit cells. Ranjan *et al.* [151] investigated the capillary pressure of several idealized geometries, such as hexagonally packed and square-packed spheres, as representations of sintered powder structures. To improve accuracy compared to simplified effective capillary pore size approximations [3], the interstitial three-dimensional static liquid–vapor interface shape was obtained by evolving the free surface toward a minimum energy configuration by a gradient-descent method [152], as shown for parallel horizontal cylinders and hexagonally packed spheres in Fig. 4.21. The maximum capillary pressure along the interface was obtained via the Young–Laplace equation and then compared for the different structures (Fig. 4.21C); evaporation characteristics were approximated by determining the percentage meniscus area below a defined thin-film thickness [151]. Nam *et al.* [49] employed a similar approach to determine the capillary pressure within an array of microposts. In addition, the permeability of the micropost array was determined by imposing a pressure gradient across the unit cell and assuming slip conditions at the fixed liquid–vapor interface. A performance parameter, K/R_{eff}, was defined to evaluate permeability versus capillary pressure as a function of micropost diameter and pitch [49].

To improve upon the simplified resistance network evaporation models for representative wick pore geometries developed earlier [151], Ranjan

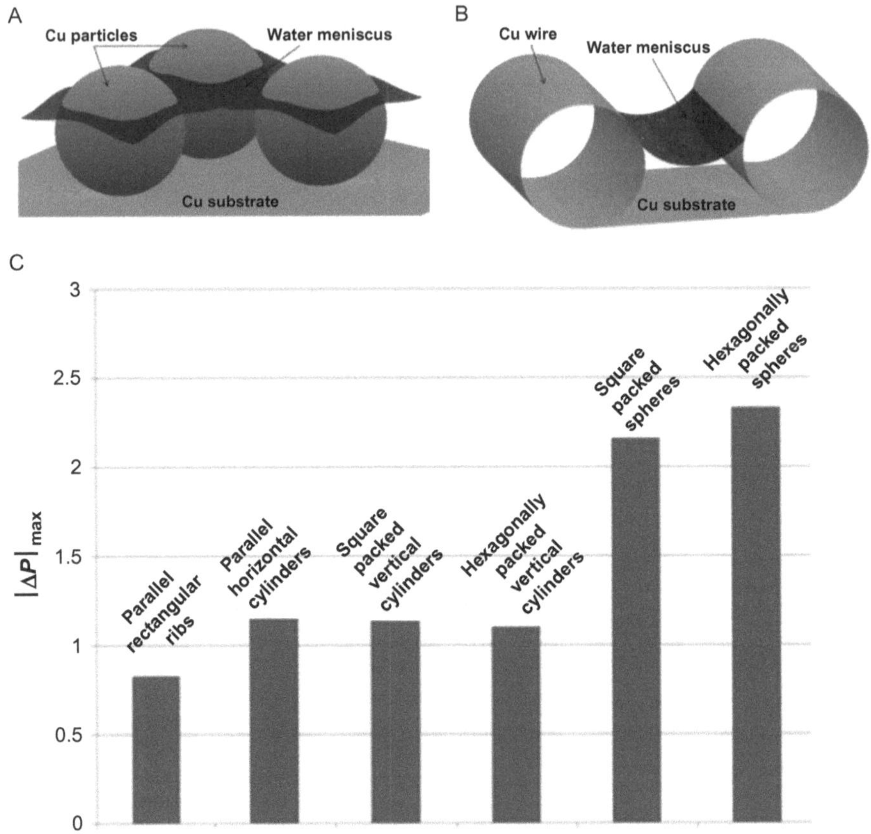

Figure 4.21 Example liquid–vapor interface shapes in (A) hexagonally packed spheres and (B) parallel horizontal cylinders obtained via surface energy minimization and (C) an assessment of the maximum nondimensional capillary pressure achieved in different topologies at any liquid volume for 64% open porosity and a fixed liquid–solid contact angle of 15° [151, *reprinted with permission from ASME*]. (For color version of this figure, the reader is referred to the online version of this chapter.)

et al. [153] formulated a numerical approach to solve three-dimensional heat transport and evaporation from the same canonical topologies. Static interface shapes were discretized for finite volume computational analysis under the prescribed boundary conditions shown in Fig. 4.22A. A constant wall superheat was imposed, and fluid was replenished at the base of the wick structure to accommodate evaporative mass loss at the interface. This model resolves the nonisothermal interface conditions that result in thermocapillary convection and nonuniform evaporative mass flux along the interface (Fig. 4.22). After validation against an analytical model for evaporation

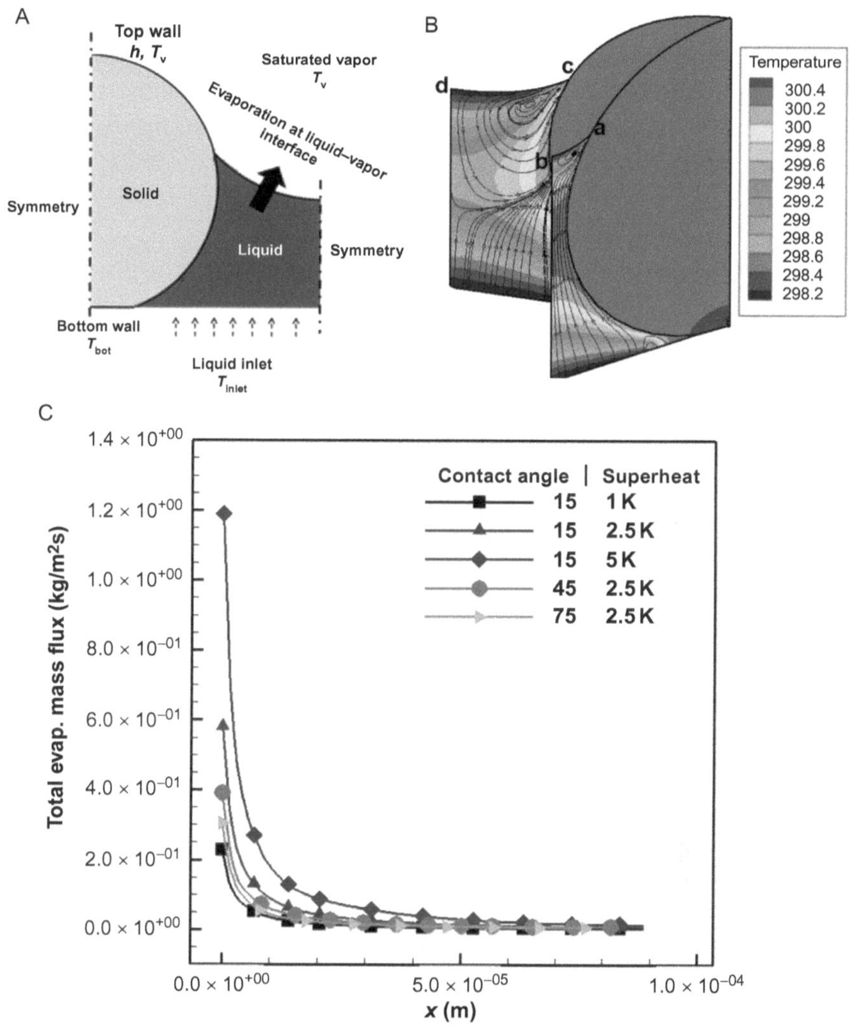

Figure 4.22 (A) Illustration of boundary conditions imposed for the evaporation model and (B) example temperature contours and particle paths formed between square-packed spheres. (C) The evaporative mass flux along the interface is shown for 200 μm spheres at a pitch of 178 μm, 56% porosity, and a saturated vapor temperature of 298 K [153, *reprinted with permission from Elsevier*]. (For color version of this figure, the reader is referred to the online version of this chapter.)

from a channel [120], a comprehensive investigation of the local and area-averaged evaporative mass flux was performed as a function of contact angle, surface superheat, pore geometry, and length scale. Similar modeling approaches that account for nonisothermal evaporation from a quasistatic liquid meniscus are presented in the literature. Dhillon *et al.* [154] investigated evaporation from micron-size rectangular channels representative of a micro-columnated porous silicon wick design. Hwang *et al.* [41] employed this local thermal nonequilibrium modeling approach to evaluate the effective thermal resistance of evaporation from a monolayer of spherical particles. Bodla *et al.* [155] described the evaporative resistance imposed by a realistic sintered powder wick via microtomography-based direct simulation.

Novel structures that enhance performance parameters of interest have been developed using these numerical methodologies for evaluation of wick capillarity, permeability, and evaporative resistance on a unit-cell basis. Ranjan *et al.* [156] computed correlations for the permeability, capillary pressure, and percentage thin-film meniscus area for cylindrical, conical, and pyramidal pillared wick structures (proposed to increase performance compared to sintered particle monolayers). For cases comparing performance at a fixed permeability, pyramidal topologies were shown to provide the highest evaporation rate due to stretching of the liquid thin-film meniscus (Fig. 4.23). When compared against spherical particle monolayers, an order of magnitude improvement in the estimated capillary limit was calculated, albeit at a moderately reduced evaporation rate [156]. Sharratt *et al.* [157] used the same modeling approach to predict the permeability and effective evaporative heat transfer coefficient for novel pillared geometries as a function of the apparent solid contact angle. Hexagonal arrays of cylindrical pillars served as the baseline for comparison against novel pie-cut cylinders and clusters of cylinders (Fig. 4.24A–C). Due to the increased flow area available in the novel geometries, permeability is increased; however, the baseline structure always had the highest permeability for a constant solid fraction [157]. In regard to the effective evaporation resistance, the pie-cut cylinders provide the most favorable trade-off between increasing liquid meniscus contact line length and decreasing thermal conductance from the substrate to the liquid meniscus. As shown in Fig. 4.24D, predictions are compared to experimental measurements for silicon-pillared wick structures.

3.3. Recent advances in ultrathin vapor chamber modeling

The review of vapor chamber and flat heat pipe modeling approaches in Section 3.1 revealed that a majority of the numerical modeling approaches do not consider the wick microstructure on flow and heat transfer. While several models [113,114] accounted for local meniscus curvature effects on

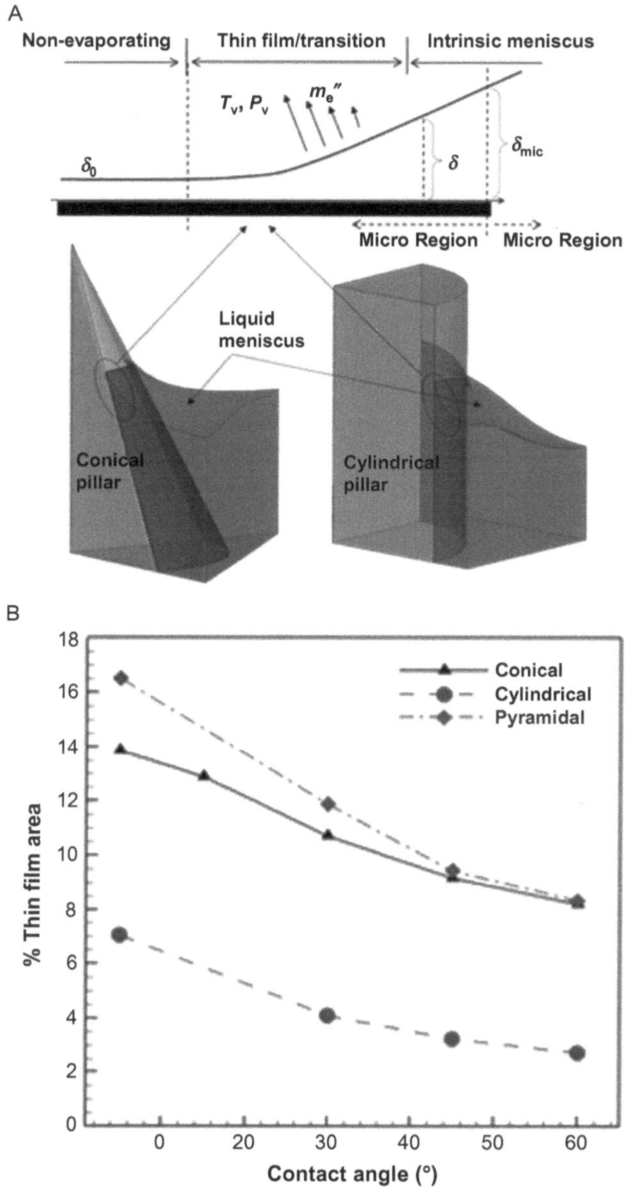

Figure 4.23 (A) Schematic diagram showing the meniscus profile obtained for conical and cylindrical pillared wick geometries and (B) comparison of computed percentage thin-film area as a function of apparent contact angle at constant permeability [156, *reprinted with permission from Elsevier*]. (For color version of this figure, the reader is referred to the online version of this chapter.)

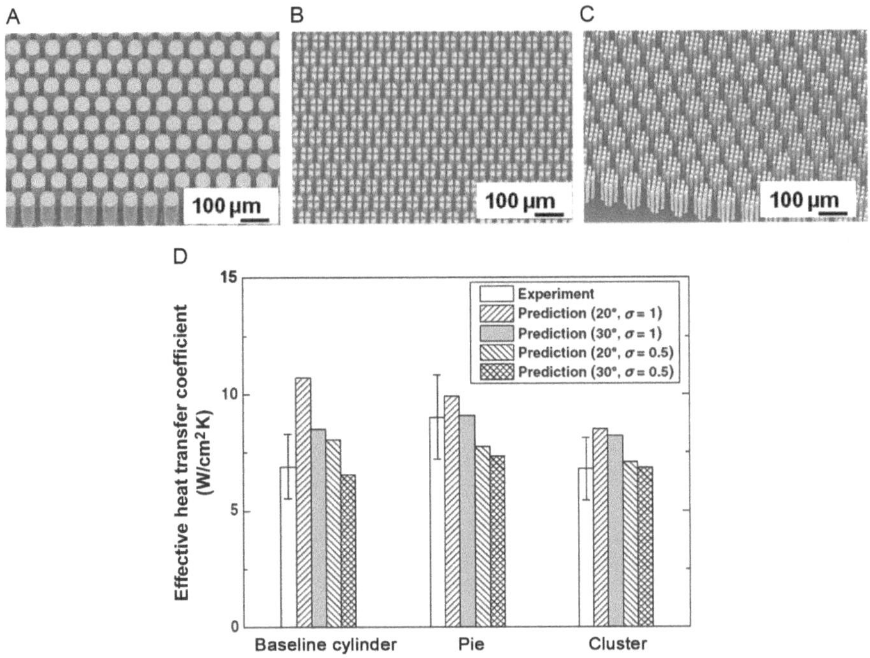

Figure 4.24 SEM images of (A) cylindrical, (B) pie-cut, and (C) clustered silicon-pillared wick structures and (D) comparison of experimentally measured and predicted effective heat transfer coefficients [157, *reprinted with permission from Elsevier*].

capillarity, the effect of local meniscus shape on evaporation was not considered. While the thermal resistance due to evaporation from the wick structure is typically small compared to conduction resistance through the wick layer, as device dimensions are reduced, and wick structure structures becomes increasingly thin and conductive [156,157], liquid–vapor interface resistance is expected to play an important role in determining overall device thermal resistance. Further, extant numerical models do not account for boiling in the wick structure despite extensive evidence of the prevalence of this heat transfer mechanism in wicks subjected to high-heat-flux hot spots (see Section 2). Recent advances in modeling approaches that address these limitations, specifically for the design and development of ultrathin vapor chamber heat spreaders for operation at high heat fluxes, are reviewed in this section.

3.3.1 Wick microstructure effects on evaporation characteristics

During vapor chamber operation, a gradient in the capillary pressure across the liquid–vapor interface is established along the device. Therefore, local meniscus curvature is highest at the evaporator section of the device, where the largest system pressure head must be sustained by capillary forces, and flattens toward the condenser section. This interface shape dictates the thickness of an extended thin-liquid film formed within the wick pores and determines the local evaporation/condensation heat transfer characteristics. To capture these effects, Ranjan *et al.* [158] developed a numerical approach wherein a macroscale device-level model was coupled with a subdevice microscale model that predicted meniscus shape and evaporation from a representative unit-cell structure.

The device-level model used by Ranjan *et al.* [158] was an adapted version of the flat heat pipe model originally developed by Vadakkan *et al.* [107]. The mass, momentum, and energy equations are solved in the solid, porous, and vapor domains to determine the temperature and pressure fields within the device. To account for phase change by evaporation/condensation, the interface temperature is first obtained from an energy balance that accounts for conduction/convection on the liquid and vapor sides of the interface. The local interface pressure is determined by the Clausius–Clapeyron equation, and the local interface mass flux can be calculated using a kinetic theory-based [118] evaporation formulation. Transient pressurization of the vapor chamber due to net mass flux across the interface is incorporated by changing the vapor density by a concomitant amount. The microscale model for evaporation from a porous unit cell is reviewed in Section 3.2.3 and is described in detail in Ref. [153]. The meniscus shape is calculated via a surface energy minimization approach, and conjugate heat transfer and evaporation are computed numerically for the static meniscus shape.

The macroscale device model computes the local interfacial pressure drop along the device; however, it assumes a flat meniscus. Therefore, Ranjan *et al.* [158] developed correction factors to the interfacial area and evaporative mass flux to account for changing liquid–vapor interface shape, thin-film evaporation, and thermocapillary convection effects. In summary, the coupling procedure first computes the flow and temperature fields using the device-level model. The microscale model is then used to develop a correlation for the local contact angle as a function of the capillary pressure difference across the interface, as obtained from the device model. For the computed contact angle, correction factors are introduced to account for the difference in evaporation mass flow from a flat versus curved interface. The mass flux and meniscus area correction ratios are obtained as a function of wick porosity and interstitial contact angle via best-fit regression of the micromodel outputs. Example correction ratios and temperature contours for a flat interface versus a wick pore are shown in Fig. 4.25.

To evaluate the effects of microscale correction ratios on device performance, Ranjan *et al.* [158] investigated the vapor chamber geometry shown in Fig. 4.26 and solved for flow and temperature in the domain using the coupled model. Details of the boundary conditions imposed and the wick, liquid, and vapor thermophysical properties can be found in Ref. [158]. As can be observed in Fig. 4.26, the correction factors are maximized at the location of highest interfacial pressure difference at the center of the heat input and decrease toward the condenser side. The correction ratios were found to be a strong function of the working fluid accommodation coefficient, leading to a maximum temperature correction due to the inclusion of the microscale meniscus effects of 16% among the cases investigated [158]. The investigation concluded

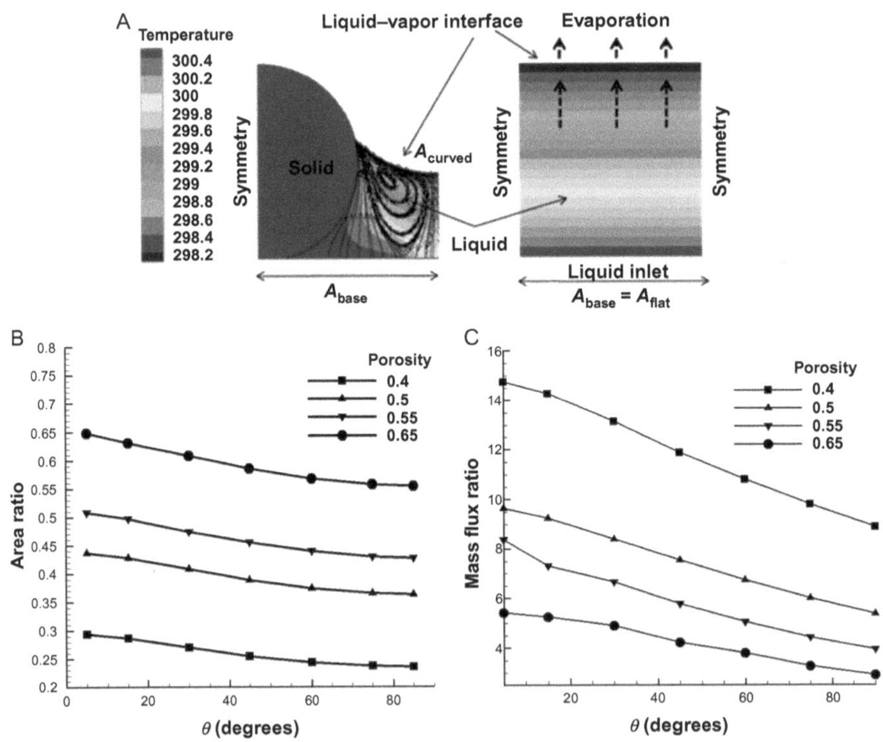

Figure 4.25 (A) Temperature contours for a liquid meniscus formed in a wick pore and under a flat meniscus with 2.5 K base superheat above the vapor temperature (298 K), and the associated (B) area and (C) mass flux correction ratios corresponding to an accommodation coefficient of 1 [158, *reprinted with permission from Elsevier*]. (For color version of this figure, the reader is referred to the online version of this chapter.)

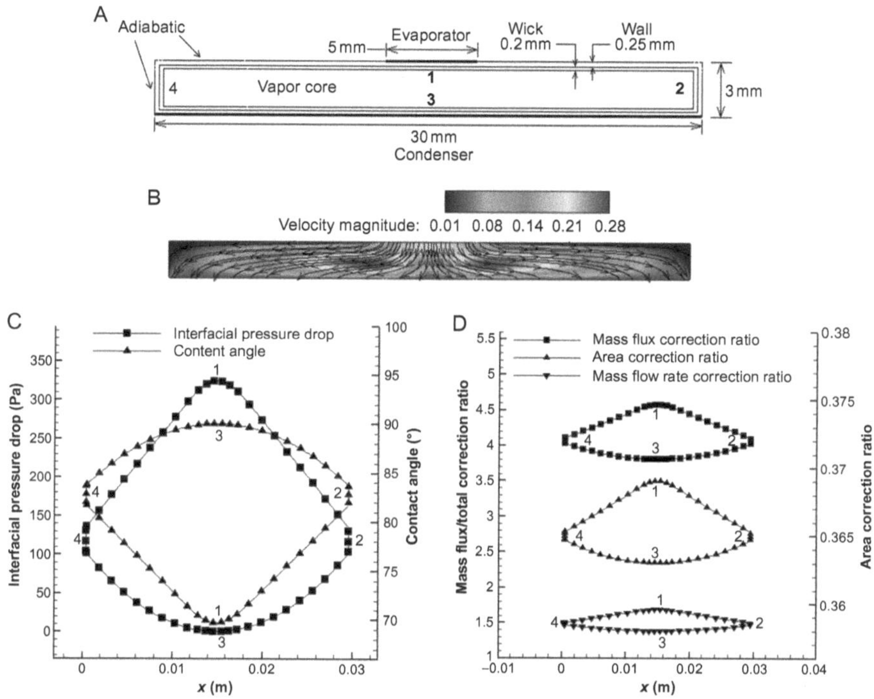

Figure 4.26 (A) Schematic diagram of the vapor chamber modeled in Ref. [158]. (B) Velocity contours are shown for an input heat flux of 10 W/cm^2 and an accommodation coefficient of 1. (C) Interfacial capillary pressure drop and (D) local contact angle mass flux, area, and mass flow rate correction ratios along the liquid–vapor interface are shown for the same case [158, *reprinted with permission from Elsevier*]. (For color version of this figure, the reader is referred to the online version of this chapter.)

that the need for a coupled model is imperative for wick thicknesses on the order of 100 μm, high wick effective thermal conductivities on the order of 100 W/mK, and low accommodation coefficients [158]. Common to these criteria is an increase in the importance of the thermal interfacial resistance associated with evaporation compared to conduction through the wick layer.

3.3.2 Boiling in the wick structure

For experimental validation, modeling tools must accurately capture the performance of ultrathin vapor chambers over the full range of operational conditions. Nucleate boiling in the wick structure has been observed at high heat fluxes under simulated conditions [37] and in testing of actual devices [79]; however, extant modeling approaches (Section 3.1) do not capture this behavior.

One potential strategy for capturing boiling effects in a device-scale vapor chamber model is to couple it to a direct simulation of bubble departure from a microporous-coated heated wall (in an analogous manner to the microscale evaporation model coupling described in Ref. [158]). As discussed by Dhir [159], measurement difficulties, prediction inaccuracies, and lack of generality associated with the effort to correlate nucleate boiling mechanisms over the last half of the twentieth century, viz., nucleation site density, bubble departure diameter, and bubble departure frequency, have spurred recent investigation of direct simulation of vapor bubble growth during nucleate boiling from flat surfaces [160–167]. Reports of similar approaches in the presence of porous structures are limited.

Ranjan *et al.* [168] investigated the formation of two-phase flow structures and vapor growth in a two-dimensional porous matrix using a validated volume-of-fluid–continuum surface force numerical model. In contrast to boiling from a smooth surface, discrete bubbles do not depart from the porous matrix and instead form continuous dendritic vapor columns due to the increased wall adhesion area that is not overcome by buoyant forces. Under these conditions, while decreasing porosity increases the phase-change heat transfer rate due to increased thin-liquid film area, the increased vapor flow pressure drop may cause dryout at a lower heat flux [168]. The formation of continuous vapor columns in a high-porosity anisotropic matrix (Fig. 4.27) mimics the mechanistic behavior proposed by the Smirnov correlation [72] evaluated in Section 2.3. A similar conduction-based equivalent fin heat transfer model was proposed by Ranjan *et al.* [168] and used to determine an optimum particle size to wick thickness ratio based on the fin efficiency, which agreed with prior experimental observations by Chien and Chang [60] and Webb [61]. Li *et al.* [169] visualized boiling from close-packed glass and copper spheres and compared the results against a volume-of-fluid numerical model of the same configuration. In both experiments and simulations, similar vapor column structures were observed at medium heat fluxes, which broke apart due to hydrodynamic instability caused by countercurrent liquid flow to the heated surface. The vapor columns only broke apart once liquid flow to the surface was insufficient to maintain new vapor mass flux to support the columns [169].

The potential for complete numerical modeling of boiling heat transfer is supported by good agreement with data obtained from surfaces with single or multiple controlled nucleation sites [161,170]; however, further computational advancements are required to predict nucleate boiling in complex three-dimensional geometries, particularly at high heat fluxes when

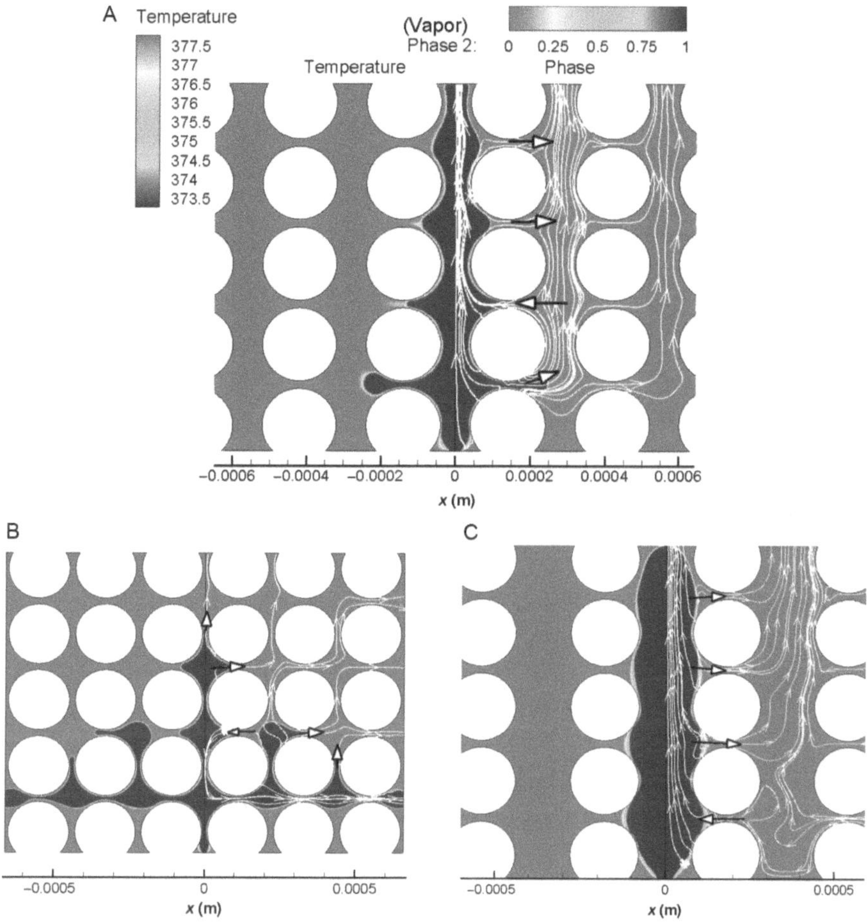

Figure 4.27 Temperature (K) and phase contours in the fluid domain for vapor generation from a heated surface covered with an anisotropic porous matrix composed of 100 μm particles. Cases shown have particle pitches of (A) 280, (B) 220, and (C) 360 μm ($\Delta T_{superheat} = 5$ K, $T_{sat} = 373$ K) [168, *reprinted with permission from IEEE*].

vapor–liquid interfaces evolve rapidly. To obtain reasonable comparison against experimental vapor chamber performance data for conditions that result in boiling in the wick structure, Ranjan *et al.* [171] adopted a semi-empirical approach to model boiling conditions at high-heat-flux inputs. The foundational numerical model was adapted from Ref. [158], as described in Section 3.3.1. In the model, nucleate boiling was assumed to occur over the spatial extent where the local wick temperature exceeded a prescribed incipience superheat. Based on the experimental observations of Weibel *et al.* [37], the conduction thermal resistance was bypassed over this area, and an empirical boiling heat transfer coefficient was instead imposed. As shown in Fig. 4.28, this approach yielded good agreement with experimental thermal resistance data obtained for a vapor chamber heat spreader [79] at heat fluxes up to 500 W/cm^2.

3.4. Design and development of ultrathin vapor chamber devices

Advances in experimental and computational wick structure characterization and device-modeling approaches have enabled the recent design, development, and optimization of ultrathin vapor chamber heat spreader devices. The need for flat, thin (~1 mm) vapor chambers with low coefficient of thermal expansion (CTE) (5–7 ppm/K) and high effective conductivity (>1000 W/m K) has recently emerged for effective proximate heat spreading from emerging high-power (>100 W/cm^2) microelectronic devices. In current state-of-the-art electronics packages, the microelectronic components are directly mounted on low-CTE solid heat spreaders. Direct mounting of microelectronic components to vapor chamber heat spreaders composed of CTE-matched materials could serve as a comparatively higher-effective-conductivity "thermal ground plane" (TGP) without compromising the package form factor or functionality. This section reviews recent developments in ultrathin vapor chamber heat spreaders with these functional performance targets set by the Defense Advanced Projects Research Agency (DARPA) and supported through significant grant funding.

3.4.1 Radio-frequency TGP

A collaborative research effort led by Raytheon Company pursued the development of a radio-frequency TGP (RFTGP) [172] composed of a copper–molybdenum–copper casing with micro-/nanostructured sintered copper powder wick structures, as depicted in Fig. 4.29A. Altman *et al.* [79] developed an experimental facility to test RFTGP devices with monolithic [37], micropatterned [43], and carbon nanotube (CNT)-coated [51] sintered powder evaporator wick structures. Performance of the 30 mm × 30 mm × 3 mm vapor chamber was directly compared to a solid copper–molybdenum heat spreader of equivalent external dimensions for a hot-spot heat input area of 5 mm × 5 mm (up to 500 W/cm^2). Incipience

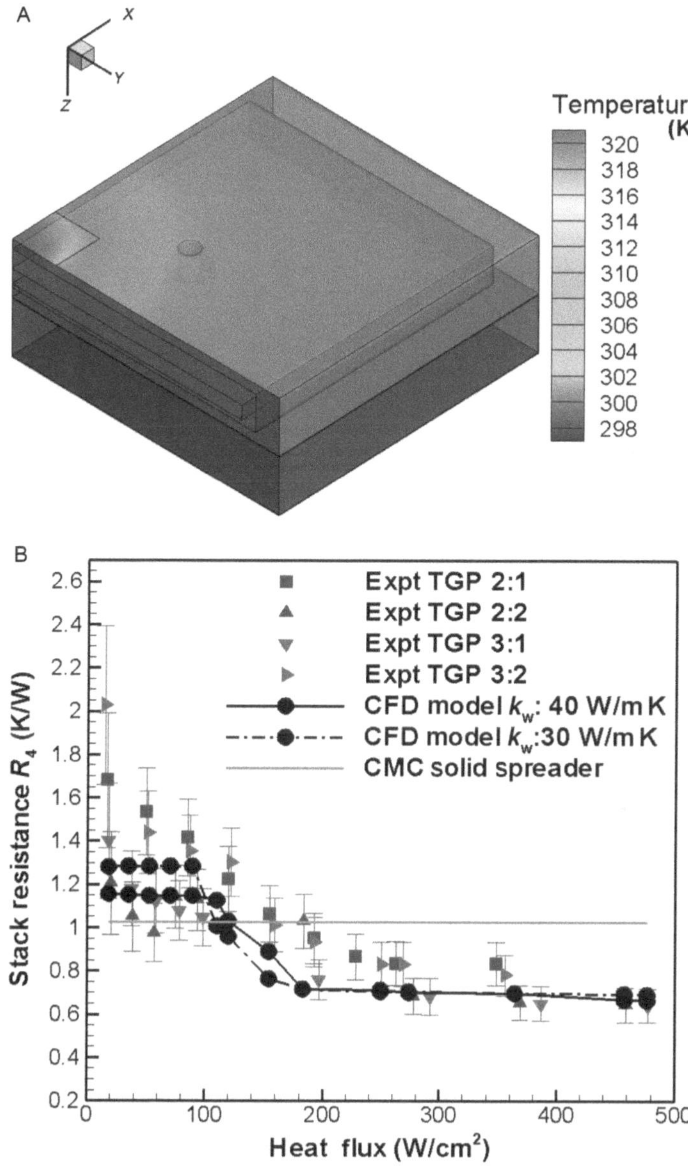

Figure 4.28 (A) Temperature contours on the outer surfaces of the thermal ground plane (TGP) vapor chamber model for an evaporator heat flux of 89 W/cm^2 and two different values of wick thermal conductivity and (B) comparison against experimental data up to 500 W/cm^2 for vapor chamber devices and a CuMoCu (CMC) solid heat spreader [171, *reprinted with permission from IEEE*]. (For color version of this figure, the reader is referred to the online version of this chapter.)

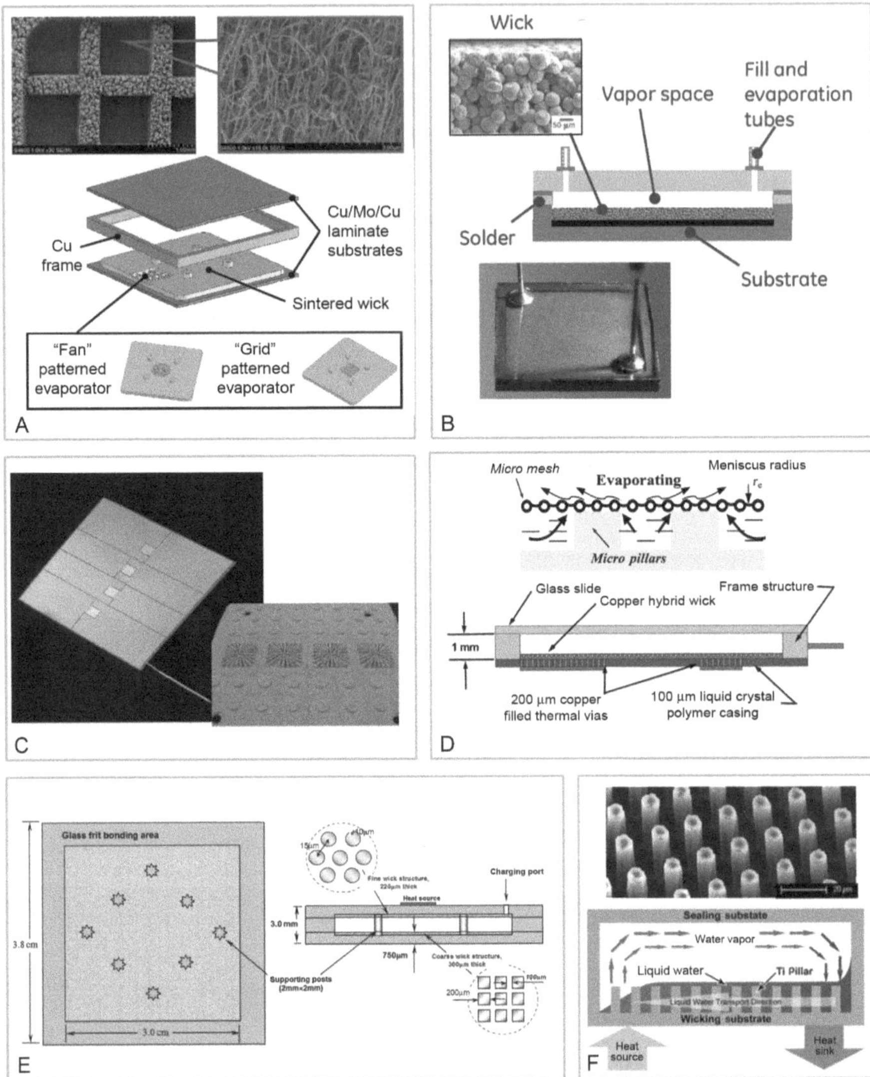

Figure 4.29 Diagrams and images of several recently developed low coefficient of thermal expansion vapor chamber thermal ground planes for direct mounting of microelectronic devices including (A) a radio-frequency thermal ground plane [79, *reprinted with permission from ASME*], (B) a micro-/nanothermal ground plane with sintered powder [173, *reprinted with permission from IEEE*], (C) an aluminum nitride vapor chamber [42, *reprinted with permission from Elsevier*], (D) a polymer-based flat heat pipe [174, *reprinted with permission from IEEE*], (E) a silicon thermal ground plane [175, *reprinted with permission from IOP Publishing*], and (F) a small-scale titanium thermal ground plane [176, *reprinted with permission from IEEE*]. (For color version of this figure, the reader is referred to the online version of this chapter.)

of boiling in the vapor chamber was shown to be the critical transition at which boiling-dominated heat transfer realized improved relative performance. Based on comparisons to direct measurement of the thermal resistance associated with boiling from monolithic sintered powder wick structures [37], it was concluded that the evaporator thermal resistance governed the overall vapor chamber thermal resistance and hydrophobic nanostructures on the internal condenser surface did not alter the overall performance. Of the various enhancement features explored, hydrophilic copper-functionalized CNTs displayed an ability to shift the critical boiling incipience transition to a lower wall superheat [79] (as further discussed in Section 4.2.2).

The measured overall vapor chamber thermal resistance was observed to be a strong function of the input heat flux and internal saturation pressure/temperature [79]. This correlation was further explored by Ranjan *et al.* [171] using the numerical vapor chamber model as described in Section 3.3.2, which was calibrated to the RFTGP experimental test data from Ref. [79]. The numerical model was used to explore the design of a 1 mm overall thickness vapor chamber device. It was observed that the thermal resistance of the vapor core (due to the saturation pressure/temperature gradient) becomes increasingly important at reduced thicknesses. Therefore, water vapor thermophysical property variations lead to nonlinear device thermal resistance behavior with linearly varying input heat flux and external condenser temperatures [171]. A parametric variation of the relative wick and vapor core thicknesses concluded that the minimum wick thickness should be selected such that a capillary limit is not reached at the desired operating heat flux in order to reduce the evaporator and vapor core thermal resistances.

To assess the improvement in package-level heat spreading from a multichip module afforded by replacing a low-CTE solid material with a vapor chamber, Altman *et al.* [177] measured the comparative thermal resistances of heat sink packages containing 53 mm × 34 mm × 1.4 mm heat spreaders. For heat dissipation from a pair of simulated 6 mm × 6 mm high-power microelectronic devices, it was shown that the package resistance was reduced by 17–26% (depending on the overall package resistance and gravitational loading) using a vapor chamber with equivalent thermal expansion characteristics and geometry as a solid copper–molybdenum heat spreader. A numerical model predicted the measured gravitational loading that induced dryout of the evaporator. Evaluation of thermal resistance in dynamic inertial loading environments at constant heat input revealed detrimental hysteresis associated with on/off cycling of high gravitational forces that induced dryout; however, the performance could be fully restored after a dryout event by power cycling below a nonzero "resetting" input heat flux that was identified.

3.4.2 Micro-/nanostructured TGP

A 30 mm × 30 mm × 3 mm vapor chamber TGP (Fig. 4.29B) was recently designed, fabricated, and tested in an effort led by researchers at GE Global Research [173,178]. Device performance is extremely sensitive to small variations in the amount of initial fluid charge at this form factor; overfilling floods the condenser and underfilling may lead to premature dryout. Hence, a fluid charging station with demonstrated accuracy of ± 2 μl was developed to evacuate and backfill prototype vapor chambers using a series of graded resolution burettes [173]. A thermal characterization facility was designed to generate 30 mm × 10 mm heat source and heat sink areas on opposite ends of the vapor chamber. An effective thermal conductivity metric was defined based on the temperature drop associated with equivalent two-dimensional heat spreading in a solid material of known conductivity, and an uncertainty analysis was used to determine the required temperature measurement accuracy [173].

To ensure device performance under adverse gravitational forces, de Bock *et al.* [178] developed device thermal performance and capillary limitation models based on an effective thermal resistance network and simplified expressions for liquid-/vapor-phase pressure drop, respectively. In order to ensure dissipation of the desired heat load against an exerted body force 10 times normal gravity, it was found that the properties of a sintered powder wick with 75 μm particle diameters [134,138] would avoid the capillary limitation [178]. The effective thermal conductivity of three prototype vapor chambers was measured as a function of the gravitational body force by mounting on a centrifuge spin table. The thermal resistance change was negligible due to the minimal additional vapor pressure drop imposed at the maximum gravitational loading. The same prototype devices were tested as a function of input heat flux and under certain operating conditions exceeded the effective thermal conductivity of copper at an adverse body force 10 times gravity [178].

3.4.3 Planar vapor chambers with hybrid evaporator wicks

Thin vapor chamber prototypes have been recently developed based on the design and testing of biporous wick structures and arterially fed evaporators as previously discussed in Sections 2.1.3 and 2.2, respectively. Prior testing of biporous sintered powder wick structures [35,36] simulated saturated vapor conditions but, unlike realistic devices, left a large open vapor space above the wick structure during boiling. Recent investigations by Catton and coworkers [179–181] attempted to reconcile this difference by testing biporous wick structures with a vapor restriction plate placed directly over the sample. In this way, the test facility provided an improved representation of vapor chamber performance based on the device thickness and lateral vapor pressure drop. Experimental results were compared directly to the thermal resistance incurred in prototype vapor chamber TGPs having the same internal wick structure and heat input area. The thermal resistance due to vapor pressure drop was significant for thin vapor chambers, and use of a vapor restriction plate in a capillary-fed evaporation/boiling test facility provided excellent agreement with vapor chamber device testing [181].

Ju *et al.* [42] explored a low-CTE vapor chamber envelope composed of aluminum nitride ceramic plates (with direct-bonded copper layers for water compatibility) spaced apart by a Kovar ring. Several wick structures were considered based on preliminary subdevice testing, viz., biporous sintered powder [36], lateral converging liquid return arteries [41], and vertical columnar arteries [66], but lateral arteries were ultimately chosen for their mechanical robustness [42]. A 100 mm × 100 mm prototype vapor chamber was constructed to accommodate an array of four 10 mm × 10 mm heat input areas for the potential thermal management of vertical-cavity surface-emitting laser arrays, as shown in Fig. 4.29C. The vapor chamber prototype was demonstrated to dissipate a total of 1500 W prior to dryout and capable of supplanting lower-reliability microchannel heat sinks currently used for this application [42].

3.4.4 Polymer-based flat heat pipe

A collaborative research effort led by the University of Colorado Boulder aimed to develop ultrathin polymer-based flat heat pipes (PFHPs) amenable to fabrication via high-volume manufacturing technologies. A hybrid (biporous) structure served as the capillary wick in a series of prototype heat pipe devices; design of this wick structure is described in Ref. [182]. The hybrid wick was composed of a fine copper mesh (intended to enhance local evaporation/condensation) that was sintered to a grooved copper surface (intended to provide liquid return at minimal flow resistance). Modeling efforts showed that this structure could provide a significant improvement in maximum heat transport capability compared to a homogeneous copper mesh [182]. A first generation 100 mm × 30 mm × 2.5 mm all-copper flat heat pipe was fabricated to demonstrate the effectiveness of the hybrid wick structure. Based on the evaporator-to-condenser temperature difference, and the device cross-sectional area, testing demonstrated that an effective thermal conductivity greater than 10,000 W/m K could be achieved for a heat input of 91.3 W over a 25 mm × 25 mm area (14.6 W/cm^2) [182].

Subsequent investigations by Oshman *et al.* [174,183] focused on development of flexible polymer-based heat pipes. A liquid-crystal polymer (LCP) casing material was chosen for chemical resistance, customizable low CTE, hermeticity, and compatibility with flexible printed circuit board manufacturing processes. In order to increase the inherently low polymer thermal conductivity, copper thermal vias were inserted into the LCP at the heat input and condenser locations. A detailed description of thermal via insertion and fabrication of the copper hybrid wick structure for LCP heat pipe walls is provided in Ref. [174]. A prototype PFHP (60 mm^3×30 mm^3×1 mm^3) was fabricated using these techniques (Fig. 4.29D). Using the same definition mentioned earlier, the maximum effective thermal conductivity was measured to be 830 W/m K over an input power range of ~3–12 W/cm^2 [174]. Using the same fabrication approaches, a 40 mm^3 × 40 mm^3 × 1 mm^3 PFHP was later fabricated with different heat input and condenser configurations [183]. For heat fluxes above ~30 W/cm^2, testing showed a favorable reduction in thermal resistance compared to pure copper.

3.4.5 Silicon TGP vapor chamber

Silicon micro heat pipes generally are described by an embedded array of discrete parallel noncircular channels that each behave as a two-phase evaporation/condensation loop to increase the inherent thermal conductivity of silicon [184–186]. Several investigations have also developed planar vapor chambers composed entirely of silicon with axially grooved wick structures [187]. Recently, researchers at Teledyne Scientific & Imaging Company developed and tested all-silicon planar vapor chambers with micropillared wick structures for spreading heat from high-flux hot spots [175,188].

Cai *et al.* [188] fabricated a flat hexagonal vapor chamber based entirely on silicon photolithography, dry etch, and wafer-bonding processes. The 2 mm thick vapor chamber had a hexagon edge length of 10 mm (total surface area of ~2 cm^2). The hexagonal shape was motivated by the ability to link together multiple "hexcell" chambers for both bulk fabrication throughput and improved spreading from an array of heat sources. A parametric optimization of the structural design was performed to ensure mechanical integrity at the internal vapor pressures encountered at 135 °C. A trade-off between reduced vapor flow area and maximum operating pressure resulted in the placement of six posts along lines bisecting the hexagon edges [188]. Operating pressure tolerance and hermitic sealing was demonstrated after bonding the upper and lower silicon chamber walls. While thermal spreading performance of a sealed device was not tested, separate capillary-fed boiling testing of the pillared silicon wick was shown to dissipate 300 W/cm^2 from a heat source area of 2 mm × 2 mm at 35 °C superheat [188].

In a later study, Cai *et al.* [175] developed a square 38 mm × 38 mm × 3 mm TGP vapor chamber (Fig. 4.29E) using several similar fabrication procedures. On the internal condenser side, a coarse-pillared silicon wick was used to facilitate liquid return to the evaporator via posts that also provided structural integrity. A finer silicon-pillared wick (10 μm diameter and 15 μm pitch) was used on the evaporator side to provide the necessary capillary pressure to sustain operation under adverse gravitational loading. A novel three-layer silicon wafer-stacking fabrication process was employed to increase device yield by reducing the required silicon wick etch depth on each wall compared to a two-wafer stack. Thermal testing of the vapor chamber was performed with heater (30 mm × 4 mm) and condenser (30 mm × 5 mm) areas at opposite ends of the vapor chamber. Performance of initial prototype devices was highly sensitive to liquid charge and noncondensable gases, leading to a large range of measured effective thermal conductivities (~900–2500 W/m K). Charge optimization led to a maximum measured device effective thermal conductivity of ~2700 W/m K [175].

3.4.6 Titanium TGP

Researchers at the University of California Santa Barbara explored fabrication of all-titanium vapor chambers [176,189]. Relative to other potential materials, titanium offers excellent corrosion resistance, is light weight, has high fracture toughness, and can be used as the substrate for microfabrication of high-aspect-ratio wick features [190].

Ding *et al.* [176] fabricated a proof-of-concept device having external dimensions of 30 mm × 30 mm × 0.6 mm for TGP applications (Fig. 4.29F). The internal wick structures were titanium pillars (10 μm diameter and 15 μm pitch) oxidized to form secondary nanostructured titania (NST) surface features (fabrication details in Ref. [176]). Transient wetting behavior of the wick structure was shown to behave in accordance with Washburn's dynamics [191], and the NST surface increased the wetting velocity. The vapor chamber was sealed along the edge by local laser welding to avoid heating the entire device to the necessary processing temperatures. By applying a fixed temperature difference between evaporator and condenser sections, the fabrication proof-of-concept device achieved a maximum effective thermal conductivity of ~350 W/m K at this form factor [176].

A large-scale titanium TGP (Ti-TGP; 300 mm × 76 mm × 4.5 mm) with 24 individual heat source mounting locations was later fabricated using the same fabrication techniques [189]. Unlike the small-scale Ti-TGP, which used an array of microfabricated pillars, the large-scale device consisted of a groove wick with NST. The groove structure was optimized to maximize the heat transport capability (by equating the capillary pressure with pressure losses) and for a device in a vertical reflux orientation. Effective device thermal conductivities of 5000–8000 W/m K were measured based on the effective working length during simulated testing of the large-scale Ti-TGP using eight independent heat sources; total heat dissipation was 500 and 1000 W at evaporator temperatures of 100 and 150 °C, respectively [189].

4. NANOSTRUCTURED CAPILLARY WICKS FOR VAPOR CHAMBER APPLICATIONS

Advances in controllable synthesis techniques continue to further enable the use of nanostructures in numerous engineering applications that exploit their tunable geometric, thermal, and mechanical properties. Nanostructures such as CNTs and metal nanowires (NW) have been evaluated for use as vapor chamber capillary wick structures owing to a number of potentially advantageous characteristics.

Conduction through the wick layer often imposes a significant thermal resistance during vapor chamber operation. The intrinsically high thermal conductivity of CNTs determined both theoretically [192,193] and experimentally [194–197] may be exploited and has previously led to a reduction in the resistance to heat flow at interfaces between components [198–201] and a development of novel composite materials with increased thermal conductivity [201–203]. The pores of nanowire arrays also have a high capillary pressure; however, their relative impermeability compared to microscale wick structures must be carefully assessed in the design process. Further, while the hydrophilicity of NWs and CNTs with water has been reported in the literature [204,205], aligned arrays of nanotubes have also been shown to behave as superhydrophobic surfaces [206]. Hence, surfactants may be used for liquid-conveying applications [207], or nanostructures may be functionalized for heat transfer applications via metallization [51], hydrochloric acid treatment [208], or ultraviolet excitation [209]. Nanostructures have a high number of pores per unit substrate area and thereby may also offer an increase in the thin-film area for evaporation. Nanostructures have been reported to improve differing aspects of the boiling process (e.g., incipience, nucleation boiling, and CHF) via CNT coating of silicon [210–212] and copper substrates [78,212] and copper nanowire [213–215] and silicon nanowire [214–216] surface coatings.

It is important to distinguish and evaluate the potential enhancement provided by nanostructured wicks during capillary-fed evaporation/boiling processes. Recent studies on the design and testing of nanostructured wicks for use in vapor chambers are discussed in this section. Two potential configurations are evaluated: (1) use of nanowire arrays as the primary wicking and evaporation structure and (2) nanostructured coating of conventional microscale wick structures.

4.1. Assessment and design of nanostructured wicks

In order to determine the viability of nanowick structures for use in vapor chambers, the morphology dependence of capillarity, permeability, and thermal resistance must be determined. Ranjan *et al.* [217] developed theoretical and numerical models to approximate these quantities for representative aligned vertical cylinders in hexagonal and square packing arrangements.

The capillary pressure was determined by obtaining the mean curvature of the liquid meniscus formed in a nanowick (as in Ref. [151]), while permeability was estimated by simulating the pressure drop associated with flow through a two-dimensional unit cell (Fig. 4.30A and B). For this analysis, wetting contact angles were assumed based on the observations of Rossi *et al.* [218] and Kim *et al.* [219], and continuum approximations for capillary dynamics [219], surface tension [220], and viscous drag [221] were shown to be justified. Properties were obtained as a function of nanowire diameters and number densities consistent with typical fabrication processes [222–224].

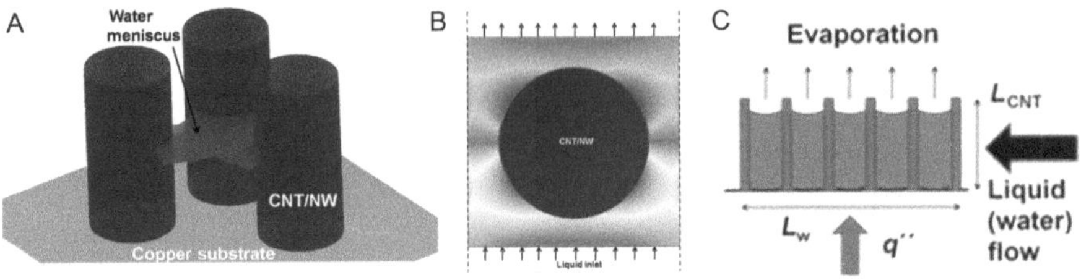

Figure 4.30 (A) Shape of the water meniscus in the nanopore formed between square-packed vertically aligned cylinders and (B) pressure contours shown in the liquid region around a cylinder (which are used to compute the nanowick permeability). (C) Two-dimensional representation of wicking length, L_w, across the nanowire array with input heat flux q'' [217, *reprinted with permission from Taylor and Francis*]. (For color version of this figure, the reader is referred to the online version of this chapter.)

4.1.1 Nanowire array wicks

Ranjan *et al.* [217] considered evaporation from a uniform nanowire array fed by capillary action. The evaporative resistance of the liquid meniscus formed in the array was computed numerically using the previously developed model discussed in Section 3.2.3 [153]. The thermal resistance of the nanowick, which is governed primarily by conduction resistance through the height of the saturated porous structure, is potentially orders of magnitude lower than typical sintered copper powder or screen wicks.

While it is clear that nanowicks may outperform conventional wick materials purely on the basis of thermal resistance, the capillary pressure generated must sustain liquid flow to the meniscus at the desired heat load. A wicking length was used to assess the feasibility of nanowick structures in this regard and was defined as the maximum length over which a given mass flow rate (i.e., heat load) can be transported through the wick structure via capillary action [217], as shown in Fig. 4.30C. Analysis as a function of nanowire density found that the maximum wicking length occurred at a nondimensional pitch of 5 due to the trade-off between capillary pressure and permeability; however, the maximum wicking lengths were only on the order of 1 cm even for modest heat loads due to the low permeability of the structure. This suggested that use of nanowick arrays over a large area on a heated smooth substrate would perform poorly [217].

Due to these inherent capillary transport limitations, Weibel *et al.* [225] proposed evaporator surfaces composed of nanowire arrays fed by interspersed conventional microscale wick structures. Design of such wicks required a study of the trade-offs between the greater permeability offered by conventional wick structures and the reduced thermal resistance offered by a nanowire array. The geometry selected for parametric investigation was a series of alternating wedges of microscale and nanoscale wick layers (Fig. 4.31). A numerical model was developed to analyze fluid flow and regions of dryout in the evaporator using estimated inputs for the capillarity, permeability, and effective thermal resistance of each region. The proposed evaporator structure was compared to a conventional homogeneous microscale wick, and thermal resistance was found to be significantly reduced when sufficiently short wicking lengths within the nanostructured regions were ensured by geometric design [225].

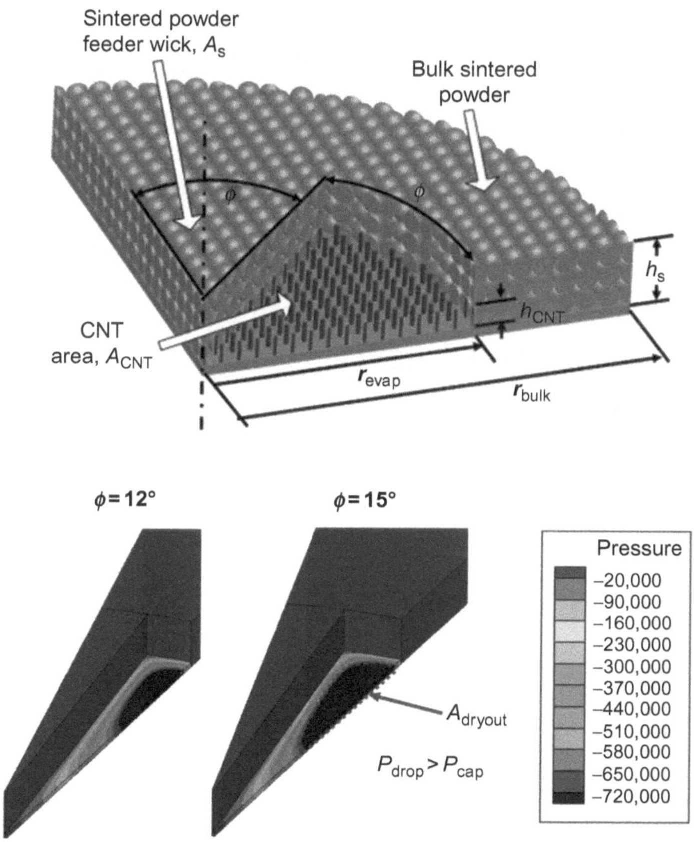

Figure 4.31 Schematic diagram of the wedge geometry chosen for the integrated evaporator wick structure and example pressure contours (in Pa) in the nanowire domain with respect to a zero pressure inlet condition for varying wedge angle, Φ. (For color version of this figure, the reader is referred to the online version of this chapter.)

4.1.2 Nanostructured coatings

An alternative wick design is to directly coat conventional wick microstructures with nanostructures in order to increase wettability and total thin-film area for enhanced evaporation heat transfer. This approach has been previously used to increase the capillarity of titanium [176] and copper [48] micropost wick structures.

Ranjan *et al.* [217] studied the case of a high-permeability sintered copper powder wick coated with nanowires and employed simplified theoretical and numerical models to estimate the potential thermal performance enhancement via nanostructuring. Two extreme cases are presented in Fig. 4.32A: (1) completely nonwetting nanowires that only serve to alter the local meniscus shape near the liquid–solid contact line formed in the microstructure and (2) nanowires with a sufficient wicking length to coat the entire microscale particle with a thin-liquid layer. An increase in the thin-film meniscus area due to the presence of nanowires is estimated based on the resolved 3D meniscus shape, and the thermal resistance is computed using a simplified network model. Figure 4.32B shows the reduction in wick thermal resistance for multiple liquid fill heights in the pore. Both wetting and nonwetting nanowires are predicted to reduce the thermal resistance (by a maximum of 14% for the most optimal configuration) [217].

4.2. Experimental evaluation of nanostructured wicks

Informed by the expected wetting behavior of low-permeability nanostructures, several novel evaporator structures composed of patterned nanowire arrays and nanostructure-coated wicks have been fabricated in the literature. In this section, several experimental studies are reviewed that evaluate the potential thermal performance enhancement provided by nanostructured wicks.

4.2.1 Nanowire array wicks

In a pair of studies conducted by Weibel *et al.* [50,226], sintered powder wick structures with an array of interspersed 1 mm × 1 mm square regions of CNTs were investigated. The sintered powder structure was composed of 100 μm copper particles, and two different wick thicknesses were evaluated, 1 mm and 200 μm. The CNTs were grown in a microwave plasma chemical vapor deposition system following deposition of metal catalyst layers (Ti/Al/Fe), where Fe provided active growth sites for the CNTs. Details of the CNT growth procedure are provided in Ref. [227]. The samples were functionalized by coating the CNTs with a thin layer of evaporated copper via physical vapor deposition, making the CNT surface hydrophilic. Images

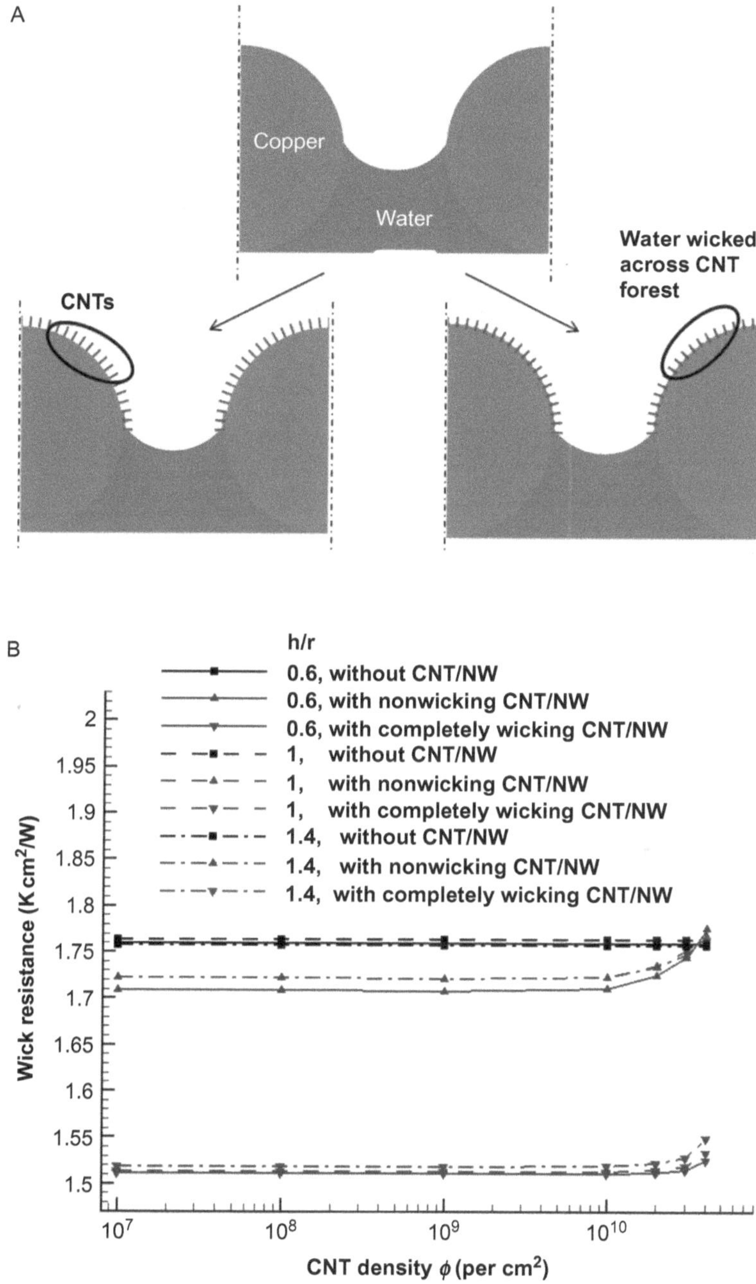

Figure 4.32 (A) Illustration of nanowire-coated spherical particle (2D representation of 3D model, not to scale) for extreme cases of completely wetting and nonwetting nanostructures and (B) thermal resistance network model results plotted versus nanowire number density for multiple liquid levels in the microscale pore [217, *reprinted with permission from Taylor and Francis*]. (For color version of this figure, the reader is referred to the online version of this chapter.)

of the copper-coated CNT structures interspersed within a 200 μm thick sintered powder sample are shown in Fig. 4.33. Samples without CNT structures were also prepared as a baseline for comparison.

Experimental evaluation of capillary-fed evaporation/boiling was performed in the experimental facility described in Section 2.1.1. For the 1 mm thick sintered powder sample, an array of 1 mm × 1 mm square recesses improved performance in the boiling regime due to the reduced resistance to vapor exiting the wick structure, as described in Section 2.2; however, addition of the CNT array did not alter performance in the boiling

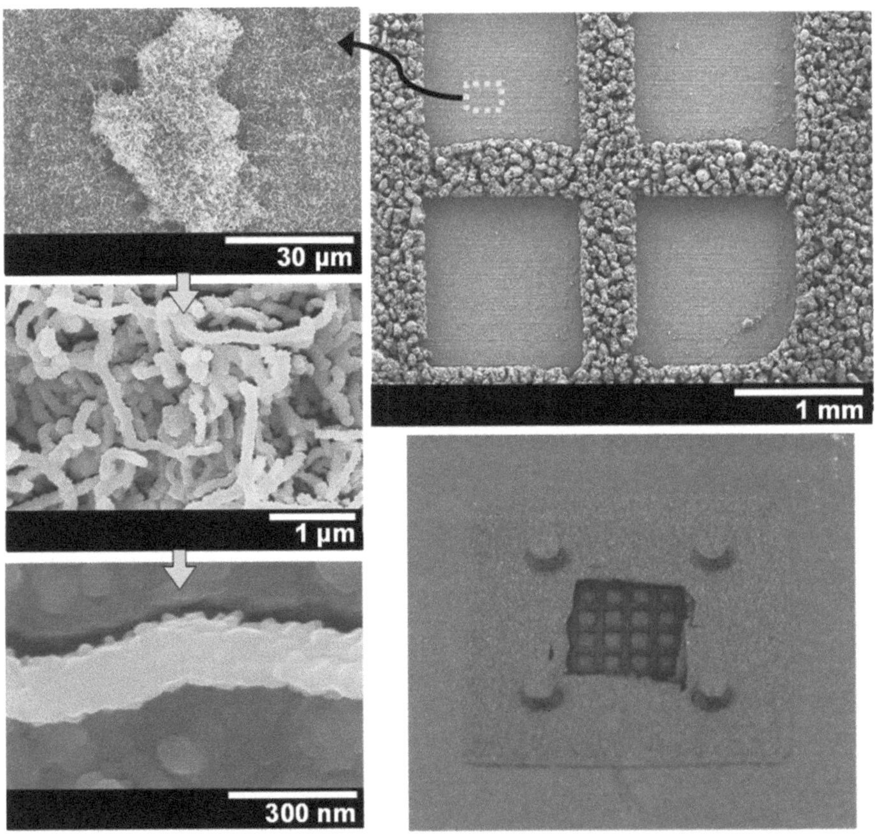

Figure 4.33 Images of a 200 μm thick sintered copper powder wick with interspersed CNT array regions [50, *reprinted with permission from IEEE*]. Low-magnification images on the right show the complete sample (bottom) and macroscale patterned features (top). The series of increasing magnification SEM images on the left show the CNT growth morphology that occurs in the patterned recesses. (For color version of this figure, the reader is referred to the online version of this chapter.)

regime because the square regions remained largely flooded during operation [226]. Conversely, for the 200 μm thick sintered powder wicks, the CNT regions were observed to form a thin-liquid film that receded at high heat fluxes during intense evaporation. The CNT array extended the dryout heat flux compared to the baseline samples [50].

Cai *et al.* [45] investigated a CNT "biwick" structure composed of a uniform 250 μm thick CNT array with parallel interspersed microgrooves. It was postulated that the nanoscale pores of the CNT array would provide a large increase in the area for thin-film evaporation and boiling heat transfer, while the groove spacing would provide area for bulk liquid supply and vapor removal. The wick structure was fabricated by using lithography processes to define the catalyst deposition area and resulting CNT growth pattern. An acid-treatment process was used to make the CNTs hydrophilic [45]. The structures tested had 100 μm wide CNT strips with 50 μm wide microgrooves. A 2 mm × 2 mm platinum heater was fabricated on the backside of the silicon growth substrate, and the capillary-fed evaporation/boiling performance was evaluated in open and saturated vapor environments. A maximum heat flux of ~600 W/cm^2 was measured at surface superheats of only 35–45 °C [45].

Subsequent studies by Cai *et al.* [46,47] investigated additional CNT biwick morphologies with parallel CNT stripes, zigzag CNT stripes, and hexagonally packed CNT clusters, as shown in Fig. 4.34. Thermal testing was performed in a saturated environment using two different heat input areas, 4 and 100 mm^2. The CNT biwick morphologies performed similarly, and a larger dependence on the heater size was noted: while maximum heat fluxes approached 1000 W/cm^2 for the 4 mm^2 heat input area, this was reduced to under 200 W/cm^2 for the 100 mm^2 heat input area across all sample morphologies tested (see Section 2.5 for additional discussion) [46,47].

Figure 4.34 CNT biwick composed of cylindrical CNT clusters, straight CNT stripes, and zigzag CNT stripes (from left to right) [47, *reprinted with permission from ASME*]. (For color version of this figure, the reader is referred to the online version of this chapter.)

4.2.2 Nanostructured coatings

Nanostructured coatings can improve the capillary-fed thermal performance of wicks by potentially increasing the wettability and surface area for evaporation. Nam *et al.* [48] evaluated the thermal performance of a nanostructured copper wick by submerging the lower edge of a sample into a pool of water and allowing liquid replenishment by capillary action to a 5 mm × 5 mm heated area located above the pool. A controlled oxidation process was used to create needlelike CuO nanostructures on top of copper microposts. The fabrication details and demonstration of superhydrophilic wetting characteristics after oxidation are described in Ref. [49]. Samples with and without CuO nanostructures were directly compared. The nanostructures provided little improvement below 25 W/cm^2, but improved capillary performance provided by the nanostructure prevented local dryout of the post surfaces and reduced the surface superheat at higher heat fluxes relative to the uncoated case. This improved capillary performance outweighed any conduction resistance added by the nanostructure layer. A 70% increase in the maximum dryout heat flux was shown for the nanostructured wick samples [48].

Kousalya *et al.* [51] explored a means of increasing the dryout heat flux by fabricating CNT on a 200 μm thick sintered copper powder wick. A physical vapor deposition process was used to coat the CNTs with copper to promote their wettability to water. Three different increasing nominal thicknesses of copper were investigated (Fig. 4.35). Unlike aligned CNT growth on a flat substrate, the randomly oriented CNTs grown on sintered powder lend themselves to a more conformal copper coating by physical vapor deposition. The nanostructured samples were compared to a bare sintered powder wick using the capillary-fed evaporation/boiling facility described in Ref. [37].

As discussed in Section 2.4, an abrupt transition from evaporation to boiling occurs at relatively large surface superheats (~10 °C) for bare sintered copper samples, resulting in a noticeable transient substrate temperature drop; however, Kousalya *et al.* [51] observed that the CNT-coated samples exhibited an earlier transition to the boiling regime. Weibel *et al.* [50] drew comparisons between the incipience behavior of bare and CNT-coated samples (fabricated using the same techniques as in Ref. [51]) using a more extensive set of 25 boiling curves. The CNT coating was able to reduce the mean surface superheat at incipience by 5.6 °C compared to uncoated samples [50]. Despite this observed behavior, conventional nucleation theory suggests that cavities formed by nanoscale pores would require very large superheats to become active due to the

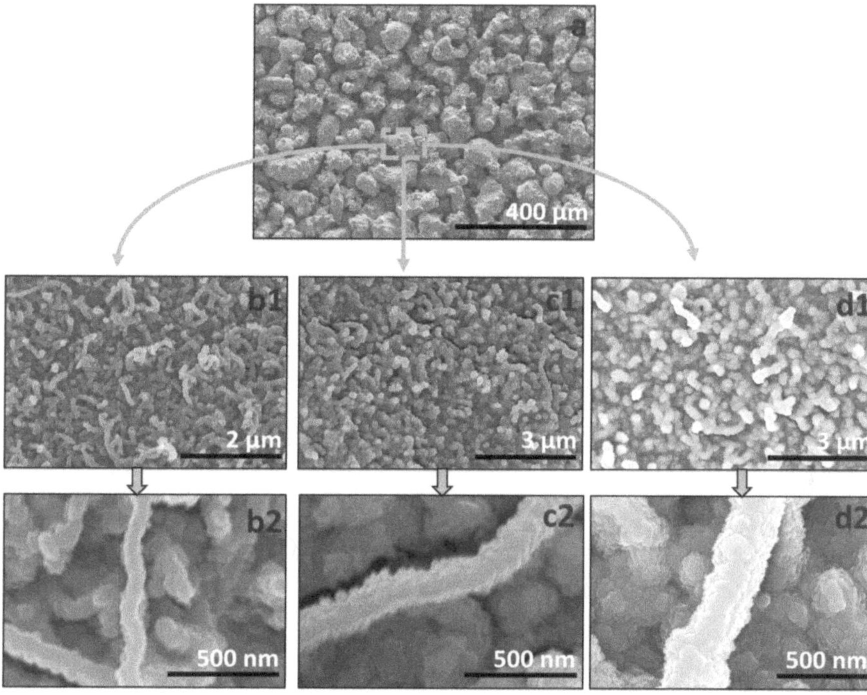

Figure 4.35 (A) A low-magnification SEM image of CNT-coated sintered powder and (B–D) medium- and high-magnification SEM images (middle and bottom rows, respectively) of CNT-coated sintered powder functionalized with increasing nominal copper coating thickness (from left to right) [51, *reprinted with permission from Elsevier*]. (For color version of this figure, the reader is referred to the online version of this chapter.)

inverse relationship between cavity radius and required activation superheat [59]. Therefore, the mechanism by which CNTs reduce the incipience superheat for capillary wicks may be attributed to: (1) an increase in the microscale thermal boundary layer as occurs in flow boiling [228] or (2) changes to the wetting characteristics of the existing microscale cavities in a manner that reduces the required superheat. For example, Li *et al.* [213] proposed that a nanorod coating increases the stability of a microcavity vapor embryo during pool boiling by feeding it with vapor trapped in the nanoscale pores.

Following boiling incipience, Kousalya *et al.* [51] observed a dryout heat flux of 437 W/cm^2 at a surface superheat of 23.3 °C for the bare sintered powder sample. For the CNT-coated samples, an increasing copper coating thickness consistently diminished the area of partial dryout visualized during testing; the maximum dryout heat flux was increased compared to the baseline for the thickest coating [51]. The authors concluded that dryout occurred due to a capillary limit because the estimated value of the maximum CHF constrained by hydrodynamic instabilities was predicted to be much higher than experimental observations. It was proposed that the CNTs functionalized with a thicker copper coating enhanced the surface wettability and thereby increased the dryout heat flux. Since a static macroscopic contact angle cannot be obtained for a porous sintered copper powder structure, a transient measurement of the dynamic contact angle during droplet imbibition was used to assess the relative wettability of the samples [51]. The surface wettability trends matched the trends in the dryout heat flux, as would be expected for a capillary-limited dryout mechanism.

5. CLOSURE

There is an immediate need for high-reliability passive heat spreading away from high-flux hot spots, which currently impose thermal limitations on a number of microelectronic systems. This need has spurred recent advances in fundamental understanding of evaporation and boiling from porous microstructures and in modeling, design, and manufacture of ultrathin vapor chamber spreaders. The major advances/developments and critical areas for further study reviewed in the foregoing are summarized here.

The thermal performance of a variety of wick microstructures has been evaluated in terms of their ability to cool a substrate by evaporation/boiling while replenishing liquid to the heat source via capillary action. This has been achieved through novel experimental facilities and has led to the identification of critical evaporation/boiling regimes and visualization of vapor formation characteristics. It is found that regimes and wick structures that increase interstitial liquid–vapor interface area for heat exchange (e.g., via discrete bubble nucleation with a high departure frequency or evaporation from continuous vapor columns) provide a significant enhancement compared to evaporation from the top of a wick structure saturated with liquid. Hence, novel heterogeneous wicks having multiple length-scale pores are proposed and shown to enhance performance by favoring such vapor removal mechanisms. A number of common trends are identified with respect to characteristic wick properties, and approximate models are developed for prediction of thermal performance; however, more knowledge of the vapor flow structures and interstitial liquid film thickness during intense evaporation/boiling is required to enable more generalized and accurate predictive methods.

Due to the severe implications on device performance, many recent investigations focused on studying the critical regime transitions during capillary-fed evaporation/boiling, viz., boiling incipience (or lack thereof) and dryout of the wick at high heat fluxes. Compared to pool boiling, even for irregular porous surface morphologies, it is observed that nucleation may be suppressed during evaporation up to a high surface superheat; this is often attributed to meniscus recession and formation of thin-liquid films in the porous structure that cannot sustain nucleation. Unfortunately, while suppression of boiling under capillary-fed conditions is observed on an anecdotal basis, prediction of inherently variable incipience criteria requires further statistically significant characterization as a function of wick parameters. Separate investigation of the maximum dissipated heat flux has revealed a strong dependence of capillary dryout on the heater size. While this general trend is anticipated, quantitative predictive methods are nonexistent. Additional investigation is required to develop methodologies for correlating and modeling the complex capillary dryout mechanisms associated with aggressive boiling in the wick structure. Generally, incorporation of nanostructures that behave as superhydrophilic coatings extends the maximum heat flux by increasing the surface wettability and reducing areas of local dryout.

From a device-modeling perspective, the importance of an accurate description of the wick properties as a function of microstructure morphology cannot be overstated. A number of novel direct numerical simulation characterization approaches have been recently developed and provide higher levels of accuracy/fidelity compared to simplified analytical approximations that are ubiquitously employed in the literature to predict effectively thermal conductivity, permeability, and capillarity as a function of wick morphology. While these approaches provide tools for characterization of both idealized and realistic structures, there is still need for process-based characterization approaches that consider the influence of actual microstructure fabrication techniques in the wick design. Transient, three-dimensional device-level models have also evolved to accommodate drastic alterations in liquid–vapor interface shape as observed during high-heat-flux operation of vapor chambers. While the potential for direct numerical simulation of vapor departure from a porous wick structure within these models is promising, further computational advancements are required to predict these phenomena in stochastic wick structures; current approaches still rely on empirical inputs to account for film evaporation or boiling behavior.

The combination of multiscale design, testing, and modeling advances has informed critical thermal transport limits in passive vapor chamber heat spreaders, which has spawned multiple strategies to break through performance barriers. A set of ultrathin vapor chambers have been demonstrated for thermal management of high-power electronic devices. Device performance trends are accurately captured by companion experimental and numerical modeling efforts, which suggest that passive cooling of millimeter-scale hot spots generating beyond 500 W/cm^2 is feasible. Further characterization and development of methodologies that accurately predict high-heat-flux operating limits as a function of wick morphology will push performance further.

ACKNOWLEDGMENTS

The authors gratefully acknowledge support for this work from industry members of the Cooling Technologies Research Center (CTRC), a National Science Foundation (NSF) Industry/University Cooperative Research Center (IUCRC) at Purdue University, and the Defense Advanced Research Project Agency (DARPA). Special thanks are extended to collaborators David Altman, Timothy Fisher, Arun Kousalya, Jayathi Murthy, Mark North, Ram Ranjan, and Kazuaki Yazawa.

REFERENCES

[1] S.W. Chi, Heat Pipe Theory and Practice: A Sourcebook, Hemisphere Publishing Corporation, Washington, DC, 1976.
[2] G.P. Peterson, Heat Pipes: Modeling, Testing, and Applications, John Wiley & Sons, New York, NY, 1994.
[3] A. Faghri, Heat Pipe Science and Technology, Taylor & Francis, Washington, DC, 1995.
[4] D. Reay, P. Kew, Heat Pipes: Theory, Design, and Applications, fifth ed., Butterworth-Heinemann, Burlington, MA, 2006.
[5] Y.H. Yau, M. Ahmadzadehtalatapeh, A review on the application of horizontal heat pipe heat exchangers in air conditioning systems in the tropics, Appl. Therm. Eng. 30 (2010) 77–84.
[6] H.N. Chaudhry, B.R. Hughes, S.A. Ghani, A review of heat pipe systems for heat recovery and renewable energy applications, Renew. Sustain. Energy Rev. 16 (2012) 2249–2259.
[7] R. Sureshkumar, S.T. Mohideen, N. Nethaji, Heat transfer characteristics of nanofluids in heat pipes: a review, Renew. Sustain. Energy Rev. 20 (2013) 397–410.
[8] S.V. Garimella, C.B. Sobhan, Recent advances in the modeling and applications of nonconventional heat pipes, in: Advances in Heat Transfer, Elsevier, Amsterdam, 2001, pp. 249–308.
[9] A. Faghri, Review and advances in heat pipe science and technology, J. Heat Transfer 134 (2012) 123001.
[10] S. Launay, V. Sartre, J. Bonjour, Parametric analysis of loop heat pipe operation: a literature review, Int. J. Therm. Sci. 46 (2007) 621–636.
[11] A. Ambirajan, A.A. Adoni, J.S. Vaidya, A.A. Rajendran, D. Kumar, P. Dutta, Loop heat pipes: a review of fundamentals, operation, and design, Heat Transfer Eng. 33 (2012) 387–405.
[12] C.B. Sobhan, R.L. Rag, G.P. Peterson, A review and comparative study of the investigations on micro heat pipes, Int. J. Energy Res. 31 (2007) 664–688.

[13] Y. Zhang, A. Faghri, Advances and unsolved issues in pulsating heat pipes, Heat Transfer Eng. 29 (2008) 20–44.
[14] R.S. Prasher, A simplified conduction based modeling scheme for design sensitivity study of thermal solution utilizing heat pipe and vapor chamber technology, J. Electron. Packag. 125 (2003) 378–385.
[15] A. Abhat, R.A. Seban, Boiling and evaporation from heat pipe wicks with water and acetone, J. Heat Transfer 96 (Ser C) (1974) 331–337.
[16] K. Nishikawa, T. Ito, K. Tanaka, Enhanced heat transfer by nucleate boiling on a sintered metal layer, Heat Transfer Jpn. Res. 8 (1979) 65–81.
[17] W. Nakayama, T. Daikoku, H. Kuwahara, T. Nakajima, Dynamic model of enhanced boiling heat transfer on porous surfaces—part I: experimental investigation, J. Heat Transfer 102 (1980) 445.
[18] A.E. Bergles, M.C. Chyu, Characteristics of nucleate pool boiling from porous metallic coatings, J. Heat Transfer 104 (1982) 279–285.
[19] N.H. Afgan, L.A. Jovic, S.A. Kovalev, V.A. Lenykov, Boiling heat transfer from surfaces with porous layers, Int. J. Heat Mass Transfer 28 (1985) 415–422.
[20] J.Y. Tsay, Y.Y. Yan, T.F. Lin, Enhancement of pool boiling heat transfer in a horizontal water layer through surface roughness and screen coverage, Heat Mass Transfer 32 (1996) 17–26.
[21] J.Y. Chang, S.M. You, Enhanced boiling heat transfer from microporous surfaces: effects of a coating composition and method, Int. J. Heat Mass Transfer 40 (1997) 4449–4460.
[22] J.W. Liu, D.J. Lee, A. Su, Boiling of methanol and HFE-7100 on heated surface covered with a layer of mesh, Int. J. Heat Mass Transfer 44 (2001) 241–246.
[23] S.G. Liter, M. Kaviany, Pool-boiling CHF enhancement by modulated porous-layer coating: theory and experiment, Int. J. Heat Mass Transfer 44 (2001) 4287–4311.
[24] K.N. Rainey, S.M. You, S. Lee, Effect of pressure, subcooling, and dissolved gas on pool boiling heat transfer from microporous surfaces in FC-72, J. Heat Transfer 125 (2003) 75–83.
[25] M.S. El-Genk, J.L. Parker, Enhanced boiling of HFE-7100 dielectric liquid on porous graphite, Energy Convers. Manag. 46 (2005) 2455–2481.
[26] J.L. Parker, M.S. El-Genk, Effect of surface orientation on nucleate boiling of FC-72 on porous graphite, J. Heat Transfer 128 (2006) 1159.
[27] C. Li, G.P. Peterson, Parametric study of pool boiling on horizontal highly conductive microporous coated surfaces, J. Heat Transfer 129 (2007) 1465.
[28] M.E. Poniewski, J.R. Thome, Nucleate Boiling on Micro-Structured Surfaces, Heat Transfer Research Inc., Lausanne–Warsaw/College Station, TX, 2008
[29] A. Brautsch, P.A. Kew, Examination and visualisation of heat transfer processes during evaporation in capillary porous structures, Appl. Therm. Eng. 22 (2002) 815–824.
[30] M.A. Hanlon, H.B. Ma, Evaporation heat transfer in sintered porous media, J. Heat Transfer 125 (2003) 644–652.
[31] T.W. Davis, S.V. Garimella, Thermal resistance measurement across a wick structure using a novel thermosyphon test chamber, Exp. Heat Transfer 21 (2008) 143–154.
[32] C. Li, G.P. Peterson, Y. Wang, Evaporation/boiling in thin capillary wicks (I)—wick thickness effects, J. Heat Transfer 128 (2006) 1312–1319.
[33] C. Li, G.P. Peterson, Evaporation/boiling in thin capillary wicks (II)—effects of volumetric porosity and mesh size, J. Heat Transfer 128 (2006) 1320–1328.
[34] M.T. North, J.H. Rosenfeld, R.M. Shaubach, Liquid film evaporation from bidisperse capillary wicks in heat pipe evaporators, in: Proceedings of the 9th International Heat Pipe Conference, Albuquerque, NM, USA, 1995.
[35] T. Semenic, Y.Y. Lin, I. Catton, D.B. Sarraf, Use of biporous wicks to remove high heat fluxes, Appl. Therm. Eng. 28 (2008) 278–283.

[36] T. Semenic, I. Catton, Experimental study of biporous wicks for high heat flux applications, Int. J. Heat Mass Transfer 52 (2009) 5113–5121.
[37] J.A. Weibel, S.V. Garimella, M.T. North, Characterization of evaporation and boiling from sintered powder wicks fed by capillary action, Int. J. Heat Mass Transfer 53 (2010) 4204–4215.
[38] J.A. Weibel, Characterization of fluid-thermal transport and boiling in micro/nano-engineered porous structures, in: PhD Thesis, Purdue University, 2012.
[39] Y. Zhao, C.-I. Chen, An investigation of evaporation heat transfer in sintered copper wicks with microgrooves, in: ASME International Mechanical Engineering Congress and Exposition, November 5–10, Chicago, IL, USA, 2006.
[40] G.S. Hwang, Y. Nam, E. Fleming, P. Dussinger, Y.S. Ju, M. Kaviany, Multi-artery heat pipe spreader: experiment, Int. J. Heat Mass Transfer 53 (2010) 2662–2669.
[41] G.S. Hwang, E. Fleming, B. Carne, S. Sharratt, Y. Nam, P. Dussinger, et al., Multi-artery heat-pipe spreader: lateral liquid supply, Int. J. Heat Mass Transfer 54 (2011) 2334–2340.
[42] Y.S. Ju, M. Kaviany, Y. Nam, S. Sharratt, G.S. Hwang, I. Catton, et al., Planar vapor chamber with hybrid evaporator wicks for the thermal management of high-heat-flux and high-power optoelectronic devices, Int. J. Heat Mass Transfer 60 (2013) 163–169.
[43] J.A. Weibel, S.V. Garimella, Visualization of vapor formation regimes during capillary-fed boiling in sintered-powder heat pipe wicks, Int. J. Heat Mass Transfer 55 (2012) 3498–3510.
[44] D. Ćoso, V. Srinivasan, M.-C. Lu, J.-Y. Chang, A. Majumdar, Enhanced heat transfer in biporous wicks in the thin liquid film evaporation and boiling regimes, J. Heat Transfer 134 (2012) 101501.
[45] Q. Cai, C.-L. Chen, Design and test of carbon nanotube biwick structure for high-heat-flux phase change heat transfer, J. Heat Transfer 132 (2010) 052403.
[46] Q. Cai, A. Bhunia, High heat flux phase change on porous carbon nanotube structures, Int. J. Heat Mass Transfer 55 (2012) 5544–5551.
[47] Q. Cai, Y.-C. Chen, Investigations of biporous wick structure dryout, J. Heat Transfer 134 (2012) 021503.
[48] Y. Nam, S. Sharratt, G. Cha, Y.S. Ju, Characterization and modeling of the heat transfer performance of nanostructured Cu micropost wicks, J. Heat Transfer 133 (2011) 101502.
[49] Y. Nam, S. Sharratt, C. Byon, S.-J. Kim, Y.S. Ju, Fabrication and characterization of the capillary performance of superhydrophilic Cu micropost arrays, J. Microelectromech. Syst. 19 (2010) 581–588.
[50] J.A. Weibel, A.S. Kousalya, T.S. Fisher, S.V. Garimella, Characterization and nano-structured enhancement of boiling incipience in capillary-fed, ultra-thin sintered powder wicks, in: 13th IEEE Intersociety Conference on Thermal and Thermomechanical Phenomena in Electronic Systems, San Diego, CA, USA, 2012.
[51] A.S. Kousalya, J.A. Weibel, S.V. Garimella, T.S. Fisher, Metal functionalization of carbon nanotubes for enhanced sintered powder wicks, Int. J. Heat Mass Transfer 59 (2013) 372–383.
[52] M.P. Mughal, O.A. Plumb, An experimental study of boiling on a wicked surface, Int. J. Heat Mass Transfer 39 (1996) 771–777.
[53] R.R. Williams, D.K. Harris, A device and technique to measure the heat transfer limit of a planar heat pipe wick, Exp. Therm. Fluid Sci. 30 (2006) 277–284.
[54] B.D. Iverson, T.W. Davis, S.V. Garimella, M.T. North, S.S. Kang, Heat and mass transport in heat pipe wick structures, J. Thermophys. Heat Transfer 21 (2007) 392–404.
[55] S.-C. Wong, Y.-H. Kao, Visualization and performance measurement of operating mesh-wicked heat pipes, Int. J. Heat Mass Transfer 51 (2008) 4249–4259.

[56] J.-H. Liou, C.-W. Chang, C. Chao, S.-C. Wong, Visualization and thermal resistance measurement for the sintered mesh-wick evaporator in operating flat-plate heat pipes, Int. J. Heat Mass Transfer 53 (2010) 1498–1506.
[57] S.-C. Wong, J.-H. Liou, C.-W. Chang, Evaporation resistance measurement with visualization for sintered copper-powder evaporator in operating flat-plate heat pipes, Int. J. Heat Mass Transfer 53 (2010) 3792–3798.
[58] C. Li, G.P. Peterson, J. Li, N. Koratkar, The visualization of thin film evaporation on thin micro sintered copper mesh screen, in: ASME Summer Heat Transfer Conference, August 10–14, Jacksonville, FL, USA, 2008.
[59] V.P. Carey, Liquid–Vapor Phase-Change Phenomena, Hemisphere Publishing Corporation, New York, NY, 1992.
[60] L.-H. Chien, C.-C. Chang, Experimental study of evaporation resistance on porous surfaces in flat heat pipes, in: 8th IEEE Intersociety Conference on Thermal and Thermomechanical Phenomena in Electronic Systems, San Diego, CA, USA, 2002.
[61] R.L. Webb, Nucleate boiling on porous coated surfaces, Heat Transfer Eng. 4 (1981) 71–82.
[62] X.L. Cao, P. Cheng, T.S. Zhao, Experimental study of evaporative heat transfer in sintered copper bidispersed wick structures, J. Thermophys. Heat Transfer 16 (2002) 547–552.
[63] S.W. Reilly, I. Catton, Utilization of advanced working fluids with biporous evaporators, J. Therm. Sci. Eng. Appl. 3 (2011) 021006.
[64] T. Semenic, Y.-Y. Lin, I. Catton, Thermophysical properties of biporous heat pipe evaporators, J. Heat Transfer 130 (2008) 022602.
[65] J. Wang, I. Catton, Evaporation heat transfer in thin biporous media, Heat Mass Transfer 37 (2001) 275–281.
[66] D.H. Min, G.S. Hwang, M. Kaviany, Multi-artery, heat-pipe spreader, Int. J. Heat Mass Transfer 52 (2009) 629–635.
[67] W.M. Rohsenow, Method of correlating heat-transfer data for surface boiling of liquids, Am. Soc. Mech. Eng. Trans. 74 (1952) 969–975.
[68] I.L. Mostinski, Application of the rule of corresponding states for calculation of heat transfer and critical heat flux, Teploenergetika 4 (1963) 66–71.
[69] M.G. Cooper, Saturation nucleate pool boiling—a simple correlation, in: Institution of Chemical Engineers Symposium Series, Leeds, England, 1984, pp. 785–793.
[70] D. Gorenflo, Pool Boiling, VDI-Verlag GmbH, Dusseldorf, Germany, 1993.
[71] S. Madhusudana Rao, A.R. Balakrishnan, Analysis of pool boiling heat transfer over porous surfaces, Heat Mass Transfer 32 (1997) 463–469.
[72] G.F. Smirnov, Approximate theory of heat transfer with boiling on surfaces covered with capillary-porous structures, Therm. Eng. (Teploenergetika) 24 (1977) 55–58.
[73] B.A. Afanas'ev, G.F. Smirnov, Investigation of heat transfer and of maximum heat fluxes with boiling in capillary-porous structures, Therm. Eng. 26 (1979) 311–313.
[74] B.A. Afanas'yev, V.V. Zrodnikov, A.L. Koba, T.F. Smirnov, A.K. Pimenov, Boiling heat transfer in slots, capillaries and in capillary-porous (wick) structures, Heat Transfer Sov. Res. 12 (1980) 70–78.
[75] S.M. You, A. Bar-Cohen, T.W. Simon, Boiling incipience and nucleate boiling heat transfer of highly wetting dielectric fluids from electronic materials, IEEE Trans. Components Hybrids Manufacturing Technol. 13 (1990) 1032–1039.
[76] B.J. Jones, J.P. McHale, S.V. Garimella, The influence of surface roughness on nucleate pool boiling heat transfer, J. Heat Transfer 131 (2009) 1–14.
[77] S.J. Reed, I. Mudawar, Elimination of boiling incipience temperature drop in highly wetting fluids using spherical contact with a flat surface, Int. J. Heat Mass Transfer 42 (1999) 2439–2454.

[78] J.P. McHale, S.V. Garimella, T.S. Fisher, G.A. Powell, Pool boiling performance comparison of smooth and sintered copper surfaces with and without carbon nanotubes, Nanoscale Microscale Thermophys. Eng. 15 (2011) 133–150.
[79] D.H. Altman, J.R. Wasniewski, M.T. North, S.S. Kim, T.S. Fisher, Development of micro/nano engineered wick-based passive heat spreaders for thermal management of high power electronic devices, in: ASME InterPACK, Portland, OR, USA, 2011, p. 52122.
[80] P.J. Marto, V.J. Lepere, Pool boiling heat transfer from enhanced surfaces to dielectric fluids, J. Heat Transfer 104 (1982) 292–299.
[81] S.M. You, T.W. Simon, A. Bar-Cohen, W. Tong, Experimental investigation of nucleate boiling incipience with a highly-wetting dielectric fluid (R-113), Int. J. Heat Mass Transfer 33 (1990) 105–117.
[82] Y.Y. Hsu, On the size range of active nucleation cavities on a heating surface, J. Heat Transfer 84 (1962) 207–213.
[83] J.H. Lienhard, V.K. Dhir, D.M. Riherd, Peak pool boiling heat-flux measurements on finite horizontal flat plates, J. Heat Transfer 95 (1973) 477–482.
[84] I.I. Gogonin, S.S. Kutateladze, Critical heat flux as a function of heater size for a liquid boiling in a large enclosure, J. Eng. Phys. 33 (1977) 1286–1289.
[85] K.-A. Park, A.E. Bergles, Effects of size of simulated microelectronic chips on boiling and critical heat flux, J. Heat Transfer 110 (1988) 728.
[86] S.H. Yang, W.-P. Baek, S.H. Chang, Pool-boiling critical heat flux of water on small plates: effects of surface orientation and size, Int. Commun. Heat Mass Transfer 24 (1997) 1093–1102.
[87] T. Chen, J.N. Chung, An experimental study of miniature-scale pool boiling, J. Heat Transfer 125 (2003) 1074–1086.
[88] M. Kaviany, Principles of Heat Transfer in Porous Media, second ed., Springer, New York, NY, 1995.
[89] D.E. Sweeney, Correlation for pressure drop in two-phase cocurrent flow in packed beds, AIChE J. 13 (1967) 663–669.
[90] V.X. Tung, V.K. Dhir, Hydrodynamic model for two-phase flow through porous media, Int. J. Multiphase Flow 14 (1988) 47–65.
[91] T. Murugesan, V. Sivakumar, Pressure drop and flow regimes in cocurrent gas–liquid upflow through packed beds, Chem. Eng. J. 88 (2002) 233–243.
[92] H. Muller-Steinhagen, M. Jamialahmadi, M.R. Izadpanah, Pressure drop, gas hold-up and heat transfer during single and two-phase flow through porous media, Int. J. Heat Fluid Flow 26 (2005) 156–172.
[93] I. Sauciuc, G. Chrysler, R. Mahajan, R. Prasher, Spreading in the heat sink base: phase change systems or solid metals? IEEE Trans. Components Packaging Technol. 25 (2002) 621–628.
[94] S.-S. Hsieh, R.-Y. Lee, J.-C. Shyu, S.-W. Chen, Analytical solution of thermal resistance of vapor chamber heat sink with and without pillar, Energy Convers. Manag. 48 (2007) 2708–2717.
[95] Z.J. Zuo, A. Faghri, A network thermodynamic analysis of the heat pipe, Int. J. Heat Mass Transfer 41 (1998) 1473–1484.
[96] Y. Cao, A. Faghri, Transient two-dimensional compressible analysis for high-temperature heat pipes with pulsed heat input, Numer. Heat Transfer A Appl. 18 (1991) 483–502.
[97] J.-Y. Chang, R.S. Prasher, S. Prstic, P. Cheng, H.B. Ma, Evaporative thermal performance of vapor chambers under nonuniform heating conditions, J. Heat Transfer 130 (2008) 121501.
[98] K. Vafai, W. Wang, Analysis of flow and heat transfer characteristics of an asymmetrical flat plate heat pipe, Int. J. Heat Mass Transfer 35 (1992) 2087–2099.

[99] N. Zhu, K. Vafai, Analytical modeling of the startup characteristics of asymmetrical flat-plate and disk-shaped heat pipes, Int. J. Heat Mass Transfer 41 (1998) 2619–2637.
[100] Y. Wang, K. Vafai, Transient characterization of flat plate heat pipes during startup and shutdown operations, Int. J. Heat Mass Transfer 43 (2000) 2641–2655.
[101] Y. Wang, K. Vafai, An experimental investigation of the thermal performance of an asymmetrical flat plate heat pipe, Int. J. Heat Mass Transfer 43 (2000) 2657–2668.
[102] Y. Wang, K. Vafai, An experimental investigation of the transient characteristics on a flat-plate heat pipe during startup and shutdown operations, J. Heat Transfer 122 (2000) 525–535.
[103] F. Lefèvre, M. Lallemand, Coupled thermal and hydrodynamic models of flat micro heat pipes for the cooling of multiple electronic components, Int. J. Heat Mass Transfer 49 (2006) 1375–1383.
[104] M. Aghvami, A. Faghri, Analysis of flat heat pipes with various heating and cooling configurations, Appl. Therm. Eng. 31 (2011) 2645–2655.
[105] N. Zhu, K. Vafai, Vapor and liquid flow in an asymmetrical flat plate heat pipe: a three-dimensional analytical and numerical investigation, Int. J. Heat Mass Transfer 41 (1998) 159–174.
[106] R. Sonan, S. Harmand, J. Pellé, D. Leger, M. Fakès, Transient thermal and hydrodynamic model of flat heat pipe for the cooling of electronics components, Int. J. Heat Mass Transfer 51 (2008) 6006–6017.
[107] U. Vadakkan, S.V. Garimella, J.Y. Murthy, Transport in flat heat pipes at high heat fluxes from multiple discrete sources, J. Heat Transfer 126 (2004) 347–354.
[108] G. Carbajal, C.B. Sobhan, G.P. Peterson, D.T. Queheillalt, H.N.G. Wadley, Thermal response of a flat heat pipe sandwich structure to a localized heat flux, Int. J. Heat Mass Transfer 49 (2006) 4070–4081.
[109] Y.-S. Chen, K.-H. Chien, C.-C. Wang, T.-C. Hung, B.-S. Pei, A simplified transient three-dimensional model for estimating the thermal performance of the vapor chambers, Appl. Therm. Eng. 26 (2006) 2087–2094.
[110] Y.-S. Chen, K.-H. Chien, T.-C. Hung, C.-C. Wang, Y.-M. Ferng, B.-S. Pei, Numerical simulation of a heat sink embedded with a vapor chamber and calculation of effective thermal conductivity of a vapor chamber, Appl. Therm. Eng. 29 (2009) 2655–2664.
[111] Y. Koito, H. Imura, M. Mochizuki, Y. Saito, S. Torii, Numerical analysis and experimental verification on thermal fluid phenomena in a vapor chamber, Appl. Therm. Eng. 26 (2006) 1669–1676.
[112] B. Xiao, A. Faghri, A three-dimensional thermal-fluid analysis of flat heat pipes, Int. J. Heat Mass Transfer 51 (2008) 3113–3126.
[113] J.-M. Tournier, M.S. El-Genk, Heat pipe transient analysis model, Int. J. Heat Mass Transfer 37 (1994) 753–762.
[114] J.A. Rice, A. Faghri, Analysis of screen wick heat pipes, including capillary dry-out limitations, J. Thermophys. Heat Transfer 21 (2007) 475–486.
[115] K. Vafai, Handbook of Porous Media, second ed., CRC Press, Boca Raton, FL, 2005.
[116] R.L. Gorring, S.W. Churchill, Thermal conductivity of heterogeneous materials, Chem. Eng. Prog. 57 (1961) 53–59.
[117] J.C. Maxwell, third ed., A Treatise on Electricity and Magnetism, vol. 1, Dover Publications, New York, 1954.
[118] R.W. Schrage, A Theoretical Study of Interface Mass Transfer, Columbia University Press, New York, NY, 1953.
[119] M. Potash Jr., P. Wayner Jr., Evaporation from a two-dimensional extended meniscus, Int. J. Heat Mass Transfer 15 (1972) 1851–1863.
[120] H. Wang, S.V. Garimella, J.Y. Murthy, Characteristics of an evaporating thin film in a microchannel, Int. J. Heat Mass Transfer 50 (2007) 3933–3942.

[121] G.P. Peterson, L.S. Fletcher, Effective thermal conductivity of sintered heat pipe wicks, J. Thermophys. 1 (1986) 343–347.
[122] N. Atabaki, B.R. Baliga, Effective thermal conductivity of water-saturated sintered powder-metal plates, Heat Mass Transfer 44 (2007) 85–99.
[123] Y.-Y. Lin, T. Semenic, I. Catton, Thermophysical properties of monoporous sintered copper, in: ASME Summer Heat Transfer Conference, San Francisco, CA, USA, 2005, pp. 17–23.
[124] S. Kirkpatrick, Percolation and conduction, Rev. Mod. Phys. 45 (1973) 574–588.
[125] J.K. Carson, S.J. Lovatt, D.J. Tanner, A.C. Cleland, Thermal conductivity bounds for isotropic, porous materials, Int. J. Heat Mass Transfer 48 (2005) 2150–2158.
[126] C. Li, G.P. Peterson, The effective thermal conductivity of wire screen, Int. J. Heat Mass Transfer 49 (2006) 4095–4105.
[127] Z. Zhao, Y. Peles, M.K. Jensen, Properties of plain weave metallic wire mesh screens, Int. J. Heat Mass Transfer 57 (2013) 690–697.
[128] J. Xu, R.A. Wirtz, In-plane effective thermal conductivity of plain-weave screen laminates, IEEE Trans. Components Packaging Technol. 25 (2002) 615–620.
[129] H. Kozai, H. Imura, Y. Ikeda, Permeability of screen wicks, JSME Int. J. II Fluid Eng. Heat Transfer Power Combust. Thermophys. Properties 34 (1991) 212–219.
[130] D.R. Adkins, R.C. Dykhuizen, Procedures for measuring the properties of heat-pipe wick materials, in: ACS Proceedings of the 28th Intersociety Energy Conversion Engineering Conference, Washington, DC, USA, 1993, pp. 911–918.
[131] R.R. Williams, D.K. Harris, Cross-plane and in-plane porous properties measurements of thin metal felts: applications in heat pipes, Exp. Therm. Fluid Sci. 27 (2003) 227–235.
[132] H. Noda, K. Yoshioka, T. Hamatake, Experimental study on the permeability of screen wicks, JSME Int. J. B Fluid Therm. Eng. 36 (1993) 357–363.
[133] D. Deng, Y. Tang, G. Huang, L. Lu, D. Yuan, Characterization of capillary performance of composite wicks for two-phase heat transfer devices, Int. J. Heat Mass Transfer 56 (2013) 283–293.
[134] H.P.J. de Bock, K. Varanasi, P. Chamarthy, T. Deng, A. Kulkarni, B.M. Rush, et al., Experimental investigation of micro/nano heat pipe wick structures, in: ASME International Mechanical Engineering Congress and Exposition, 2008, Boston, MA, USA, 2008, pp. 991–996.
[135] H. Imura, H. Kozai, Y. Ikeda, Effective pore radius of screen wicks, Heat Transfer Eng. 15 (1994) 24–32.
[136] E.W. Washburn, The dynamics of capillary flow, Phys. Rev. 17 (1921) 273–283.
[137] B. Holley, A. Faghri, Permeability and effective pore radius measurements for heat pipe and fuel cell applications, Appl. Therm. Eng. 26 (2006) 448–462.
[138] P. Chamarthy, H.P.J. de Bock, B. Russ, S. Chauhan, B. Rush, S.E. Weaver, et al., Novel fluorescent visualization method to characterize transport properties in micro/nano heat pipe wick structures, in: ASME International Technical Conference and Exhibition on Packaging and Integration of Electronic and Photonic Microsystems, San Francisco, CA, USA, 2009, pp. 419–425.
[139] C. Byon, S.J. Kim, Capillary performance of bi-porous sintered metal wicks, Int. J. Heat Mass Transfer 55 (2012) 4096–4103.
[140] M.R.S. Shirazy, L.G. Fréchette, Capillary and wetting properties of copper metal foams in the presence of evaporation and sintered walls, Int. J. Heat Mass Transfer 58 (2013) 282–291.
[141] R. Xiao, R. Enright, E.N. Wang, Prediction and optimization of liquid propagation in micropillar arrays, Langmuir 26 (2010) 15070–15075.
[142] H.K. Dhavaleswarapu, P. Chamarthy, S.V. Garimella, J.Y. Murthy, Experimental investigation of steady buoyant-thermocapillary convection near an evaporating meniscus, Phys. Fluids 19 (2007) 082103.

[143] H. Wang, S.V. Garimella, J.Y. Murthy, An analytical solution for the total heat transfer in the thin-film region of an evaporating meniscus, Int. J. Heat Mass Transfer 51 (2008) 6317–6322.
[144] H.K. Dhavaleswarapu, S.V. Garimella, J.Y. Murthy, Microscale temperature measurements near the triple line of an evaporating thin liquid film, J. Heat Transfer 131 (2009) 1–7.
[145] H.K. Dhavaleswarapu, J.Y. Murthy, S.V. Garimella, Numerical investigation of an evaporating meniscus in a channel, Int. J. Heat Mass Transfer 55 (2012) 915–924.
[146] C.P. Migliaccio, H.K. Dhavaleswarapu, S.V. Garimella, Temperature measurements near the contact line of an evaporating meniscus V-groove, Int. J. Heat Mass Transfer 54 (2011) 1520–1526.
[147] H. Wang, Z. Pan, S.V. Garimella, Numerical investigation of heat and mass transfer from an evaporating meniscus in a heated open groove, Int. J. Heat Mass Transfer 54 (2011) 3015–3023.
[148] C.P. Migliaccio, S.V. Garimella, Evaporative heat and mass transfer from the free surface of a liquid wicked into a bed of spheres, Int. J. Heat Mass Transfer 54 (2011) 3440–3447.
[149] K.K. Bodla, J.A. Weibel, S.V. Garimella, Advances in fluid and thermal transport property analysis and design of sintered porous wick microstructures, J. Heat Transfer 135 (6) (2013) 061202.
[150] K.K. Bodla, J.Y. Murthy, S.V. Garimella, Direct simulation of thermal transport through sintered wick microstructures, J. Heat Transfer 134 (2011) 012602.
[151] R. Ranjan, J.Y. Murthy, S.V. Garimella, Analysis of the wicking and thin-film evaporation characteristics of microstructures, J. Heat Transfer 131 (2009) 1–11.
[152] K.A. Brakke, The surface evolver, Exp. Math. 1 (1992) 141–165.
[153] R. Ranjan, J.Y. Murthy, S.V. Garimella, A microscale model for thin-film evaporation in capillary wick structures, Int. J. Heat Mass Transfer 54 (2011) 169–179.
[154] N.S. Dhillon, J.C. Cheng, A.P. Pisano, Heat transfer due to microscale thin film evaporation from the steady state meniscus in a coherent porous silicon based micro-columnated wicking structure, in: ASME International Mechanical Engineering Congress and Exposition, Denver, CO, USA, 2011.
[155] K.K. Bodla, J.Y. Murthy, S.V. Garimella, Evaporation analysis in sintered wick microstructures, Int. J. Heat Mass Transfer 61 (2013) 729–741.
[156] R. Ranjan, A. Patel, S.V. Garimella, J.Y. Murthy, Wicking and thermal characteristics of micropillared structures for use in passive heat spreaders, Int. J. Heat Mass Transfer 55 (2012) 586–596.
[157] S. Sharratt, C. Peng, Y.S. Ju, Micro-post evaporator wicks with improved phase change heat transfer performance, Int. J. Heat Mass Transfer 55 (2012) 6163–6169.
[158] R. Ranjan, J.Y. Murthy, S.V. Garimella, U. Vadakkan, A numerical model for transport in flat heat pipes considering wick microstructure effects, Int. J. Heat Mass Transfer 54 (2011) 153–168.
[159] V.K. Dhir, Mechanistic prediction of nucleate boiling heat transfer—achievable or a hopeless task? J. Heat Transfer 128 (2006) 1.
[160] G. Son, V.K. Dhir, N. Ramanujapu, Dynamics and heat transfer associated with a single bubble during nucleate boiling on a horizontal surface, J. Heat Transfer 121 (1999) 623–631.
[161] G. Son, N. Ramanujapu, V.K. Dhir, Numerical simulation of bubble merger process on a single nucleation site during pool nucleate boiling, J. Heat Transfer 124 (2001) 51–62.
[162] D. Gerlach, G. Tomar, G. Biswas, F. Durst, Comparison of volume-of-fluid methods for surface tension-dominant two-phase flows, Int. J. Heat Mass Transfer 49 (2006) 740–754.

[163] A. Mukherjee, S.G. Kandlikar, Numerical study of single bubbles with dynamic contact angle during nucleate pool boiling, Int. J. Heat Mass Transfer 50 (2007) 127–138.
[164] C. Kunkelmann, P. Stephan, CFD simulation of boiling flows using the volume-of-fluid method within OpenFOAM, Numer. Heat Transfer A Appl. 56 (2009) 631–646.
[165] C. Kunkelmann, P. Stephan, Numerical simulation of the transient heat transfer during nucleate boiling of refrigerant HFE-7100, Int. J. Refrig. 33 (2010) 1221–1228.
[166] W. Lee, G. Son, J.J. Jeong, Numerical analysis of bubble growth and departure from a microcavity, Numer. Heat Transfer B Fundam. 58 (2010) 323–342.
[167] B.A. Nichita, I. Zun, J.R. Thome, A level set method coupled with a volume of fluid method for modeling of gas–liquid interface in bubbly flow, J. Fluids Eng. 132 (2010) 081302.
[168] R. Ranjan, J.Y. Murthy, S.V. Garimella, Bubble dynamics during capillary-fed nucleate boiling in porous media, in: 13th IEEE Intersociety Conference on Thermal and Thermomechanical Phenomena in Electronic Systems, 2012, San Diego, CA, USA, 2012, pp. 1114–1126.
[169] C.H. Li, T. Li, P. Hodgins, G.P. Peterson, Characteristics of pool boiling bubble dynamics in bead packed porous structures, J. Heat Transfer 133 (2011) 031004.
[170] A. Mukherjee, V.K. Dhir, Study of lateral merger of vapor bubbles during nucleate pool boiling, J. Heat Transfer 126 (2005) 1023–1039.
[171] R. Ranjan, J.Y. Murthy, S.V. Garimella, D.H. Altman, M.T. North, Modeling and design optimization of ultrathin vapor chambers for high heat flux applications, IEEE Trans. Components Packaging Manufacturing Technol. 2 (2012) 1465–1479.
[172] D.H. Altman, J.A. Weibel, S.V. Garimella, T.S. Fisher, J.H. Nadler, M.T. North, Thermal ground plane vapor chamber heat spreaders for high power packaging density electronic systems, Electron. Cooling (2012) 20–27.
[173] H.P.J. de Bock, S. Chauhan, P. Chamarthy, S.E. Weaver, T. Deng, F.M. Gerner, et al., On the charging and thermal characterization of a micro/nano structured thermal ground plane, in: 12th IEEE Intersociety Conference on Thermal and Thermomechanical Phenomena in Electronic Systems, Las Vegas, NV, USA, 2010.
[174] C. Oshman, B. Shi, C. Li, R. Yang, Y.C. Lee, G.P. Peterson, et al., The development of polymer-based flat heat pipes, J. Microelectromech. Syst. 20 (2011) 410–417.
[175] Q. Cai, B.-C. Chen, C. Tsai, Design, development and tests of high-performance silicon vapor chamber, J. Micromech. Microeng. 22 (2012) 035009.
[176] C. Ding, G. Soni, P. Bozorgi, B.D. Piorek, C.D. Meinhart, N.C. MacDonald, A flat heat pipe architecture based on nanostructured titania, J. Microelectromech. Syst. 19 (2010) 878–884.
[177] D.H. Altman, A. Gupta, T.E. Dubrowski Jr., D.J. Sharar, N.R. Jankowski, M.T. North, Analysis and characterization of thermal expansion-matched wick-based multi-chip passive heat spreaders in static and dynamic environments, in: ASME InterPACK, Burlingame, CA, USA, 2013, p. 73087.
[178] H.P.J. de Bock, S. Chauhan, P. Chamarthy, C. Eastman, S. Weaver, B.P. Whalen, et al., Development and experimental validation of a micro/nano thermal ground plane, in: Proceedings of the ASME/JSME 8th Thermal Engineering Joint Conference, Honolulu, HI, USA, 2011.
[179] S.W. Reilly, I. Catton, Characterization of vapor escape restriction in biporous wicks with monolayers for thermal ground plane optimization, in: Proceedings of the ASME International Mechanical Engineering Congress & Exposition, November 13–19, Lake Buena Vista, FL, USA, 2009.
[180] S.W. Reilly, L. Amouzegar, I. Catton, Comparison of vacuum chamber tested biporous wicks with thermal ground plane testing, in: Proceeding of the ASME International Heat Transfer Conference, August 8–13, Washington, DC, USA, 2010.

[181] I. Catton, S. Reilly, L. Amouzegar, Advances in biporous wick design and testing for thermal ground planes, Front. Heat Pipes 1 (2010) 013001.
[182] X. Dai, L. Tran, F. Yang, B. Shi, R. Yang, Y.C. Lee, et al., Characterization of hybrid-wicked copper heat pipe, in: Proceedings of the ASME/JSME 8th Thermal Engineering Joint Conference, Honolulu, HI, USA, 2011.
[183] C. Oshman, Q. Li, L.-A. Liew, R. Yang, Y.C. Lee, V.M. Bright, Flat polymer heat spreader with high aspect ratio micro hybrid wick operating under adverse gravity, in: Proceedings of the ASME International Mechanical Engineering Congress and Exposition, Denver, CO, USA, 2011.
[184] G.P. Peterson, A.B. Duncan, M.H. Weichold, Experimental investigation of micro heat pipes fabricated in silicon wafers, J. Heat Transfer 115 (1993) 751–756.
[185] M.L. Berre, S. Launay, V. Sartre, M. Lallemand, Fabrication and experimental investigation of silicon micro heat pipes for cooling electronics, J. Micromech. Microeng. 13 (2003) 436.
[186] S. Launay, V. Sartre, M. Lallemand, Experimental study on silicon micro-heat pipe arrays, Appl. Therm. Eng. 24 (2004) 233–243.
[187] C. Gillot, Y. Avenas, N. Cezac, G. Poupon, C. Schaeffer, E. Fournier, Silicon heat pipes used as thermal spreaders, IEEE Trans. Components Packaging Technol. 26 (2003) 332–339.
[188] Q. Cai, B.-C. Chen, C. Tsai, C. Chen, Development of scalable silicon heat spreader for high power electronic devices, J. Therm. Sci. Eng. Appl. 1 (2010) 041009.
[189] M. Sigurdson, Y. Liu, P. Bozorgi, D. Bothman, N. MacDonald, C. Meinhart, A large scale titanium thermal ground plane, Int. J. Heat Mass Transfer 62 (2013) 178–183.
[190] M.F. Aimi, M.P. Rao, N.C. MacDonald, A.S. Zuruzi, D.P. Bothman, High-aspect-ratio bulk micromachining of titanium, Nat. Mater. 3 (2004) 103–105.
[191] N. Srivastava, C. Din, A. Judson, N.C. MacDonald, C.D. Meinhart, A unified scaling model for flow through a lattice of microfabricated posts, Lab Chip 10 (2010) 1148–1152.
[192] S. Berber, Y.-K. Kwon, D. Tomanek, Unusually high thermal conductivity of carbon nanotubes, Phys. Rev. Lett. 84 (2000) 4613–4616.
[193] J. Che, T. Cagin, W.A. Goddard III., Thermal conductivity of carbon nanotubes, Nanotechnology 11 (2000) 65–69.
[194] P. Kim, L. Shi, A. Majumdar, P.L. McEuen, Thermal transport measurements of individual multiwalled nanotubes, Phys. Rev. Lett. 87 (2001) 215502, 1–4.
[195] A.E. Aliev, M.H. Lima, E.M. Silverman, R.H. Baughman, Thermal conductivity of multi-walled carbon nanotube sheets: radiation losses and quenching of phonon modes, Nanotechnology 21 (2010) 1–11.
[196] E. Pop, D. Mann, Q. Wang, K. Goodson, H. Dai, Thermal conductance of an individual single-wall carbon nanotube above room temperature, Nano Lett. 6 (2006) 96–100.
[197] M.S. Dresselhaus, P.C. Eklund, Phonons in carbon nanotubes, Adv. Phys. 49 (2000) 705–814.
[198] P.B. Amama, B.A. Cola, T.D. Sands, X. Xu, T.S. Fisher, Dendrimer-assisted controlled growth of carbon nanotubes for enhanced thermal interface conductance, Nanotechnology 18 (2007) 385303.
[199] B.A. Cola, J. Xu, C. Cheng, X. Xu, T.S. Fisher, H. Hu, Photoacoustic characterization of carbon nanotube array thermal interfaces, J. Appl. Phys. 101 (2007) 054313–054313-9.
[200] H. Wang, J. Feng, X. Hu, K.M. Ng, Synthesis of aligned carbon nanotubes on double-sided metallic substrate by chemical vapor deposition, J. Phys. Chem. C 111 (2007) 12617–12624.
[201] Q. Ngo, B.A. Cruden, A.M. Cassell, G. Sims, M. Meyyappan, J. Li, et al., Thermal interface properties of Cu-filled vertically aligned carbon nanofiber arrays, Nano Lett. 4 (2004) 2403–2407.

[202] M.J. Biercuk, M.C. Llaguno, M. Radosavljevic, J.K. Hyun, A.T. Johnson, J.E. Fischer, Carbon nanotube composites for thermal management, Appl. Phys. Lett. 80 (2002) 2767–2769.
[203] C.H. Liu, H. Huang, Y. Wu, S.S. Fan, Thermal conductivity improvement of silicone elastomer with carbon nanotube loading, Appl. Phys. Lett. 84 (2004) 4248–4250.
[204] Y. Gogotsi, J.A. Libera, A. Güvenç-Yazicioglu, C.M. Megaridis, In situ multiphase fluid experiments in hydrothermal carbon nanotubes, Appl. Phys. Lett. 79 (2001) 1021–1023.
[205] J.-G. Fan, D. Dyer, G. Zhang, Y.-P. Zhao, Nanocarpet effect: pattern formation during the wetting of vertically aligned nanorod arrays, Nano Lett. 4 (2004) 2133–2138.
[206] L. Feng, S. Li, Y. Li, H. Li, L. Zhang, J. Zhai, et al., Super-hydrophobic surfaces: from natural to artificial, Adv. Mater. 14 (2002) 1857–1860.
[207] J.J. Zhou, F. Noca, M. Gharib, Flow conveying and diagnosis with carbon nanotube arrays, Nanotechnology 17 (2006) 4845–4853.
[208] Q. Cai, C.-L. Chen, Design and test of carbon nanotube biwick structure for high-heat-flux phase change heat transfer, J. Heat Transfer 132 (2010) 1–8.
[209] A.S. Kousalya, C.N. Hunter, S.A. Putnam, T. Miller, T.S. Fisher, Photonically enhanced flow boiling in a channel coated with carbon nanotubes, Appl. Phys. Lett. 100 (2012) 071601–071601-4.
[210] H.S. Ahn, N. Sinha, M. Zhang, D. Banerjee, S. Fang, R.H. Baughman, Pool boiling experiments on multiwalled carbon nanotube (MWCNT) forests, J. Heat Transfer 128 (2006) 1335–1342.
[211] S. Launay, A.G. Fedorov, Y. Joshi, A. Cao, P.M. Ajayan, Hybrid micro-nano structured thermal interfaces for pool boiling heat transfer enhancement, Microelectron. J. 37 (2006) 1158–1164.
[212] S. Ujereh, T. Fisher, I. Mudawar, Effects of carbon nanotube arrays on nucleate pool boiling, Int. J. Heat Mass Transfer 50 (2007) 4023–4038.
[213] C. Li, Z. Wang, P.-I. Wang, Y. Peles, N. Koratkar, G.P. Peterson, Nanostructured copper interfaces for enhanced boiling, Small 4 (2008) 1084–1088.
[214] R. Chen, M.-C. Lu, V. Srinivasan, Z. Wang, H.H. Cho, A. Majumdar, Nanowires for enhanced boiling heat transfer, Nano Lett. 9 (2009) 548–553.
[215] Z. Yao, Y.-W. Lu, S.G. Kandlikar, Effects of nanowire height on pool boiling performance of water on silicon chips, Int. J. Therm. Sci. 50 (2011) 2084–2090.
[216] M.-C. Lu, R. Chen, V. Srinivasan, V.P. Carey, A. Majumdar, Critical heat flux of pool boiling on Si nanowire array-coated surfaces, Int. J. Heat Mass Transfer 54 (2011) 5359–5367.
[217] R. Ranjan, S.V. Garimella, J.Y. Murthy, K. Yazawa, Assessment of nanostructured capillary wicks for passive two-phase heat transport, Nanoscale Microscale Thermophys. Eng. 15 (2011) 179–194.
[218] M.P. Rossi, H. Ye, Y. Gogotsi, S. Babu, P. Ndungu, J.-C. Bradley, Environmental scanning electron microscopy study of water in carbon nanopipes, Nano Lett. 4 (2004) 989–993.
[219] B.M. Kim, S. Sinha, H.H. Bau, Optical microscope study of liquid transport in carbon nanotubes, Nano Lett. 4 (2004) 2203–2208.
[220] C. Fradin, A. Braslau, D. Luzet, D. Smilgies, M. Alba, N. Boudet, et al., Reduction in the surface energy of liquid interfaces at short length scales, Nature 403 (2000) 871–874.
[221] J.H. Walther, T. Werder, R.L. Jaffe, P. Koumoutsakos, Hydrodynamic properties of carbon nanotubes, Phys. Rev. E (2004) 062201.
[222] Z.P. Huang, J.W. Xu, Z.F. Ren, J.H. Wang, M.P. Siegal, P.N. Provencio, Growth of highly oriented carbon nanotubes by plasma-enhanced hot filament chemical vapor deposition, Appl. Phys. Lett. 73 (1998) 3845–3847.

[223] J. Li, C. Papadopoulos, J.M. Xu, M. Moskovits, Highly-ordered carbon nanotube arrays for electronics applications, Appl. Phys. Lett. 75 (1999) 367–369.
[224] Y. Tu, Z.P. Huang, D.Z. Wang, J.G. Wen, Z.F. Ren, Growth of aligned carbon nanotubes with controlled site density, Appl. Phys. Lett. 80 (2002) 4018–4020.
[225] J.A. Weibel, S.V. Garimella, J.Y. Murthy, D.H. Altman, Design of integrated nanostructured wicks for high-performance vapor chambers, IEEE Trans. Components Packaging Manufacturing Technol. 1 (2011) 859–867.
[226] J.A. Weibel, S.S. Kim, T.S. Fisher, S.V. Garimella, Carbon nanotube coatings for enhanced capillary-fed boiling from porous microstructures, Nanoscale MicroscaleThermophys. Eng. 16 (2012) 1–17.
[227] R.K. Garg, S.S. Kim, D.B. Hash, J.P. Gore, T.S. Fisher, Effects of feed gas composition and catalyst thickness on carbon nanotube and nanofiber synthesis by plasma enhanced chemical vapor deposition, J. Nanosci. Nanotechnol. 8 (2008) 3068–3076.
[228] N. Singh, V. Sathyamurthy, W. Peterson, J. Arendt, D. Banerjee, Flow boiling enhancement on a horizontal heater using carbon nanotube coatings, Int. J. Heat Fluid Flow 31 (2010) 201–207.

CHAPTER FIVE

Applications of Nanomaterials in Solar Energy and Desalination Sectors

Khalil Khanafer*, **Kambiz Vafai**†
*Department of Biomedical Engineering, Frankel Vascular Mechanics Laboratory, University of Michigan, Ann Arbor, Michigan, USA
†Mechanical Engineering Department, University of California, Riverside, California, USA

Contents

Abstract

This work provides an overview of the use of nanomaterials in solar energy and desalination sectors. Nanotechnology has received considerable attention in the past few years due to availability of new structures at nanoscales with potential applications in various industrial applications especially in the energy field. This work offers the most recent advances of nanotechnology in thermal storage systems, photovoltaic systems, and solar desalination. With the application of nanomaterials, photovoltaic solar cells are increasing their efficiency while reducing the production costs of electricity and manufacturing. According to the US Department of Energy, few power-generating technologies have as little environmental impact as photovoltaic solar panels. Photovoltaic systems generate considerably smaller amount of harmful air emissions (at least 89%) per kilowatt hour than conventional fossil fuel-fired technologies.

1. INTRODUCTION

Nanotechnology, a term normally used to describe materials and phenomena at a nanoscale, has been widely used in various engineering and scientific applications. A great deal of interest has been directed at the use

Advances in Heat Transfer, Volume 45
ISSN 0065-2717
http://dx.doi.org/10.1016/B978-0-12-407819-2.00005-0

of nanotechnology and nanomaterials in the energy sector. Nanotechnology has the potential to develop new industries that contribute to a sustainable economic growth. Moreover, nanotechnology has been used in many applications intended to provide cleaner and more efficient energy supplies and uses. According to the "Roadmap Report Concerning the Use of nanomaterials in the Energy Sector" [1], nanomaterials can play an important role in various domains of the energy sector, namely, energy conversion (e.g., solar cells, fuel cells, and thermoelectric devices), energy storage (e.g., rechargeable batteries and supercapacitors), and energy saving (e.g., insulation such as aerogels and smart glazes and efficient lightning such as light-emitting diode and organic light-emitting diode).

Recent advances in nanotechnology have led to the development of an innovative class of heat-transfer fluids (HTFs) (nanofluids) created by dispersing nanoparticles (10–50 nm) in traditional HTFs [2]. Nanofluids show the potential to significantly increase heat-transfer rates in a variety of areas such as industrial cooling applications, nuclear reactors, transportation industry (automobiles, trucks, and airplanes), microelectromechanical systems, electronics and instrumentation, and biomedical applications (nanodrug delivery, cancer therapeutics, and cryopreservation) [3]. Possible improved thermal conductivity translates into higher energy efficiency, better performance, and lower operating costs. A significant number of research work associated with heat-transfer enhancement using nanofluids both experimentally and theoretically have been conducted by many researchers [4–17]. Figure 5.1 shows the rapid growth of nanofluid research in recent years. It is estimated that more than 2000 articles related to nanofluids have been published in the literature. Furthermore, several review papers on nanofluids have also been published. The potential market of nanofluids for heat-transfer applications is estimated by Commissariat à l'énergie atomique (CEA—France) to be over 2 billion dollars per year worldwide [18].

While different studies have shown that nanofluids demonstrate higher heat-transfer enhancement than those of base fluids, conflicting results on nanofluid performance have also been reported [6]. A variety of thermal conductivity enhancement ratios were reported for various particle diameter sizes D_p and volume fractions φ_p [6]. Table 5.1 shows a comparison of the experimental thermal conductivity enhancements of metallic and non-metallic nanofluids cited in the literature. Recently, Khanafer and Vafai [6] presented a critical synthesis of the variants within the thermophysical

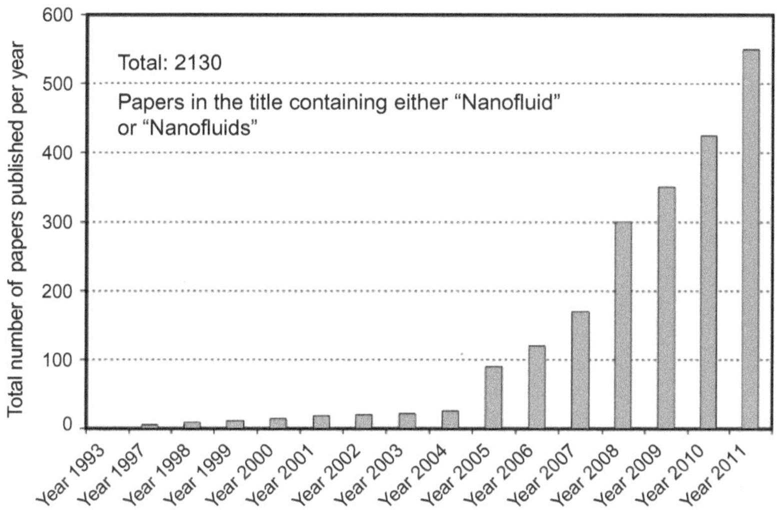

Figure 5.1 Total number of papers published per year containing the term nanofluid or nanofluids between 1993 and 2011. (For color version of this figure, the reader is referred to the online version of this chapter.)

properties of nanofluids. The authors demonstrated that the experimental results for the effective thermal conductivity and viscosity reported by several authors are in disagreement. Correlations for effective thermal conductivity and viscosity were synthesized and developed in their study in terms of pertinent physical parameters based on the reported experimental data as shown in Table 5.2.

Contradictory results were also reported in the literature regarding natural convection heat-transfer enhancement using nanofluids. The conclusions for both experimental and analytical investigations are still in disagreement. Analytical studies show an increase in heat transfer with an increase in the volume fraction of nanoparticles, which is not in agreement with experimental results [4,6]. Since the Rayleigh number, the ratio of buoyant to the viscous forces, represents a significant parameter in natural convection processes, a comparison of nanofluid Rayleigh number to the base fluid Rayleigh number at various volume fractions and temperatures was rigorously highlighted by Khanafer and Vafai [6] for Al_2O_3–water nanofluid.

Figure 5.2 shows that the ratio of the Rayleigh number of nanofluid to that of the base fluid decreases with an increase in the Al_2O_3 volume

Table 5.1 Comparison of the experimental thermal conductivity enhancements of metallic and nonmetallic nanofluids cited in the literature

References	Base fluid	Particle	D_p (nm)	φ_p (%)	Thermal conductivity enhancement
[17]	Water	Al_2O_3	38.4	4	9% (21 °C), 16% (36 °C), 24% (51 °C)
[17]	Water	CuO	28.6	4	14% (21 °C), 26% (36 °C), 36% (51 °C)
[19]	Water	Al_2O_3	131	4	24% (51 °C)
[20]	Water	Al_2O_3	13	4.3	32.4% (31.8 °C)
[20]	Water	Al_2O_3	13	4.3	29.6% (46.8 °C)
[20]	Water	Al_2O_3	13	4.3	26.2% (66.8 °C)
[20]	Water	SiO_2	12	2.3	1.1% (31.8 °C)
[20]	Water	SiO_2	12	2.3	1% (46.8 °C)
[21]	Water	CuO	23.6	3.4	12%
[21]	EG[a]	CuO	23.6	4	23%
[21]	Water	Al_2O_3	38.4	4.3	11%
[21]	EG[a]	Al_2O_3	38.4	5	19%
[22]	Water	CuO	23	9.7	34%
[22]	EG[a]	CuO	23	14.8	54%
[22]	Water	Al_2O_3	28	5.5	16%
[22]	EG[a]	Al_2O_3	28	5	24.5%
[23]	Water	Al_2O_3	11	1	14.8% (70 °C)
[23]	Water	Al_2O_3	47	1	10.2% (70 °C)
[23]	Water	Al_2O_3	150	1	4.8% (60 °C)
[23]	Water	Al_2O_3	47	4	28.8% (70 °C)
[23]	Water	Al_2O_3	47	1	3% (21 °C)
[24]	Water	CuO	29	6	36% (28.9 °C)
[24]	Water	CuO	29	6	50% (31.3 °C)
[24]	Water	Al_2O_3	36	6	28.2%
[24]	Water	Al_2O_3	47	6	26.1%

Table 5.1 Comparison of the experimental thermal conductivity enhancements of metallic and nonmetallic nanofluids cited in the literature—cont'd

References	Base fluid	Particle	D_p (nm)	φ_p (%)	Thermal conductivity enhancement
[25]	Water	Al_2O_3	20	5	15%
[26]	Water	Al_2O_3	11	5	8%
[26]	Water	Al_2O_3	20	5	7%
[26]	Water	Al_2O_3	40	5	10%
[27]	Water	Cu	100	7.5	78%
[28]	Water	Au	10–20	0.03	21%
[28]	Water	Ag	60–80	0.001	17%

[a]EG, ethylene glycol.

fraction. Higher volume fractions of the solid nanoparticles cause an increase in the viscous force of nanofluids, which consequently suppresses heat transfer. Moreover, Fig. 5.2A shows the effect of varying particle diameter on the Rayleigh number ratio. As the particle diameter increases, the ratio of the Rayleigh numbers decreases because the effective thermal conductivity of nanofluids decreases and the kinematic viscosity increases. However, the rate of increase of the kinematic viscosity of the nanofluid with the particle size is larger than the resulting decrease of the effective thermal conductivity. This may provide a physical reason for the reduction of natural convection heat-transfer enhancement with an increase in the volume fraction of nanoparticles at room temperature.

The effect of varying the temperature of nanofluids and volume fraction on the ratio of Rayleigh numbers is illustrated in Fig. 5.2B for nanoparticle diameter of 36 nm. Figure 5.2B shows that the ratio of nanofluid Rayleigh number, to the base fluid Rayleigh number, increases with an increase in the temperature. Moreover, this ratio is higher for volume fraction of 1% compared to 4% for various temperatures. This is because the kinematic viscosity and the effective thermal conductivity of nanofluids increase with an increase in the volume fraction of nanoparticles. For volume fraction of 1%, Fig. 5.2B shows an interesting result associated with the fact that the nanofluid Rayleigh number is smaller than the Rayleigh number of a water base below 31 °C. For temperatures greater than 31 °C, Fig. 5.2B shows that

Table 5.2 Summary of the correlations synthesized and developed by Khanafer and Vafai [6] based on the reported experimental data

Physical properties	Room temperature	Temperature-dependent
Density	$\rho_{eff}=(1-\varphi_p)\rho_f+\varphi_p\rho_p$	Al_2O_3–water $\rho_{eff}=1001.064+2738.6191\varphi_p-0.2095T$ $0\le\varphi_p\le0.04, 5\le T(^\circ C)\le 40$
Specific heat	$c_{eff}=\frac{(1-\varphi_p)\rho_f c_f+\varphi_p\rho_p c_p}{\rho_{eff}}$	N/A
Thermal expansion coefficient	$\beta_{eff}=\frac{(1-\varphi_p)(\rho\beta)_f+\varphi_p(\rho\beta)_p}{\rho_{eff}}$ $\beta_{eff}=(1-\varphi_p)\beta_f+\varphi_p\beta_p$	Al_2O_3–water $\beta_{eff}=\left(-0.479\varphi_p+9.3158\times10^{-3}T-\frac{4.7211}{T^2}\right)\times10^{-3}$ $0\le\varphi_p\le0.04, 10^\circ C\le T\le 40^\circ C$
Viscosity	N/A	Al_2O_3–water $\mu_{eff}=-0.4491+\frac{28.837}{T}+0.574\varphi_p-0.1634\varphi_p^2$ $+23.053\frac{\varphi_p^2}{T^2}+0.0132\varphi_p^3-2354.735\frac{\varphi_p}{T^3}$ $+23.498\frac{\varphi_p^2}{d_p^2}-3.0185\frac{\varphi_p^3}{d_p^2},$ $1\%\le\varphi_p\le9\%, 20\le T(^\circ C)\le70, 13\,nm\le d_p\le131\,nm$
Thermal conductivity	Al_2O_3–water and CuO–water $\frac{k_{eff}}{k_f}=1.0+1.0112\varphi_p+2.4375\varphi_p\left(\frac{47}{d_p(nm)}\right)$ $-0.0248\varphi_p\left(\frac{k_p}{0.613}\right)$	Al_2O_3–water $\frac{k_{eff}}{k_f}=0.9843+0.398\varphi_p^{0.7383}\left(\frac{1}{d_p(nm)}\right)^{0.2246}\left(\frac{\mu_{eff}(T)}{\mu_f(T)}\right)^{0.0235}$ $-3.9517\frac{\varphi_p}{T}+34.034\frac{\varphi_p^2}{T^3}+32.509\frac{\varphi_p}{T^2}$ $0\le\varphi_p\le10\%, 11\,nm\le d\le150\,nm, 20^\circ C\le T\le70^\circ C$

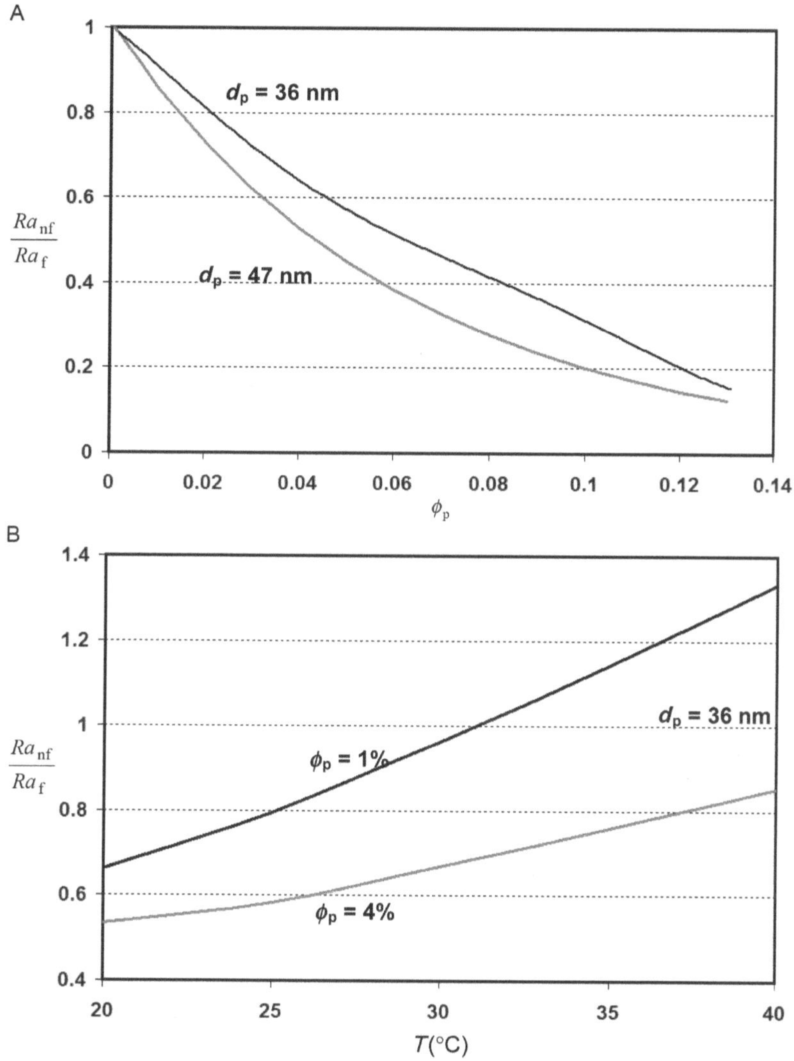

Figure 5.2 Effect of volume fraction and temperature on the ratio of the Rayleigh numbers for different particle diameters (Al_2O_3–water nanofluid); (A) effect of volume fraction at room temperature on the Rayleigh number ratio; (B) effect of temperature on the Rayleigh number ratio. *Reprinted from Khanafer and Vafai [6] with permission from Elsevier.* (For color version of this figure, the reader is referred to the online version of this chapter.)

a nanofluid Rayleigh number is higher than that of the base water. Hence, nanofluids may exhibit natural convection heat-transfer enhancement at high temperatures. This is associated with the behavior of the kinematic viscosity and the thermal diffusivity for both a nanofluid and a water base at various temperatures, which can be seen in Fig. 5.3. Whereas Fig. 5.2A and B shows that natural convection heat transfer not only is exclusively characterized by the effective thermal conductivity of nanofluids but also depends on the viscosity of nanofluids.

Another importance of nanoparticle application can be found in boiling heat-transfer processes. Boiling heat transfer plays an important role in a range

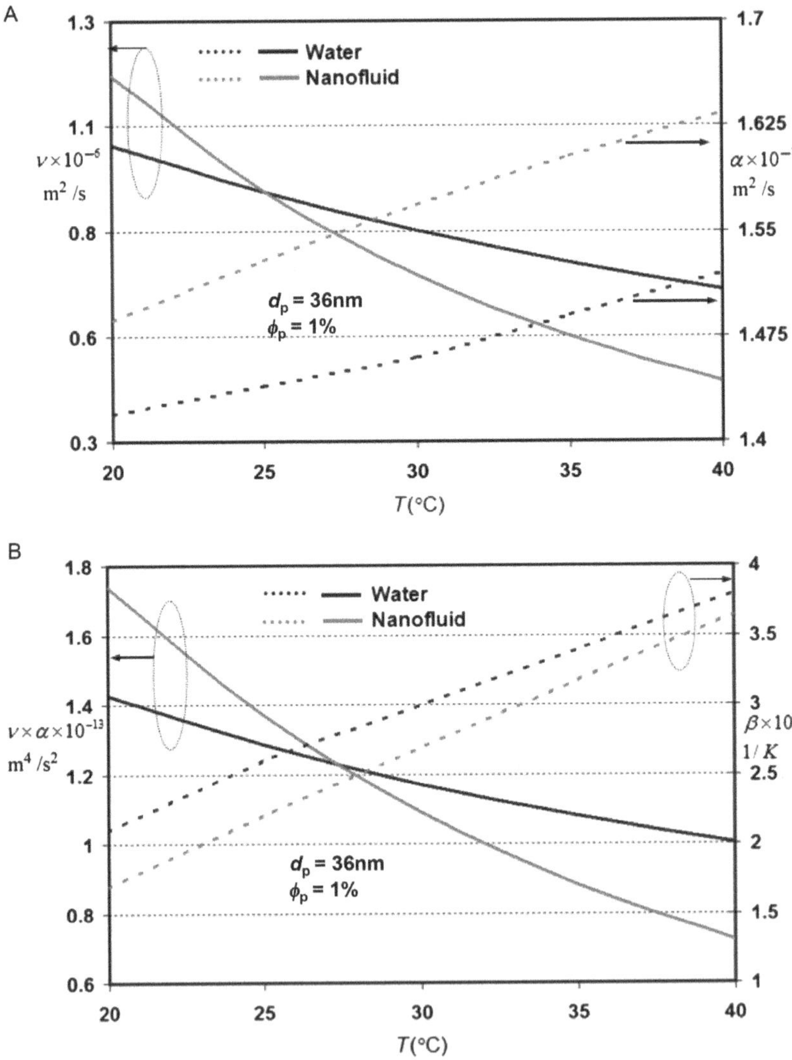

Figure 5.3 Effect of varying temperature on the thermophysical properties (A) Al_2O_3–water nanofluid; (B) water. (For color version of this figure, the reader is referred to the online version of this chapter.)

of technological and industrial applications such as refrigeration, heat exchangers, cooling of high-power electronics, and nuclear reactors. The application of nanofluids in enhancing boiling heat-transfer characteristics is of great importance [29–31]. Many experimental investigations on the nucleate pool boiling and critical heat flux (CHF) characteristics of nanofluids have been carried out in the literature [32–44]. Conflicting results on the effect of nanoparticles on the nucleate boiling heat-transfer rate and CHF were reported. For example, Das *et al.* [15,16] conducted an experimental study on pool boiling characteristics of Al_2O_3–water nanofluids on smooth and roughened heating surfaces for various particle concentrations. The authors showed that nanoparticles degraded the boiling performance with increasing particle concentration. You *et al.* [33] found that nucleate boiling heat-transfer coefficients remained unchanged with the addition of Al_2O_3 nanoparticles compared with water. Contrary to the aforementioned findings, Wen and Ding [36] showed that alumina nanofluids (particle sizes of 10–50 nm) can significantly enhance boiling heat transfer. The enhancement in the boiling heat-transfer coefficient increased with increasing particle concentration up to ~40% at a particle loading of 1.25% by weight.

Most CHF experimental investigations using nanofluids have shown CHF enhancement under pool boiling conditions [33,34,39,40]. You *et al.* [33] studied experimentally the effect of Al_2O_3 nanoparticles on CHF of water in pool boiling. Their results demonstrated that the CHF increased dramatically (~200%) compared to that of pure water. Kim *et al.* [39] carried out an experimental study on the CHF characteristics of nanofluids in pool boiling. Their results illustrated that the CHF of nanofluids containing TiO_2 or Al_2O_3 was enhanced up to 100% over that of pure water. Vassallo *et al.* [41] experimentally illustrated an increase in the CHF (up to 60%) for both nano- and microsolutions (silica–water) at the same concentration (0.5% volume fraction) compared to the base water. Figure 5.4 shows a comparison of CHF enhancements between experimental results for various volume concentrations, nanoparticle material, and particle diameter. In addition, a summary of research investigations on nucleate pool boiling heat-transfer coefficients (BHT) and CHF of nanofluids is presented in Table 5.3.

Compared to the studies on the enhanced thermal characteristics of nanofluids, the optical and radiative properties of nanofluids have received much less attention. Recently, several researchers have addressed usage of nanofluids in thermal storage and solar thermal collectors. The addition of small particles causes scattering of the incident radiation allowing higher

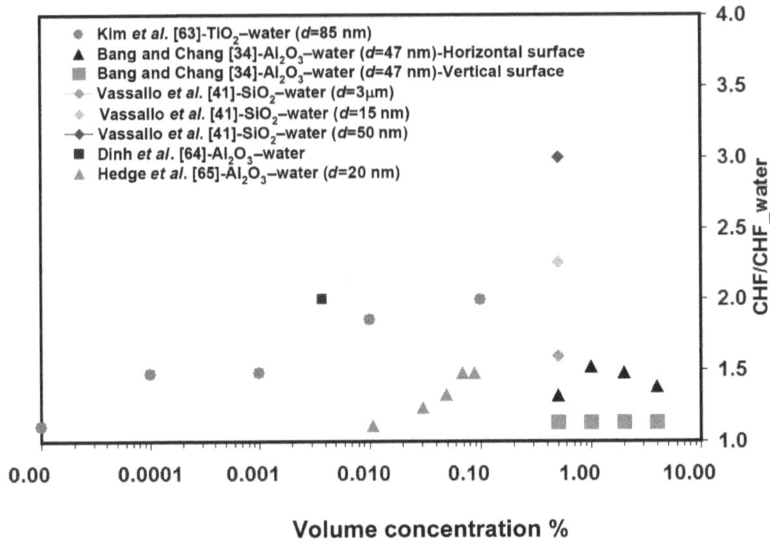

Figure 5.4 Comparison of CHF enhancements between experimental results for various volume concentrations, nanoparticles materials, and nanoparticles diameter. (For color version of this figure, the reader is referred to the online version of this chapter.)

levels of absorption within the fluid [51–53]. The optical properties of the effective fluid are highly dependent on the particle shape, particle size, and the optical properties of the base fluid and particles themselves [53]. The aim of this chapter is to help identify the potential role of nanoparticles in solar energy and desalination sectors. Nanotechnology-based nanoparticles can be used to develop new industries based on cost-effective and cost-efficient economies leading to a sustainable economic growth. As such, the drivers and requirements for solar and desalination sectors using nanoparticles are examined. This work provides an overview of the contribution of nanotechnology in these sectors towards more sustainable ways to store energy.

2. SOLAR ENERGY

This section deals with the use of nanoparticles in various energy processes that engage the use of solar radiation as an energy source. This energy source can be used in thermal energy storage (TES), direct absorption in a solar collector, photovoltaic (PV) technology, solar desalination, etc.

Table 5.3 Summary of research studies on nucleate pool boiling heat-transfer coefficient (BHT) and CHF of nanofluids

References	Nanofluids	Remarks
[15,16]	Al_2O_3–water	• BHT degradation
[45]	ZrO_2–water	• BHT enhancement at low volume fraction of nanoparticles (<0.07%) • BHT degradation (>0.07%)
[33]	Al_2O_3–water	• No change in BHT coefficient • CHF enhancement up to 200%
[34]	Al_2O_3–water	• BHT degradation • CHF enhancement up to 32%
[37]	γ-Al_2O_3–water	• BHT enhancement up to 40%
[38]	Carbon nanotube (CNT)–deionized water	• Both BHT and CHF enhancement • Decrease in pressure, increase in BHT, and CHF enhancement
[46]	Al_2O_3–water, TiO_2–water	• BHT enhancement for both TiO_2 and Al_2O_3
[43]	TiO_2–water Al_2O_3–water	• CHF enhancement up to 100%
[44]	TiO_2–water	• CHF enhancement up to 200%
[41]	SiO_2–water	• No change in BHT coefficient • CHF enhancement up to 60%
[47]	SiO_2–water (also in salt and strong electrolyte solution)	• CHF enhancement: three times greater than pure water
[48]	SiO_2–water	• CHF enhancement: 50% with no nanoparticle deposition on wire
[49]	Al_2O_3–water Bismuth oxide (Bi_2O_3)–water	• CHF enhancement: up to 50% for Al_2O_3 and 33% for Bi_2O_3
[50]	Al_2O_3–water, CuO–water, and diamond–water	• BHT degradation • CHF enhancement: increases with nanoparticle concentration until reaches an asymptotic value

2.1. Thermal energy storage systems

The demand for energy and electricity increases as the global economy continues to grow. However, higher fuel prices, lack of grid infrastructure investment, safety issues with nuclear power plants, and the desire to minimize CO_2 emissions that cause global warming are some of the reasons for replacing fossil fuels with renewable energy. Renewable energy production is irregular and the power output depends on weather and location. Fossil fuel generation can be turned on or off as demand requires, while shifting to renewable energy generation needs management of demand and supply. As such, TES systems are essential to store the generated renewable energy. Thus, TES enables the electric grid to overcome the intermittent power output of renewable energy, keeping the electric grid stable and reliable.

The efficiency and reliability of solar thermal energy conversion systems depend significantly on the specific heat of the HTF and on the operating temperature of the TES systems. The operating temperature of a conventional TES system is restricted to 400 °C due to the limitation of the materials used in TES systems such as mineral oil and fatty acids [54]. Molten salt has been recently used in concentrated solar power (CSP) facilities because it is stable at very high temperatures, that is, exceeding 600 °C [55,56], and can store more heat than the synthetic oil used in the CSP and therefore produces electricity even after the sun has gone down. Typical molten salt materials include alkali–carbonate, alkali–nitrate, alkali–chloride, or their eutectic [56]. The use of molten salt as a HTF in solar plants increases the Rankine cycle efficiency of the power steam turbine (from 54% at 400 °C to 63% at 560 °C [56]) and may reduce the physical size of the thermal storage system for a given capacity. In addition, molten salt is cheap and more environmentally safe than the present HTF [54]. The major challenge of molten salt is its high freezing point, leading to complications related to freeze protection in the solar field. In addition, molten salts exhibit poor thermophysical properties (e.g., specific heat capacity ~1.55 J/gK at 350 °C and thermal conductivity ~1 W/mK, while the specific heat of water is 4.2 J/gK at room temperature) that may increase the size requirement of TES.

Several papers have been published in the literature dealing with enhancing the thermophysical properties of base fluids by adding nanoparticles. For example, Khanafer and Vafai [6] presented a critical synthesis of the variants within the thermophysical properties of nanofluids. Correlations for the effective thermal conductivity and viscosity were synthesized and developed in terms of pertinent physical parameters based on the reported experimental data. The majority of studies have shown that the thermal conductivity of nanofluid increases with the addition of nanoparticles while specific heat capacity decreases [57–65]. Khanafer and Vafai [6] have demonstrated analytically and verified experimentally [57] that the addition of nanoparticles decreases the specific heat capacity of nanofluid at room temperature. Vajjha and Das [58] experimentally illustrated that the specific heat value of the nanofluid increases moderately with an increase in temperature. However, the specific heat decreases substantially with an increase in particle volumetric concentration. This study confirms that a nanofluid exhibits a lower specific heat than a base fluid; an illustration of this finding is presented in Fig. 5.5.

There are some studies in the literature that show an increase in the specific heat with the addition of nanoparticles [56,66–69]. Nelson *et al.* [66] reported that the specific heat of nanofluids (exfoliated graphite nanoparticle fibers suspended in polyalphaolefin at mass concentrations of 0.6% and 0.3%) was found to be 50% higher compared with pure polyalphaolefin. Shin and Banerjee [56] conducted an experimental study showing the effect of dispersing silica nanoparticles (1% by weight) for enhancing the specific heat

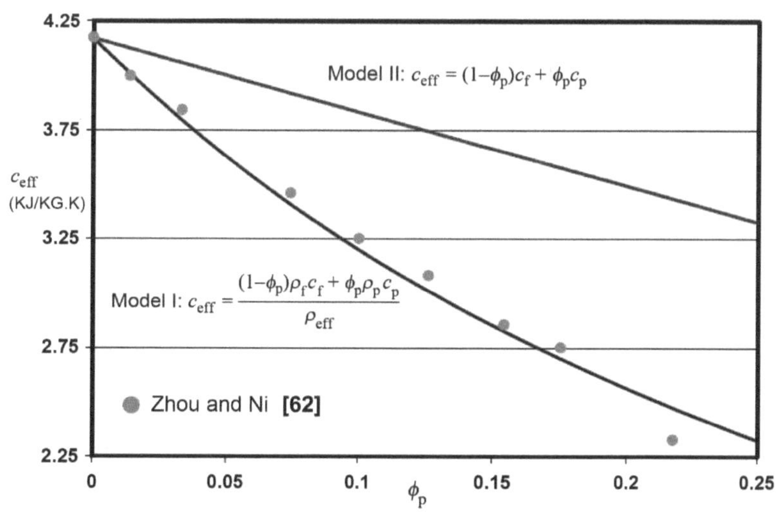

Figure 5.5 Comparison of the heat capacity of Al_2O_3–water nanofluid obtained by models I and II and the experimental data of Zhou and Ni [57]. *Reprinted from Khanafer and Vafai [6] with permission from Elsevier.* (For color version of this figure, the reader is referred to the online version of this chapter.)

capacity of the eutectic of lithium carbonate and potassium carbonate (62:38 ratio). A differential scanning calorimeter instrument was used to measure the specific heat of the molten salt eutectic after addition of nanoparticles. They found that the specific heat of the nanofluid was enhanced by 19–24%.

Shin and Banerjee [56] claimed that this finding is important in enhancing the stability and performance of solar thermal plants. The application of high-temperature nanofluids in the form of molten salts doped with nanoparticles in thermal storage systems is essential for continuous operation of solar thermal power plants. The anomalous enhancement of specific heat capacity of this new class [56,67,68] of nanofluids (molten salts doped with nanoparticles) can help to decrease the cost and size of TES and increase the operating temperature of the commercial solar towers from 400 [54] to 500–600 °C, which results in better thermal efficiency of the overall system. The sensible heat–thermal energy storage systems depend substantially on the specific heat and the operating temperature. The amount of energy stored in a TES system can be written as

$$Q(T) = M_s \int_{T_L}^{T_H} C_p(T) \mathrm{d}T \tag{5.1}$$

where M_s is the mass of the working fluid in the TES, $C_p(T)$ is the temperature-dependent specific heat capacity, and T_C and T_H are the lowest and highest operating temperatures, respectively.

Shin and Banerjee [68] proposed three thermal mechanisms to explain the abnormal enhancement of the specific heat capacity. These mechanisms include (1) higher specific heat capacity of nanoparticles compared with the bulk value of the base fluid, (2) fluid–solid interaction energy, and (3) "layering" of liquid molecules at the surface to form a semisolid layer. These mechanisms appear to be valid for other nanofluids reported in the literature in addition to molten salts. However, enhancements such as these have not been universal for other types of nanofluids. Therefore, more experimental and theoretical studies need to be conducted in order to explain the anomalous behavior between the specific heat capacity values of molten salt doped with nanoparticles to other nanofluids.

The mechanisms proposed by Shin and Banerjee [68] were similar to the mechanisms proposed by Keblinski *et al.* [70] to explain the thermal conductivity enhancement of nanofluids. Recently, Tiznobaik and Shin [71] dispersed four different-sized silicon dioxide nanoparticles (5, 10, 30, and 60 nm in diameter) in a molten salt eutectic (lithium carbonate and

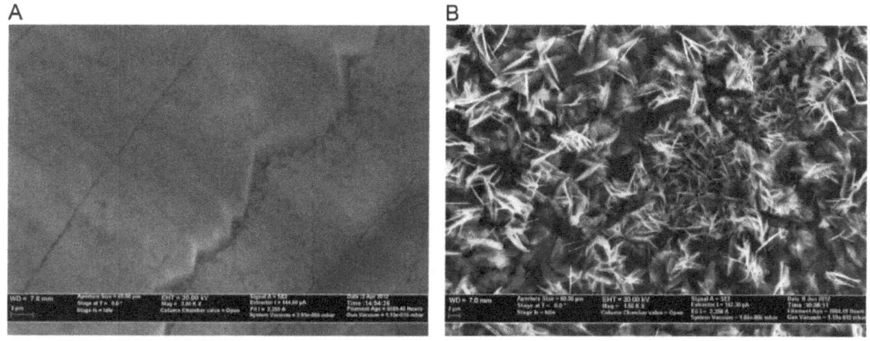

Figure 5.6 (A) Scanning electron micrograph (SEM) of pure eutectic mixture after testing and (B) scanning electron micrograph (SEM) of nanomaterial (30 nm) after testing. Special needlelike structures are formed all over the nanomaterials. *Reprinted from Tiznobaik and Shin [71] with permission from Elsevier.*

potassium carbonate, 62:38 by molar ratio) to obtain high-temperature operating fluids. These authors showed a 25% enhancement in the specific heat of nanomaterials regardless of the size of the embedded nanoparticles. The authors attributed this enhancement to the formation of needlelike structures (very large specific surface area) induced by the addition of nanoparticles, which can be seen in Fig. 5.6.

2.2. Direct absorption solar collectors

Flat-plate solar collectors are extensively used to harness solar energy. They absorb radiation through a black absorbing surface and transfer energy to the working fluid flowing through it. The performance of these collectors depends on a number of aspects, such as climatological and micro-climatological factors, geographic factors, geometry, and orientation of the collector [72]. Due to the shortcomings of the flat-plate black-surface absorbers (such as relatively high heat losses, corrosion effects, and limitations on incident flux density [72]), different concepts were proposed in the literature to allow the working fluid to directly absorb the incident radiation. The use of black liquids [73] and particles mixed with a gaseous working fluid [74–76] is one notable example. Typical working fluids used in the solar thermal collectors exhibited relatively low absorptive properties over the solar spectrum [77]. Therefore, nanoparticles were utilized in the solar energy applications to enhance the absorption properties of the base fluid. The addition of nanoparticles results in scattering of the incident solar energy within the working fluid and consequently increases the absorption within inside it [78]. The optical properties of the effective working fluid are highly dependent on the particle shape, particle size, and the optical properties of the base fluid and particles themselves [53]. Minardi and Chuang [73] presented experimental performance data for a low-flux black liquid (a suspension of micron-sized carbonaceous particles in shellac) collector used for hot water heating and compared it with flat-plate solar collectors. They found that solar radiation could be absorbed directly by the black liquid with minimal losses to other structures within the collector. Bertocchi *et al.* [74] conducted an experimental evaluation of a nonisothermal high-temperature solar particle receiver. Gas heating experiments were conducted with four different working gases using micron-sized spherical particles (600 nm diameter).

Use of micron-sized particles in the working fluids presents various operational challenges. Micron-sized particles have a tendency to settle rather than remaining suspended in the working fluid; hence, their distribution is highly nonuniform. Furthermore, they can lead to clogging of pumps and valves used in the overall system. These difficulties can be alleviated by use of nanoparticles in liquid (nanofluid). Nanofluids have the potential of improving the thermal and radiative properties of the working fluid. The significance of using nanoparticles on the thermal properties of the working fluid was recently shown by Khanafer and Vafai [6]. The authors presented a critical analysis of the variants within the thermophysical properties of nanofluids.

The application of nanofluids as a working fluid for solar collectors is a relatively new concept. Tyagi *et al.* [79] investigated theoretically the feasibility of using a nanofluid, a mixture of water and aluminum nanoparticles, as an absorbing medium for a low-temperature (<100 °C) direct absorption solar collector (DASC). The effects of absorption and scattering within the nanofluid were considered in that study. The authors showed that the presence of nanoparticles increased the absorption of incident radiation by more than nine times over that of pure water. Moreover, the efficiency of a DASC using a nanofluid as the working fluid was found to be up to 10% higher than that of a flat-plate collector under similar operating conditions [79]. Taylor *et al.* [78] analyzed theoretically the applicability of nanofluids in high-flux solar collectors. The authors showed that efficiency improvements on the order of 5–10% were possible with a nanofluid receiver.

Otanicar *et al.* [80] reported experimental results on solar collectors based on nanofluids made from a variety of nanoparticles (carbon nanotubes, graphite, and silver). Their results showed that the efficiency of a direct absorption solar collector was improved by 5% when utilizing nanofluids as the absorption mechanism. Moreover, the authors reported that the size, shape, and volume fraction of nanoparticles have a significant effect on broadening the spectral absorption of solar energy throughout the working fluid. This expansion allows the nanofluid to absorb a larger portion of the spectrum. This can readily be seen in Fig. 5.7. Taylor *et al.* [81] presented measurement and modeling techniques for determining the optical properties of nanofluids. The results of that study showed that the Maxwell–Garnett effective medium approach does not correctly predict the extinction coefficient for nanofluids.

One can note from the foregoing discussion that the researchers have focused their interest on the radiative properties of nanoparticles in liquid suspensions due to their potential applications in maximizing the amount of solar absorption. However, the volume fraction of nanoparticles must be chosen carefully to achieve the optimum thermal and optical characteristics of nanofluids. For high volume fraction of nanoparticles, the incoming sunlight will be absorbed in a thin layer near the surface of the receiver where the energy is easily lost to the environment. On the other hand, if the volume fraction of nanoparticles is low, the nanofluid will not absorb all of the incoming solar radiation. Another challenge associated with

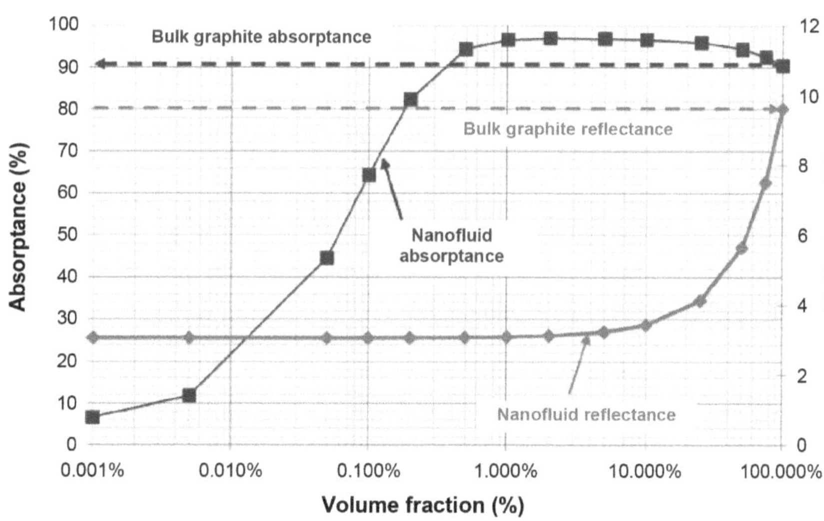

Figure 5.7 Theoretical benefit of volumetric absorption when utilizing a 30 nm graphite nanofluid in comparison to conventional area-based absorption. *Reprinted from* Otanicar et al. *[80] with permission from American Institute of Physics.* (For color version of this figure, the reader is referred to the online version of this chapter.)

nanoparticles is to achieve an even distribution of the absorbed solar energy within the nanofluid. Even distribution of the absorbed heat results in a uniform temperature profile within the fluid and consequently eliminates peak temperature at surfaces exposed to ambient temperature (i.e., minimizing heat loss at the boundaries). Therefore, for an optimum performance of a solar thermal collector, solar radiation should be absorbed within a small wavelength range ($0.25\ \mu m < \lambda < 2.5\ \mu m$) and converted directly to heat inside the working fluid to minimize the heat losses and the effect of fouling and pumping cost.

2.3. Photovoltaic technology

Photovoltaic systems produce clean, reliable energy without consuming fossil fuels and can be used in a wide variety of applications such as building-integrated photovoltaic systems (i.e., photovoltaic materials that are used to replace conventional building materials in parts of the building envelope such as the roof, skylights, or facades) [82–84], solar-powered cars, backup systems for critical equipment, PV-powered reverse osmosis (PV-RO) in remote areas [85–89], and stand-alone devices (e.g., water pumps, parking meters, emergency telephones, and traffic signs). However, as with all energy sources, there are potential environmental, health, and safety hazards associated with the full product life cycle of photovoltaics. The most important concerns are associated with the use of harmful chemicals in the manufacturing phase of the solar cells (e.g., crystalline silica dust). Improper disposal of solar panels at the end of their life cycle also presents similar concerns. It is possible to mitigate these risks with effective regulations by manufacturers and operators. According to the US Department of Energy, few power-generating technologies have as little environmental impact as photovoltaic solar panels. PV systems generate significantly fewer harmful air emissions (at least 89%) per kilowatt hour than conventional fossil fuel-fired technologies [90].

The solar photovoltaic industry started modestly and has been growing rapidly to where the total global capacity by the end of 2011 was 69.7 GW [91]. Figure 5.8 illustrates that the world solar PV capacity (grid-connected) has increased significantly from 5.4 GW in 2005 to 69.7 GW in 2011 [91].

Due to rapid advances in PV technology and the increase in manufacturing volume, the price of PV modules/MW has fallen by 60% from 2008 to 2011 according to estimates by Bloomberg New Energy Finance. Moreover, crystalline silicon photovoltaic cell prices have fallen from $76.67/watt in 1977 to an estimated $.74/watt in 2013 [92]. This is seen as evidence supporting Swanson's law, an observation similar to the famous Moore's law, that states that solar cell prices fall 20% for every doubling of industry capacity. This trend is depicted in Fig. 5.9.

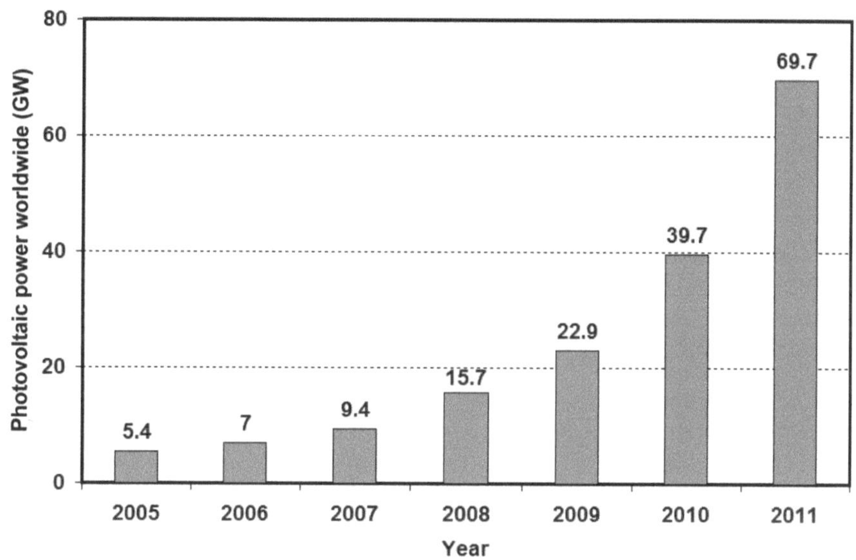

Figure 5.8 Worldwide photovoltaic power capacity connected to the power grid from 2005 to 2011 [91]. (For color version of this figure, the reader is referred to the online version of this chapter.)

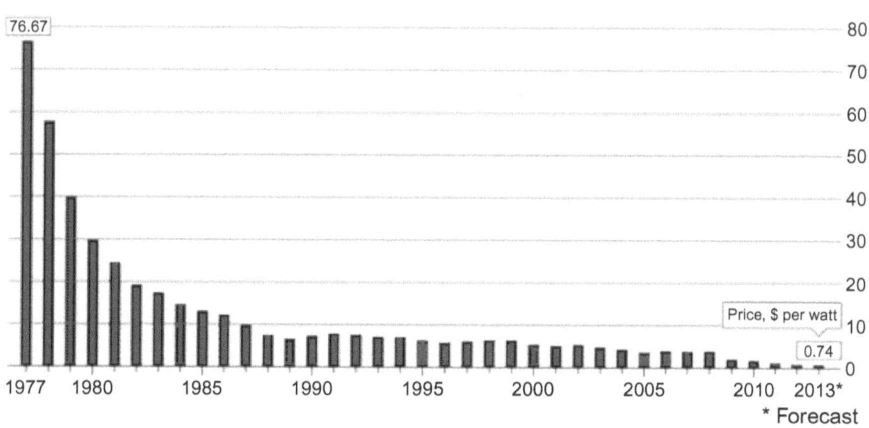

Figure 5.9 Swanson's law: price of crystalline silicon photovoltaic cell per watt ($/watt) over time. *Source: Bloomberg New Energy Finance.* (For color version of this figure, the reader is referred to the online version of this chapter.)

Silicon-based PVs currently comprise the great majority of the PV solar cell market (thick cells of around 150–300 nm made of crystalline silicon). The bandgap for silicon is about 1.1 eV. This technology was considered the first model of photovoltaic cells that accounts for more than 86% of the global solar cell market [66]. The second generation of photovoltaic materials was based on adding thin-film layers of semiconductor materials (1–2 nm) to achieve efficiencies greater than 12% [93,94]. There are a number of disadvantages for silicon-based PVs including high-temperature production requirements, low conversion efficiencies, and limitations in solar-grade feedstock availability [94,95]. Moreover, thin-film solar cells (e.g., Cu(InGa)Se2) also require high-temperature deposition methods [56].

To overcome some of the aforementioned limitations, nanoscale materials can be used in PV cells. Recently, thin-film solar cells enhanced by nanoparticles have attracted much attention of the scientific community [96]. The addition of nanoparticles in PV systems may enhance the effective optical path, resulting in the highest possible solar energy absorption. The use of nanocrystal quantum dots has led to thin-film solar cells based on a silicon or conductive transparent oxide substrate [97].

Quantum dot solar cells are an emerging area in solar cell research that uses quantum dots as an absorbing photovoltaic material, as opposed to well-known bulk materials such as silicon and copper indium gallium selenide. Quantum dots have bandgaps that are tunable over a wide range of energy levels by changing the quantum dot size. This is in contrast to other materials, where the bandgap is fixed by the choice of material composition. This property makes quantum dots attractive for multijunction solar cells, where a variety of different bandgap materials are used to improve efficiency by harvesting select portions of the solar spectrum.

2.4. Desalination

To protect the environment and to make seawater a more sustainable potable water source, renewable energy and more energy-efficient desalting systems should be used for desalting seawater. PV-powered reverse osmosis is considered one of the most promising forms of renewable energy-powered desalination systems, especially when it is used in remote areas. Therefore, small-scale PV-powered reverse osmosis (PV-RO) has received much attention in recent years and numerous demonstration systems have been built [98]. Since brackish water (water that has more salinity than freshwater, but less than seawater) has a much lower osmotic pressure than seawater, its desalination requires much less energy and therefore much smaller PV arrays in the case of PV-RO. Many brackish-water PV-RO systems have been installed in different parts of the world [85–89,98,99]. Lamei *et al.* [100] discussed electricity price at which solar energy can be considered economical to be used for RO desalination. They proposed an equation to estimate the unit production costs of RO desalination plants using PV solar energy based on current and future PV module prices. Kershman *et al.* [101] studied an experimental plant for seawater reverse osmosis (SWRO) desalination powered from renewable energy sources. The reverse osmosis desalination plant used both wind energy conversion and photovoltaic power generation while being integrated into a grid-connected power supply to provide power recovery. Darwish *et al.* [102] examined the feasibility of using renewable energy resources, such as solar and wind energies, to run the SWRO desalting plants and selecting the suitable renewable energy for pumping the seawater. To protect the environment and to make the desalted seawater (DW) more sustainable as a potable water source, they recommended the use of renewable energy and more energy-efficient desalting methods.

New methodologies that combine solar desalination and nanotechnology are essential. Ling and Chung [103] proposed a potentially integrated forward osmosis–ultrafiltration (FO–UF) system for water reuse and desalination with the aid of superhydrophilic nanoparticles. The system uses FO as the semipermeable membrane to reject salts and the UF membranes to regenerate the draw solutes. The authors present the proposed FO–UF integrated system, using superhydrophilic nanoparticles as draw solutes, as a promising technology to desalinate both seawater and brackish water and to reclaim water from wastewater.

Recently, researchers have attempted to develop a new way to use sunlight to produce steam, and other vapors, without directly heating an entire container of fluid to its boiling point. This new technology uses nanoparticles to more effectively produce steam by allowing the surfaces of nanoparticles to serve as boiling nucleation sites and not directly heating the fluid to its boiling point [104]. The nanoparticle surfaces absorb light energy and become elevated in temperature to a point beyond the normal boiling point of the fluid.

3. CONCLUSIONS

Nanotechnology-based nanomaterials are receiving considerable attention these days in various areas of the energy sector. However, the main challenges facing nanomaterials in the energy sector are the enhancement of efficiency, reliability, safety, and product life in addition to the reduction of costs. Promising applications of nanomaterials can be found in areas such as photovoltaics (solar cells), thermal storage, hydrogen conversion (fuel cells), and solar desalination. With the application of nanomaterials, PV solar cells are experiencing an increase in efficiency while simultaneously reducing the production costs of electricity and manufacturing.

REFERENCES

[1] Department of Energy in U.S. Hydrogen Storage Technologies Roadmap. http://www1.eere.energy.gov/vehiclesandfuels/pdfs/program/hydrogen_storage_roadmap.pdf, 2005.

[2] S.U.S. Choi, Nanofluids: From vision to reality through research, J. Heat Transf. 131 (2009) 1–9.

[3] K.V. Wong, O. Leon, Applications of nanofluids: Current and future, Adv. Mech. Eng. 2010 (2010) 1–11.

[4] K. Khanafer, K. Vafai, M. Lightstone, Buoyancy-driven heat transfer enhancement in a two-dimensional enclosure utilizing nanofluids, Int. J. Heat Mass Transf. 46 (2003) 3639–3653.

[5] A.R.A. Khaled, K. Vafai, Heat transfer enhancement through control of thermal dispersion effects, Int. J. Heat Mass Transf. 48 (2005) 2172–2185.

[6] K. Khanafer, K. Vafai, A critical synthesis of thermophysical characteristics of nanofluids, Int. J. Heat Mass Transf. 54 (2011) 4410–4428.

[7] A. Ali, K. Vafai, A.-R.A. Khaled, Comparative study between parallel and counter flow configurations between air and falling film desiccant in the presence of nanoparticle suspensions, Int. J. Energy Res. 27 (2003) 725–745.

[8] M. Shafahi, V. Bianco, K. Vafai, O. Manca, Thermal performance of flat-shaped heat pipes using nanofluids, Int. J. Heat Mass Transf. 53 (2010) 1438–1445.

[9] M. Shafahi, V. Bianco, K. Vafai, O. Manca, An investigation of the thermal performance of cylindrical heat pipes using nanofluids, Int. J. Heat Mass Transf. 53 (2010) 376–383.

[10] J.A. Eastman, S.U.S. Choi, S. Li, L.J. Thompson, S. Lee, Enhanced thermal conductivity through the development of nanofluids, in: S. Komarneni, J.C. Parker, H.J. Wollenberger (Eds.), 1996 Fall meeting of the Materials Research Society (MRS), Boston, USA, 1996.

[11] J.A. Eastman, S.U.S. Choi, S. Li, W. Yu, L.J. Thompson, Anomalously increased effective thermal conductivities of ethylene glycol-based nanofluids containing copper nanoparticles, Appl. Phys. Lett. 78 (2001) 718–720.

[12] S.P. Jang, S.U.S. Choi, Role of Brownian motion in the enhanced thermal conductivity of nanofluids, Appl. Phys. Lett. 84 (2004) 4316–4318.

[13] S. Lee, S.U.S. Choi, Application of metallic nanoparticle suspensions in advanced cooling systems, in: 1996 International Mechanical Engineering Congress and Exhibition, Atlanta, USA, 1996.

[14] B.X. Wang, L.P. Zhou, X.F. Peng, A fractal model for predicting the effective thermal conductivity of liquid with suspension of nanoparticles, Int. J. Heat Mass Transf. 46 (2003) 2665–2672.

[15] S.K. Das, N. Putra, W. Roetzel, Pool boiling characteristics of nanofluids, Int. J. Heat Mass Transf. 46 (2003) 851–862.

[16] S.K. Das, N. Putra, W. Roetzel, Pool boiling of nanofluids on horizontal narrow tubes, Int. J. Multiphase Flow 29 (2003) 1237–1247.

[17] S.K. Das, N. Putra, P. Thiesen, W. Roetzel, Temperature dependence of thermal conductivity enhancement for nanofluids, J. Heat Transf. 125 (2003) 567–574.
[18] CEA, Nanofluids for Heat Transfer Applications, Marketing Study Unit, CEA, France, 2007.
[19] N. Putra, W. Roetzel, S.K. Das, Natural convection of nanofluids, Heat Mass Transf. 39 (2003) 775–784.
[20] H. Masuda, A. Ebata, K. Teramae, N. Hishinuma, Alteration of thermal conductivity and viscosity of liquid by dispersing ultra-fine particles (dispersion of c-Al_2O_3, SiO_2 and TiO_2 ultra-fine particles), Netsu Bussei 4 (1993) 227–233.
[21] S. Lee, S.U.S. Choi, S. Li, J.A. Eastman, Measuring thermal conductivity of fluids containing oxide nanoparticles, ASME J. Heat Transf. 121 (1999) 280–289.
[22] X. Wang, X. Xu, S.U.S. Choi, Thermal conductivity of nanoparticles–fluid mixture, J. Thermophys. Heat Transf. 13 (1999) 474–480.
[23] C.H. Chon, K.D. Kihm, S.P. Lee, S.U.S. Choi, Empirical correlation finding the role of temperature and particle size for nanofluid (Al_2O_3) thermal conductivity enhancement, Appl. Phys. Lett. 87 (2005) 153107.
[24] C.H. Li, G.P. Peterson, The effect of particle size on the effective thermal conductivity of Al_2O_3-water nanofluids, J. Appl. Phys. 101 (2007) 044312.
[25] X. Zhang, H. Gu, M. Fujii, Effective thermal conductivity and thermal diffusivity of nanofluids containing spherical and cylindrical nanoparticles, J. Appl. Phys. 100 (2006) 1–5.
[26] E.V. Timofeeva, A.N. Gavrilov, J.M. McCloskey, Y.V. Tolmachev, Thermal conductivity and particle agglomeration in alumina nanofluids: Experiment and theory, Phys. Rev. E. 76 (2007) 061203.
[27] Y.M. Xuan, Q. Li, Heat transfer enhancement of nanofluids, Int. J. Heat Fluid Flow 21 (2000) 58–64.
[28] H.E. Patel, S.K. Das, T. Sundararajan, Thermal conductivities of naked and monolayer protected metal nanoparticle based nanofluids: Manifestation of anomalous enhancement and chemical effects, Appl. Phys. Lett. 83 (2003) 2931–2933.
[29] R.A. Taylor, P.E. Phelan, Pool boiling of nanofluids. Comprehensive review of existing data and limited new data, Int. J. Heat Mass Transf. 52 (2009) 5339–5347.
[30] H.S. Ahn, H. Kim, H. Jo, S. Kang, W. Chang, M.H. Kim, Experimental study of critical heat flux enhancement during forced convective flow boiling of nanofluid on a short heated surface, Int. J. Multiphase Flow 36 (5) (2010) 375–384.
[31] J. Lee, I. Mudawar, Assessment of the effectiveness of nanofluids for single-phase and two-phase heat transfer in micro-channels, Int. J. Heat Mass Transf. 50 (3–4) (2007) 452–463.
[32] G. Prakash Narayan, K.B. Anoop, S.K. Das, Mechanism of enhancement/deterioration of boiling heat transfer using stable nanoparticle suspensions over vertical tubes, J. Appl. Phys. 102 (2007) 074317.
[33] S.M. You, J.H. Kim, K.H. Kim, Effect of nanoparticles on critical heat flux of water in pool boiling heat transfer, Appl. Phys. Lett. 83 (2003) 3374–3376.
[34] I.C. Bang, S.H. Chang, Boiling heat transfer performance and phenomena of Al_2O_3-water nanofluids from a plain surface in a pool, Int. J. Heat Mass Transf. 48 (2005) 2407–2419.
[35] J.P. Tu, N. Dinh, T. Theofanous, An experimental study of nanofluid boiling heat transfer, in: Proceedings of 6th International Symposium on Heat Transfer, Beijing, China, 2004.
[36] D.S. Wen, Y.L. Ding, Experimental investigation into the pool boiling heat transfer of aqueous based γ-alumina nanofluids, J. Nanopart. Res. 7 (2005) 265–274.
[37] D.S. Wen, Y.L. Ding, R.A. Williams, Pool boiling heat transfer of aqueous based TiO_2 nanofluids, J. Enhanc. Heat Transf. 13 (2006) 231–244.

[38] Z.H. Liu, J.G. Xiong, R. Bao, Boiling heat transfer characteristics of nanofluids in a flat heat pipe evaporator with micro-grooved heating surface, Int. J. Multiphase Flow 33 (2007) 1284–1295.
[39] M.H. Kim, J.B. Kim, H.D. Kim, Experimental studies on CHF characteristics of nanofluids at pool boiling, Int. J. Multiphase Flow 33 (2007) 691–706.
[40] S.J. Kim, T. McKrell, J. Buongiorno, L.W. Hu, Experimental study of flow critical heat flux in alumina–water, zinc-oxide–water and diamond-water nanofluids. J. Heat Transf. 131 (4) (Feb 13, 2009) 7 pages, 043204, http://dx.doi.org/10.1115/1.3072924.
[41] P. Vassallo, R. Kumar, S.D. Amico, Pool boiling heat transfer experiments in silica–water nanofluids, Int. J. Heat Mass Transf. 47 (2004) 407–411.
[42] C.H. Li, B.X. Wang, X.F. Peng, Experimental investigations on boiling of nano-particle suspensions, in: 2003 Boiling Heat Transfer Conference, Jamaica, USA, 2003.
[43] S.J. Kim, I.C. Bang, J. Buongiorno, L.W. Hu, Effects of nanoparticle deposition on surface wettability influencing boiling heat transfer in nanofluids, Appl. Phys. Lett. 89 (2006) 153107.
[44] S.J. Kim, I.C. Bang, J. Buongiorno, L.W. Hu, Surface wettability change during pool boiling of nanofluids and its effect on critical heat flux, Int. J. Heat Mass Transf. 50 (2007) 4105–4116.
[45] M. Chopkar, A.K. Das, I. Manna, P.K. Das, Pool boiling heat transfer characteristics of ZrO_2-water nanofluids from a flat surface in a pool, Heat Mass Transf. 44 (8) (2007) 999–1004.
[46] Y. Ding, H. Chen, L. Wang, et al., Heat transfer intensification using nanofluids, J. Part. Powder 25 (2007) 23–36.
[47] D. Milanova, R. Kumar, Role of ions in pool boiling heat transfer of pure and silica nanofluids, Appl. Phys. Lett. 87 (23) (2005) 233107.
[48] D. Milanova, R. Kumar, Heat transfer behaviour of silica nanoparticles in pool boiling experiment, J. Heat Transf. 130 (4) (2008) 1–6.
[49] M. Golubovic, H.D.M. Hettiarachchi, W.M. Worek, W.J. Minkowycz, Nanofluids and critical heat flux, experimental and analytical study, Appl. Therm. Eng. 29 (2009) 1281–1288.
[50] S.M. Kwark, R. Kumar, G. Moreno, J. Yoo, S.M. You, Pool boiling characteristics of low concentration nanofluids, Int. J. Heat Mass Transf. 53 (2010) 972–981.
[51] D. Chicea, Coherence light scattering on nanofluids: Computer simulation results, Appl. Opt. 47 (2008) 1434–1442.
[52] R.S. Prasher, P.E. Phelan, Modeling of radiative and optical behavior of nanofluids based on multiple and dependent scattering theories, Paper No. IMECE2005-80302, ASME International Mechanical Engineering Congress & Exposition, Orlando, FL, November, 2005.
[53] N.G. Khlebtsov, L.A. Trachuk, A.G. Mel'nikov, The effect of the size, shape and structure of metal nanoparticles on the dependence of their optical properties on the refractive index of a disperse medium, Opt. Spectrosc. 98 (2005) 77–83.
[54] D. Kearney, U. Hermann, P. Nava, B. Kelly, Assessment of a molten salt heat transfer fluid in a parabolic trough solar field, J. Sol. Energy Eng. 125 (2003) 170–176.
[55] G. Janz, C. Allen, N. Bansal, R. Murphy, R. Tomkins, Physical properties data compilations relevant to energy storage, U. S. Dept. of Commerce, National Bureau of Standards, Washington, DC, 1979.
[56] D. Shin, D. Banerjee, Effects of silica nanoparticles on enhancing the specific heat capacity of carbonate salt eutectic (work in progress), Int. J. Struct. Changes Solids Mech. Appl. 2 (2010) 25–31.
[57] S.Q. Zhou, R. Ni, Measurement of the specific heat capacity of water-based Al_2O_3 nanofluid, Appl. Phys. Lett. 92 (2008) 093123.

[58] R.S. Vajjha, D.K. Das, Specific heat measurement of three nanofluids and development of new correlations, J. Heat Transf. 131 (2009) 071601.
[59] B.-X. Wang, L.-P. Zhou, X.-F. Peng, Surface and size effects on the specific heat capacity of nanoparticles, Int. J. Thermophys. 27 (2006) 139–151.
[60] B.C. Pak, Y.I. Cho, Hydrodynamic and heat transfer study of dispersed fluids with submicron metallic oxide particles, Exp. Heat Transf. 11 (1999) 151–170.
[61] H. O'Hanley, J. Buongiorno, T. McKrell, H. Lin-wen, Measurement and model validation of nanofluid specific heat capacity with differential scanning calorimetry. Adv. Mech.Eng. 2012 (2012), http://dx.doi.org/10.1155/2012/181079, ID 181079.
[62] L.-P. Zhou, B.-X. Wang, X.-F. Peng, X.-Z. Du, Y.-P. Yang, On the specific heat capacity of CuO nanofluid. Adv. Mech. Eng. 2010 (2010), http://dx.doi.org/10.1155/2010/172085,ID 172085.
[63] A. Mohebbi, Prediction of specific heat and thermal conductivity of nanofluids by a combined equilibrium and non-equilibrium molecular dynamics simulation, J. Molecular Liquids 175 (2013) 51–58.
[64] Y.R. Sekhar, K.V. Sharm, Study of viscosity and specific heat capacity characteristics of water-based Al_2O_3 nanofluids at low particle concentrations. J. Experimental Nanoscience, 2013, http://dx.doi.org/10.1080/17458080.2013.796595.
[65] H. Tiznobaik, D. Shin, Experimental validation of enhanced heat capacity of ionic liquid-based nanomaterials, Applied Physics Letters 102 (2013) 173906–173906-3.
[66] I.C. Nelson, D. Banerjee, P. Rengasamy, Flow loop experiments using polyalphaolefin nanofluids, J. Thermophys. Heat Transf. 23 (2009) 752–761.
[67] D. Shin, D. Banerjee, Enhancement of specific heat capacity of high-temperature silica-nanofluids synthesized in alkali chloride salt eutectics for solar thermal-energy storage applications, Int. J. Heat Mass Transf. 54 (2011) 1064–1070.
[68] D. Shin, D. Banerjee, Enhanced specific heat of silica nanofluid, J. Heat Transf. 133 (2011) 024501.
[69] D. Malik, Evaluation of composite alumina nanoparticle and nitrate eutectic materials for use in concentrating solar power plants (Thesis), Texas A&M, 2010.
[70] P. Keblinski, S.R. Phillpot, S.U.S. Choi, J.A. Eastman, Mechanisms of heat flow in suspensions of nano-sized particles (nanofluids), Int.J. Heat Mass Transf. 45 (2002) 855–863.
[71] H. Tiznobaik, D. Shin, Enhanced specific heat capacity of high-temperature molten salt-based nanofluids, Int. J. Heat Mass Transf. 57 (2013) 542–548.
[72] C.U. Okujagu, S.K. Adjepong, Performance of a simple flat-plate collector at an equatorial location, Sol. Wind Technol. 6 (1989) 283–289.
[73] J.E. Minardi, H.N. Chuang, Performance of a "black" liquid flat-plate, solar collector, Sol. Energy 17 (1975) 179–183.
[74] R. Bertocchi, J. Karni, A. Kribus, Experimental evaluation of a non-isothermal high temperature solar particle receiver, Energy 29 (2004) 687–700.
[75] M. Abdelrahman, P. Fumeaux, P. Suter, Study of solid-gas suspensions used for direct absorption of concentrated solar radiation, Sol. Energy 22 (1979) 45–48.
[76] A.J. Hunt, A New Solar Thermal Receiver Utilizing a Small Particle Heat Exchanger, in: 13th Intersociety Energy Conversion Engineering Conference, Boston, MA, August 5-10, Lawrence Berkeley National Laboratory, Berkeley, California, 1979.
[77] T.P. Otanicar, P.E. Phelan, J.S. Golden, Optical properties of liquids for direct absorption solar thermal energy systems, Sol. Energy 83 (2009) 969–977.
[78] R.A. Taylor, P.E. Phelan, T.P. Otanicar, C.A. Walker, M. Nguyen, Applicability of nanofluids in high flux solar collectors, J. Renew. Sust. Energ 3 (2011) 023104.
[79] H. Tyagi, P. Phelan, R. Prasher, Predicted efficiency of a low-temperature nanofluid-based direct absorption solar collector, J. Sol. Energy Eng. 131 (2009) 041004.
[80] T.P. Otanicar, P.E. Phelan, R.S. Prasher, G. Rosengarten, R.A. Taylor, Nanofluid-based direct absorption solar collector, J. Renew. Sust. Energ. 2 (2010) 033102.

[81] R.A. Taylor, P.E. Phelan, T.P. Otanicar, R. Adrian, R. Prasher, Nanofluid optical property characterization: Towards efficient direct absorption solar collectors, Nanoscale Res. Lett. 6 (2011) 225–235.
[82] M. Ordenes, D.L. Marinoski, P. Braun, R. Ruther, The impact of building integrated photovoltaics on the energy demand of multi-family dwellings in Brazil, Energy Build. 39 (2007) 629–642.
[83] X. Xu, S. Van Dessel, Evaluation of an active building envelope window-system, Energy Build. 40 (2008) 168–174.
[84] T.T. Chow, J.W. Hand, P.A. Strachan, Building-integrated photovoltaic and thermal applications in a subtropical hotel building, Appl. Therm. Eng. 23 (2003) 2035–2049.
[85] S. Kehal, Reverse Osmosis Unit of 0.85 m^3/h Capacity Driven by Photovoltaic Generator in South Algeria, in: Proceedings of the New Technologies for the Use of Renewable Energy Sources in Water Desalination Conference, Session II, Athens, Greece, 1991, pp. 8–16.
[86] E. Tzen, K. Perrakis, Prefeasibility study for small Aegean islands, RENA-CT94–0063, CRES, Greece, 1996.
[87] M. Thomson, D. Infield, A photovoltaic powered seawater reverse-osmosis system without batteries, Desalination 153 (2002) 1–8.
[88] A. Hafez, S. El-Manharawy, Economics of seawater RO desalination in the Red Sea region, Egypt. Part 1: A case study, Desalination 153 (2003) 335–347.
[89] P. Finken, K. Korupp, Water Desalination Plants Powered by Wind Generators and Photovoltaic Systems, in: Proceedings of the New Technologies for the Use of Renewable Energy Sources in Water Desalination Conference, Session II, Athens, Greece, 1991, pp. 65–98.
[90] V. Fthenakis, H.C. Kim, E. Alsema, Emissions from photovoltaic life-cycles, Environ. Sci. Technol. 42 (2008) 2168–2174.
[91] EPIA, Global Market Outlook for Photovoltaics until 2016, 2012, p. 9,11,12,64.
[92] G. Carr, Sunny uplands: Alternative energy will no longer be alternative, The Economist (2012).
[93] E. Serrano, G. Rus, J. Garcı´a-Martı´nez, Nanotechnology for sustainable energy, Renew. Sust. Energ. Rev. 13 (2009) 2373–2384.
[94] S. Hegedus, A. Luque, Status, trends, challenges and the bright future of solar electricity from photovoltaics, in: A. Luque, S. Hegedus (Eds.), Handbook of Photovoltaic Science and Engineering, John Wiley & Sons Ltd, The Atrium, Southern Gate, Chichester, West Sussex PO19 8SQ, England, 2003, pp. 25–67.
[95] P. Woditsch, W. Koch, Solar grade silicon feedstock supply for PV industry, Sol. Energy Mater. Sol. Cells 72 (2002) 11–26.
[96] Y. Akimov, W.S. Koh, S.Y. Sian, S. Ren, Nanoparticle-enhanced thin film solar cells: Metallic or dielectric nanoparticles? Appl. Phys. Lett. 96 (2010) 073111.
[97] R.T. Ross, A.J. Nozik, Efficiency of hot-carrier solar energy converters, J. Appl. Phys. 53 (1982) 3813–3818.
[98] A. Al-Karaghouli, D. Renne, L.L. Kazmerski, Technical and economic assessment of photovoltaic-driven desalination systems, Renew. Energy 35 (2010) 323–328.
[99] K. Perrakis, E. Tzen, P. Baltas. PV-RO desalination plant in Morocco, JOR3-CT95-0066, CRES, Greece.
[100] A. Lamei, P. Van der Zaag, E. Von Munch, Impact of solar energy cost on water production cost of seawater desalination plants in Egypt, Energy Policy 36 (2008) 1748–1756.
[101] S.A. Kershman, J. Rheinlander, H. Gabler, Seawater reverse osmosis powered Tom renewable energy sources—hybrid wind/photovoltaic/grid power supply for small-scale desalination in Libya, Desalination 153 (2002) 17–23.

[102] M. Darwish, R. Mohtar, Y. Elgendy, M. Chmeissani, Desalting seawater in Qatar by renewable energy: A feasibility study, Desalin. Water Treat. 47 (2012) 279–294.
[103] M.M. Ling, T.S. Chung, Desalination process using super hydrophilic nanoparticles via forward osmosis integrated with ultrafiltration regeneration, Desalination 278 (2011) 194–202.
[104] O. Neumann, A.S. Urban, J. Day, S. Lal, P. Nordlander, N.J. Halas, Solar vapor generation enabled by nanoparticles, ACS Nano 7 (2013) 42–49.

AUTHOR INDEX

Note: Page numbers followed by "*f*" indicate figures and "*t*" indicate tables.

T

SUBJECT INDEX

Note: Page numbers followed by "*f*" indicate figures and "*t*" indicate tables.

D

E

R

S

T

U

V

Kolibri 360 GmbH
Ferdinand-Jühlke-Straße 7
99095 Erfurt, Deutschland
produktsicherheit@kolibri360.de